Learning
Trigonometry by
Problem Solving

Problem Solving in Mathematics and Beyond

Print ISSN: 2591-7234
Online ISSN: 2591-7242

Series Editor: Dr. Alfred S. Posamentier
Distinguished Lecturer
New York City College of Technology - City University of New York

There are countless applications that would be considered problem solving in mathematics and beyond. One could even argue that most of mathematics in one way or another involves solving problems. However, this series is intended to be of interest to the general audience with the sole purpose of demonstrating the power and beauty of mathematics through clever problem-solving experiences.

Each of the books will be aimed at the general audience, which implies that the writing level will be such that it will not engulfed in technical language — rather the language will be simple everyday language so that the focus can remain on the content and not be distracted by unnecessarily sophiscated language. Again, the primary purpose of this series is to approach the topic of mathematics problem-solving in a most appealing and attractive way in order to win more of the general public to appreciate his most important subject rather than to fear it. At the same time we expect that professionals in the scientific community will also find these books attractive, as they will provide many entertaining surprises for the unsuspecting reader.

Published

Vol. 23   *Learning Trigonometry by Problem Solving*
          by Alexander Rozenblyum and Leonid Rozenblyum

Vol. 22   *Mathematical Labyrinths. Pathfinding*
          by Boris Pritsker

Vol. 21   *Adventures in Recreational Mathematics: Selected Writings on*
          Recreational Mathematics and its History
          (In 2 Volumes)
          by David Singmaster

Vol. 20   *X Games in Mathematics: Sports Training That Counts!*
          by Tim Chartier

For the complete list of volumes in this series, please visit www.worldscientific.com/series/psmb

Problem Solving in Mathematics and Beyond Volume **23**

Learning Trigonometry by Problem Solving

Alexander Rozenblyum

The City University of New York, USA

Leonid Rozenblyum

World Scientific

NEW JERSEY · LONDON · SINGAPORE · BEIJING · SHANGHAI · HONG KONG · TAIPEI · CHENNAI · TOKYO

Published by

World Scientific Publishing Co. Pte. Ltd.
5 Toh Tuck Link, Singapore 596224
USA office: 27 Warren Street, Suite 401-402, Hackensack, NJ 07601
UK office: 57 Shelton Street, Covent Garden, London WC2H 9HE

Library of Congress Cataloging-in-Publication Data
Names: Rozenblyum, Alexander, author. | Rozenblyum, Leonid, author.
Title: Learning trigonometry by problem solving / Alexander Rozenblyum,
    City University of New York, USA, Leonid Rozenblyum.
Description: New Jersey : World Scientific, [2021] | Series: Problem solving in mathematics
    and beyond, 2591-7234 ; volume 23 | Includes bibliographical references and index.
Identifiers: LCCN 2020051832 | ISBN 9789811231209 (hardcover) |
    ISBN 9789811232848 (paperback) | ISBN 9789811231216 (ebook for institutions) |
    ISBN 9789811231223 (ebook for individuals)
Subjects: LCSH: Trigonometry--Problems, exercises, etc.
Classification: LCC QA537 .R69 2021 | DDC 516.24--dc23
LC record available at https://lccn.loc.gov/2020051832

British Library Cataloguing-in-Publication Data
A catalogue record for this book is available from the British Library.

For any available supplementary material, please visit
https://www.worldscientific.com/worldscibooks/10.1142/12130#t=suppl

Desk Editors: George Vasu/Tan Rok Ting

Typeset by Stallion Press
Email: enquiries@stallionpress.com

Preface

In this book, trigonometry is presented mainly through the solution of specific problems. The purpose of the problems is two-fold. As in more conventional texts, the problems are meant to help the reader consolidate their knowledge of the subject. In addition, they perform another, no less important function. Namely, they serve to motivate and provide context for the concepts, definitions, and results as they are presented. The theory is developed as necessarily arising from considerations of specific problems, making it much more natural. For example, the definitions of trigonometric functions are introduced not just as abstract ratios of the sides in a right triangle, but as convenient and useful tools while solving practical problems. This approach allows the reader to more easily understand and absorb the meaning of abstract mathematical concepts. In this way, it enables a more active mastery of the subject, directly linking the results of the theory with their applications. Some historical notes are also embedded in select chapters.

Most of the problems are solved directly within the text. Some are given to be independently solved by the reader. Additional problems are offered to deepen the knowledge gained. These problems appear at the end of the chapters as exercises. Solutions to all independent problems and exercises are given at the end of the book.

The problems in the book are selected from a variety of disciplines, such as physics, medicine, architecture, and so on. They include solving triangles, trigonometric equations, and their applications. Taken together,

the problems cover the entirety of material contained in a standard trigonometry course which is studied in high school and college. We have
taken care to address many typical misconceptions and errors that are
made when solving problems.

The book is intended for a wide range of audiences, including math
instructors, high school and undergraduate students, and is well-suited for
independent study. This book can be used not only as a textbook, but also as
a book of exercises. It can serve as a guide to lead an enrichment class with
emphasis on fundamental understanding rather than mere memorization.

Understanding this book requires very little prior knowledge of mathematics: the ability to perform simple algebraic transformations, some
knowledge of plane geometry, the ability to solve linear and quadratic
equations, and familiarity with their graphs.

We've also added some interesting, in our opinion, entertainment
problems. To solve them, no special knowledge is required. The reader
only needs to apply some creativity, imagination, and a bit of logic. While
they are not directly related to the subject of the book, they reflect its spirit.
We hope that the inclusion of these problems will not distract the reader
from the main subject matter. These puzzles will contribute to a more
lighthearted reading of the material and will also be useful for gymnastics
and training of the mind and, as a result, will hopefully please all lovers of
mathematics. Below is one such entertainment problem.

Problem. Two books stand on a bookshelf next to each other as shown:

The first book contains 250 pages and the second 300 pages. One
of the most curious and prolific readers is a bookworm — a sneaky little

insect that likes to chew through books. It chewed these books, starting with the first page of book 1 and ending with the last page of book 2. Through how many pages did the worm chew? Let's assume the pages of the book are single sided — meaning one page per sheet of paper.

Hint: Don't rush to answer.
Solution is given at the end of the book.

The problem shown on the Cover Page is presented in Chapter 18, Problem 18.7.

About the Authors

Alexander Rozenblyum, currently, an Associate Professor of Mathematics at New York City College of Technology, CUNY, where he has taught a broad spectrum of courses from developmental to upper-level for the past 20 years. Graduated summa cum laude from Belarusian University, Belarus, USSR with a Master's degree in mathematics. Worked as a researcher at the Institute of Land Reclamation and Water Management in Minsk, Belarus, creating mathematical models of water flow in drainage systems. Earned a Ph.D. in mathematics from Belarusian University. Then moved to Grodno, Belarus, to take up a position as a professor of mathematics at Grodno University. Earned a doctorate degree in mathematics from St. Petersburg University, Russia. Research interests are in the theory of special functions of mathematical physics and the theory of representations of groups, and contemporary methods of teaching in mathematics. Alexander is the author of more than 50 scientific and methodological publications in mathematics, including a research monograph "Multidimensional special functions in the theory of group representations" (co-authored with his brother, Leonid). In his leisure time, Alexander likes to listen to classical music (he graduated from music school, class of violin) and play table tennis.

Leonid Rozenblyum, graduated summa cum laude from Belarusian University, Belarus, USSR, earning a Master's degree in mathematics. Worked as a Leading Software Specialist in the Computer Center of Minsk Tractor Works, Belarus. Earned a Ph.D. degree at Belarusian University. Worked as a Visiting Professor at the Institute for Qualification Improvement, Minsk, Belarus, teaching subjects in computer science. Published more than 30 Scientific and Technical Publications in mathematics and computer science. The areas of scientific research include elaboration of algorithms, mathematical simulation, design and development of database management systems, theory of special functions of mathematical physics, representations of groups, and contemporary methods of teaching in mathematics. He published (together with his brother, Alexander) the research monograph "Multidimensional special functions in the theory of group representation." He has about 20 Certificates of Innovation in Information Technology. In the USA, Leonid worked as a scientific and software engineer in such companies as IBM, Bio-Rad Laboratories, Oracle Corporation. He retired from the position of Principal Software Engineer, Oracle Corporation, Redwood City, CA. Leonid enjoys listening to classical music (he graduated from music school, class of violin) and likes to play table tennis.

Contents

Preface   v

About the Authors   ix

Chapter 1   Introduction to Trigonometric Functions   1

Chapter 2   Solving Right Triangles   17

Chapter 3   Trigonometric Properties in Right Triangles   33

Chapter 4   General Definitions of Trigonometric Functions   47

Chapter 5   Basic Properties of General Trigonometric Functions.
             Methods for Reducing to Acute Angles   63

Chapter 6   Solving Oblique Triangles: Law of Sines   77

Chapter 7   Solving Oblique Triangles: Law of Cosines   93

Chapter 8   Addition and Double Angle Formulas   105

Chapter 9   Product, Sum, and Difference of Trigonometric
             Functions   121

Chapter 10   Radian Measure   133

Chapter 11   Graphs of Sine and Cosine   149

Chapter 12   Sinusoidal Functions   167

Chapter 13   Applications of Sinusoids   181

Chapter 14   Graphs of Tangent and Cotangent   191

Chapter 15 Basic Sine Equation and Inverse Sine Function   205

Chapter 16 Inverse Cosine and Tangent. Basic Trigonometric
            Equations for Cosine and Tangent   221

Chapter 17 Trigonometric Identities, Inequalities, and Equations 235

Chapter 18 Applications to Geometry   247

Chapter 19 Historical Background of Complex Numbers.
            Cardano's Formula   265

Chapter 20 Definition and Properties of Complex Numbers   279

Appendix: Two Remarkable Triangles   297

Summary of Results   303

Answers and Solutions to Problems and Exercises   313

Index   365

Chapter 1

Introduction to Trigonometric Functions

Every scientific theory is created to solve a certain range of problems. For trigonometry, at its initial stage, such problems are finding distances to objects that are not reachable for direct measurement. These could be widths of rivers, or heights of mountains or buildings. Fortunately, it turned out that it is possible to measure angles to such objects. The main idea of trigonometry is to take advantage of these data. It is done by establishing a relation between angles and distances. Using this relation, it is possible to calculate the needed distances without having to measure them directly. The tools that link angles to distances are trigonometric functions. They are introduced in this chapter while solving practical problems below.

If you are curious, the term *trigonometry* consists of two parts: *trigono* and *metry*. These words come from the Ancient Greek *trigonon* and *metron*. The word *trigonon*, in turn, can be divided into *tri + gonon* (meaning three + angles), so *trigonon* is a triangle. The word *metron* means a measure and the entire term *trigonometry* literally means the measuring (sides and angles) in a triangle.

From this origin of the word, it may seem that trigonometry arose from the need of measuring plots of land or heights of different objects on Earth. However, trigonometry was born, one might say, not on Earth, but in the skies: it arose from the need of astronomy by ancient Egyptians and Babylonians at around 3000 BC in determining the distances to celestial bodies. And only later did trigonometry "descend to the ground," and was

1

used in geodesy and architecture, such as in the construction of Egyptian pyramids. We will say more about the history of trigonometry throughout this book.

Following the initial definition of trigonometry as the science of calculating sides and angles in triangles, we'll start our discussion with a problem of measuring the height of a tree. In doing this, we will rediscover trigonometric functions and their applications. Let's begin.

Problem 1.1. On a beautiful sunny day, we were asked to measure the height of our apple tree. We don't have any special tools other than a simple tape measure, and we're not going to climb the tree. What can we do? Before reading the solution, try to invent your own method.

Solution. In the given conditions, the tree casts a shadow, and we can measure it. Let's denote the tree shadow as D and the tree's height as H.[1] Using a tape measure, we determine that $D = 9.43$ m. Then we stick a small vertical wand next to the tree, and measure its length h and shadow d: $h = 7$ cm, $d = 8.2$ cm. We can draw the following figure:

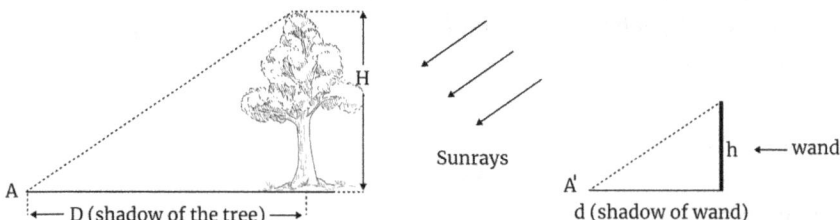

Since sunrays are parallel, $\angle A = \angle A'$.

To use these data for finding the tree's height H, we take advantage of the concept of **similarity**. Informally speaking, two figures are similar if they have the same shape but may have different sizes. In other words, figures are similar if one is a reduced or an enlarged copy of the other. Think of similar figures as though you are looking at them through a magnifying or a diminishing glass. What you'll see through the glass is an object that is similar to the original one. The main property of similar figures is that

---

[1] Throughout the book, we will use the same notation when referring to line segments as well as their lengths. The same idea will apply to angles and their values.

their corresponding sides are proportional. It means that if, for example, one side of the first figure is 3 times larger than the corresponding side of the second figure, then all other sides from the first figure will also be 3 times larger. In contrast to the sides, the angle values remain unchanged when using a magnifying or diminishing glass.

If two figures are triangles, they are similar if their corresponding angles are equal. Look at these triangles:

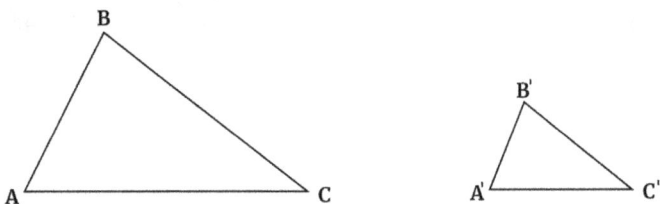

Here, the angles in individual triangles can be different, but angles A, B, and C from the bigger triangle are equal to the corresponding angles A', B' and C' from the smaller one: $A = A'$, $B = B'$ and $C = C'$, so these triangles are similar.

For triangles to be similar, it is enough that only two angles from one triangle are equal to the corresponding two angles from the other. In this case the third angles will be equal automatically, because the sum of all three angles in any triangle is 180°. In particular, two right triangles are similar if they have just one equal acute angle.[2]

In our problem about the height of the tree, two right triangles drawn for the tree and the wand are similar because angles A and A' are equal. Therefore, we can use the proportionality property of similarity: $\frac{H}{h} = \frac{D}{d}$. From here, $H = \frac{D \cdot h}{d}$. Substituting values $h = 7$ cm, $d = 8.2$ cm, and $D = 9.43$ m into the last formula, we get:

$$H = \frac{9.43 \times 7}{8.2} = 8.05 \, \text{m}.$$

---

[2] Strictly speaking, instead of "equal" we should use the term "congruent." The term "equal angles" means that we are dealing with the same angle, while the term "congruent angles" means that angles may belong to different figures (or different parts of the figure), but these angles have the same measurement. However, here and throughout the text, we will often use the term "equal."

Problem solved: we were able to find the height of the tree to be 8.05 m without climbing it and using only a tape measure. ■

Now consider the case when the weather changes and the problem becomes more complicated.

Problem 1.2. Again, we need to measure the tree's height, but this time the weather is cloudy.

Solution. Since the tree's shadow is not seen, we need to invent another approach. Consider the following figure:

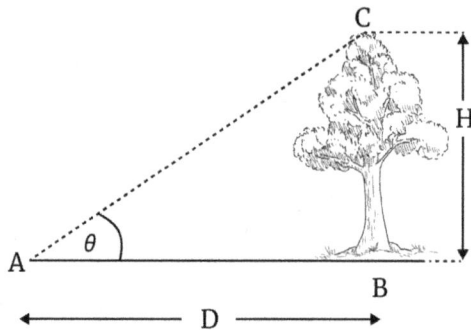

Figure 1.

Assume we are standing at point A on the ground at some distance D from the tree. Distance D is easy to measure, because nothing prevents us from doing it. However, this information is not enough to determine the height H. Let's see what else we can measure. At first glance, it looks like nothing more. This is true if we use a tape measure only, and we don't want to climb the tree.

To solve the problem, we will use one more device. Nowadays it is called a **clinometer**. This is an optical device that can measure angles. It is widely used in industry, in particular in land surveying and mapping, construction, and even in artillery. In ancient times such a device was called an **astrolabe**. Schematically, it looks like a pipe with a protractor attached to it. Astrolabes were used by early navigators in determining the latitude by measuring the angle of the North Star (Polaris) above the horizon:

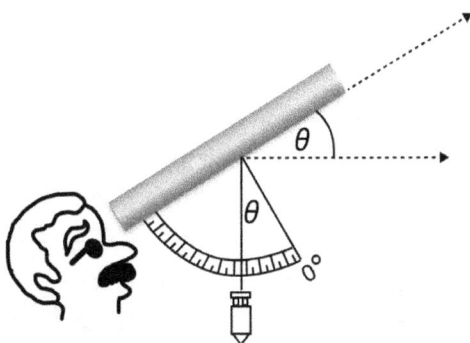

In trigonometry, it is very common to use Greek letters for angles. In the above figure, we used the letter θ (theta, the eighth letter in the Greek alphabet). Let's assume that in addition to a tape measure, we also have a simple astrolabe. Using it, we measure angle CAB from point A to the top point C of the tree. Now we have two measurements: distance D to the tree and the angle $\theta = \angle CAB$. At this point we can again use the concept of similarity of triangles. We draw a right triangle $A'B'C'$ of any convenient size with exactly the same angle θ. Then we measure its legs d and h:

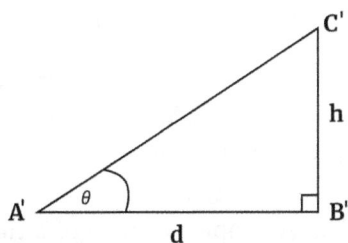

Since $\triangle ABC$ in Figure 1 is similar to $\triangle A'B'C'$, their sides are proportional:

$$\frac{H}{h} = \frac{D}{d}.$$

From here we get a formula for the tree's height H, the same as in Problem 1.1:

$$H = \frac{D \cdot h}{d}. \quad \blacksquare$$

Problem 1.3. We need to measure the heights of all the trees in some grove.

Solution. You may be wondering why we are considering such a problem once again, since it is quite similar to the previous one. Why can't we just use the same method to solve it? Yes, in principle, we can draw a figure with a triangle for each tree, measure its sides h and d from the figure, and calculate the tree's height H by the formula: $H = \dfrac{D \cdot h}{d}$. However, such a procedure in practice requires a lot of effort and time, especially if the number of trees is large. Fortunately, there is a better method. To find it, let's rewrite the above formula as $H = D \cdot \dfrac{h}{d}$ and look carefully at the ratio $\dfrac{h}{d}$.

The most important observation here is that this ratio does not depend on the specific values of h and d. In fact, this ratio depends only on the value of the angle θ. In other words, if we have two right triangles of different sizes, but with an identical angle θ, then the ratio h/d will be the same for both of them. Thus, this ratio is fully determined by the angle θ for **any size** of a triangle.

We get the following relation: for each angle θ there is a corresponding number, which is the ratio h/d. Such a correspondence is called a **function**. In general, a function can be considered as some transformer: it takes a number as an input and calculates (assigns or produces) output. The input number is called the **argument**, and the output number is called the **value** of the function. A simple example of a function is $y = x^2$. This function takes any number x as the argument, squares it, and thereby gets the value of the function y. For example, if the input is the number 3, then the function $y = x^2$ produces the output number $3^2 = 9$.

Our function takes the angle θ as an argument and transforms it into the value h/d. Such a function is given a special name **tangent,** and it is denoted as $\tan \theta$.

Look again at the right triangle with angle θ and legs h and d:

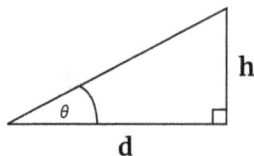

Leg h is called the **opposite** to the angle θ, while leg d is called the **adjacent** one. So, the tangent of angle θ is defined as a ratio of the opposite leg to the adjacent leg:

$$\tan\theta = \frac{\text{opposite leg}}{\text{adjacent leg}} = \frac{h}{d}.$$

Using tangent, we can rewrite the formula for the height of a tree in the form:

$$H = D \tan\theta.$$

But what is the benefit of using the function $\tan\theta$ instead of the h/d ratio? Looking at a function as an independent mathematical object allows us to develop methods for calculating its values for various angles regardless of specific triangles. Having such values of $\tan\theta$ for different angles θ, we can use them easily and quickly in the formula $H = D\tan\theta$ without making any effort to draw triangles and calculate the h/d ratio every time.

The tangent function is one of the so-called **trigonometric functions**. Trigonometry studies the various properties of tangent and some other trigonometric functions (in short, trig functions), develops methods for calculating their values, and applies them to solve numerous problems of nature, science and practical life. Before the era of calculators, special tables of trig functions were built. The work on the creation of such tables was undertaken several thousand years ago. A Greek astronomer and mathematician Hipparchus of Nicaea (190 BC?–127 BC?) created the first known tables of trig functions. As the main mathematical object, he used a chord of a circle as a function of corresponding central angle (or arc). He made calculations of chords at increments 1/48 of a circle that correspond to increments of 7°30′. In the 2nd century AD, an Egyptian astronomer, mathematician and geographer Claudius Ptolemy (100 AD–170 AD) created even more detailed tables by calculating the values of trig functions at 30′ increments. Ptolemy is also known as the author of the geocentric model (Earth-centered) of the universe, the so-called Ptolemaic system (this system turned out to be inadequate). In the next chapters we will create our own tables of trig functions for some specific angles.

In our days, scientific calculators permit easy calculation of trig functions. For example, to get the value of tan 35° (tangent of a 35° angle), you can use the button "tan" on a calculator. The calculator will show that tan 35° ≈ 0.7.

Returning to Problem 1.3, we get the following new method of calculating the height H of a tree. We stand at some distance D from a tree and measure the angle θ to the top of the tree. Then, using a calculator, we obtain the value for tan θ. Finally, we multiply it by D, and get the answer $H = D \tan \theta$. As you see, this method is much simpler than drawing triangles on paper and measuring their sides. ■

As we mentioned, tangent is not the only trig function. The problem below will lead us to the discovery of other trig functions.

Problem 1.4. Trees A, B, and C are separated by cliffs as shown in the following figure:

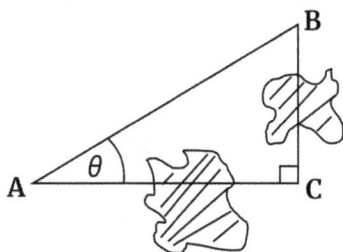

Here shaded areas show cliffs. The distance between A and B is 50 m, and the angle θ is 23°. We need to find the distances BC and AC.

Solution. Let's start with the same method as we used in Problem 1.2. We draw a right $\triangle A'B'C'$ with the same angle θ and any suitable size:

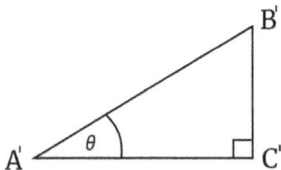

Then we measure $A'B'$, $B'C'$, and $A'C'$. Since $\triangle ABC$ and $\triangle A'B'C$ are similar (they have the same angle θ), we can set up proportions:

$$\frac{AB}{A'B'} = \frac{BC}{B'C'} = \frac{AC}{A'C'}.$$

From here,

$$BC = AB \cdot \frac{B'C'}{A'B'} \text{ and } AC = AB \cdot \frac{A'C'}{A'B'}.$$

Formally speaking, we found formulas for distances BC and AC, and can stop here. However, each time we use them, we would have to draw a similar triangle and measure its sides. To avoid such procedures, we can use the fact that the ratios $B'C'/A'B'$ and $A'C'/A'B'$ depend only on angle θ regardless of the individual sizes $B'C'$, $A'B'$, and $A'C'$. Therefore, we can interpret both ratios $B'C'/A'B'$ and $A'C'/A'B'$ as two new functions of θ, which create a correspondence between argument θ and the values $B'C'/A'B'$ and $A'C'/A'B'$. A function that produces the ratio $B'C'/A'B'$ (the ratio of the opposite leg to the hypotenuse) is called **sine** and is denoted as $\sin \theta$.

A function that produces the ratio $A'C'/A'B'$ (the ratio of the adjacent leg to the hypotenuse) is called **cosine** and is denoted as $\cos \theta$. Using these functions, we can rewrite formulas for BC and AC in terms of sine and cosine:

$$BC = AB \sin \theta, AC = AB \cos \theta.$$

Scientific calculators have buttons "sin" and "cos" to get values of these functions. Let's use them for a given angle $\theta = 23°$: $\sin 23° \approx 0.39$, $\cos 23° \approx 0.92$. From here we get the final answer:

$$BC = 50 \sin 23° = 19.54 \text{ m}, AC = 50 \cos 23° = 46.03 \text{ m}. \blacksquare$$

The term "sine" comes from a series of language transformations (and mistranslations). It is believed that this word originated from the word "half-chord" in Sanskrit, a language of ancient India. In the figure below, you can see a unit circle (a circle with a radius equal to 1), a central angle θ, and a chord AB:

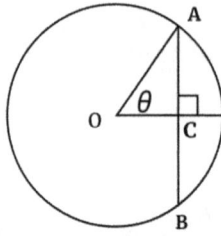

The sine of angle θ is AC/OA. Since the radius OA is 1, the sine of angle θ is AC, which is the half of chord AB. The word "half-chord" was later incorrectly translated into Arabic as "breast." This in turn was interpreted as a "curve." The last word was translated into Latin as "sinus." The abbreviations sin, cos and tan were first used by a French mathematician Albert Girard (1595–1632).

Problem 1.5. Prove that the area of a triangle (not necessarily a right one) is equal to $\frac{1}{2}ab\sin\theta$, where angle θ is between sides a and b.

Solution. Let h be a height to the base b, and S is the area of a triangle:

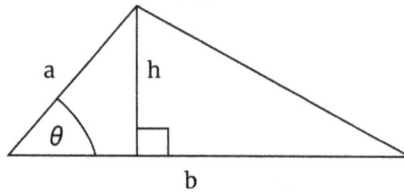

We know from geometry that $S = \frac{1}{2}bh$. By the definition of sine, $\sin\theta = \frac{h}{a}$. From here, $h = a\sin\theta$. Substituting the last expression into the formula for S, we get

$$S = \frac{1}{2}ab\sin\theta. \tag{1}$$

Note. We considered the case when the angle θ is acute. Let's take a look at an obtuse angle

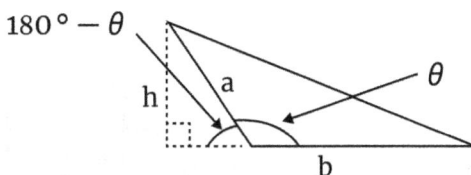

Here, $h = a \sin (180° - \theta)$ and $S = \dfrac{1}{2} ab \sin (180° - \theta)$. Later, in Chapter 5, after we define trig functions for arbitrary angles, we will show that $\sin (180° - \theta) = \sin \theta$. Thus, formula (1) is universal. For now, we just accept it. ■

As we saw, when solving the above problems, we discovered and used three trig functions: $\tan \theta$, $\sin \theta$, $\cos \theta$. Using notation c for the hypotenuse, a for the leg opposite to the angle θ, and b for the leg adjacent to θ, we can present these functions as ratios:

$$\sin \theta = \frac{a}{c}, \quad \cos \theta = \frac{b}{c}, \quad \tan \theta = \frac{a}{b}.$$

With three sides a, b, and c, you can create exactly six combinations of ratios: a/b, a/c, b/c, b/a, c/a, and c/b. Each of them depends only on angle θ and, therefore, represents a certain function. The first three ratios we already used. The three remaining, b/a, c/a, and c/b correspond to the trig functions with names **cotangent**, **cosecant**, and **secant**. The following notations are used:

$$\cot \theta = \frac{b}{a}, \quad \csc \theta = \frac{c}{a}, \quad \sec \theta = \frac{c}{b}.$$

So, the entire set of trig functions consists of **six functions**, which are listed in the table below:

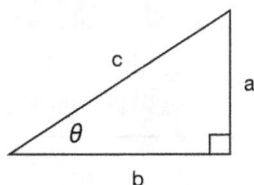

Function Name	Function Notation	Definition
Sine	$\sin\theta$	$\dfrac{a}{c}$
Cosine	$\cos\theta$	$\dfrac{b}{c}$
Tangent	$\tan\theta$	$\dfrac{a}{b}$
Cotangent	$\cot\theta$	$\dfrac{b}{a}$
Secant	$\sec\theta$	$\dfrac{c}{b}$
Cosecant	$\csc\theta$	$\dfrac{c}{a}$

Functions sine, cosine, and tangent are the most frequently used trig functions. They are called **basic** trig functions. Let's present verbal definitions of the basic functions:

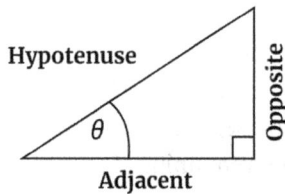

Sine definition

Sine of angle θ is the ratio of the opposite leg to the hypotenuse:
$$\sin\theta = \frac{\text{opposite}}{\text{hypotenuse}}$$

Cosine definition

Cosine of angle θ is the ratio of the adjacent leg to the hypotenuse:
$$\cos\theta = \frac{\text{adjacent}}{\text{hypotenuse}}$$

Tangent definition

Tangent of angle θ is the ratio of the opposite leg to the adjacent one:

$$\tan \theta = \frac{\text{opposite}}{\text{adjacent}}$$

Since these trig functions are so fundamental and we will actively use them throughout this book, we recommend that you pause here for a few minutes and try to memorize the definitions. A widely used "mnemonic device" (a word or a phrase that makes memorizing information easier) is the word "SohCahToa." Here's how it helps:

Soh ... Sine = Opposite/Hypotenuse
Cah ... Cosine = Adjacent/Hypotenuse
Toa ... Tangent = Opposite/Adjacent

Exercises for Solving on Your Own

Exercise 1.0. Both figures in the picture below consist of the same four components. However, the bottom figure contains an empty cell. How can this be?

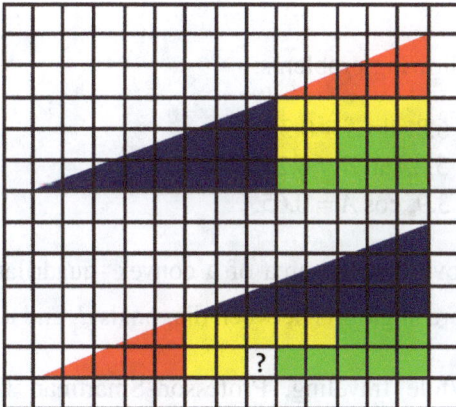

Exercise 1.1. Michelle and Nick are walking on the beach. The length of Michelle's shadow is 108 cm, and Nick's is 104 cm. Michelle is 161 cm tall. How tall is Nick?

Exercise 1.2. For the right triangle *KLM*,

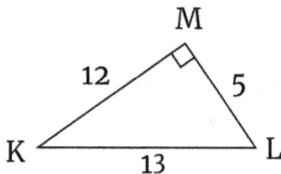

find: (a) sin *K*, (b) cos *K*, (c) tan *L*.

Exercise 1.3. Let *ABC* be a right triangle with hypotenuse *c* and legs *a* and *b*:

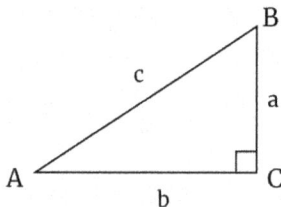

Solve the following problems:

(a) Find *a*, if $c = 6.8$, sin $A = 0.35$.
(b) Find *b*, if $a = 5.2$, tan $B = 6.25$.
(c) Find *c*, if $b = 3.9$, cos $A = 0.65$.

Exercise 1.4. Prove that the area of a convex[3] quadrilateral is equal to $\frac{1}{2} d_1 d_2 \sin \theta$, where angle θ is between diagonals d_1 and d_2.

Exercise 1.5. While traveling, Professor Smartman finds himself at the shore of a river he needs to cross to get to the other side. The professor

---

[3] A plane figure is called convex if the line segment connecting any two points of the figure lies inside the figure.

can swim, but since the river's current is so strong, he will get carried away. Thankfully, he has a crossbow with an arrow to which a rope can be tied. On the opposite shore a sturdy tree grows near the water. The professor can shoot an arrow at this tree and thus fasten the rope to the other side. Then, holding on to the rope, he will be able to swim across the river. The problem here is to determine if his rope is long enough — or in other words, measure the distance to the tree. As an experienced traveler, Professor Smartman carries with him a tape measure, a device for measuring angles, as well as a scientific calculator. How can he measure the distance to the tree?

Before you continue reading, try to come up with your own idea.

The professor did the following (for numerical calculations, he used the data below): First, he stood close to the water and measured the angle to the top of the tree. Let the angle be 43°. Then he moved a little off the shore (say, 10 m), and measured the angle again. Let this angle be 32°. Using these data, determine if 30 m of the rope is sufficient to reach the tree (the distance to the tree is equal to the width of the river).

Entertainment Problem

Problem E1.1. Eli purchased a construction set, which contained 100 sticks of different lengths. The instruction on the box mentions that you can make a triangle from any three sticks. Eli decided to check this statement by making some triangles. The sticks are in the ascending order of length. What is the smallest number of checks Eli needs to make to prove or disprove the statement on the box?

Chapter 2

Solving Right Triangles

In the previous chapter, we introduced trigonometric functions. In this chapter, we consider applications of these functions as well as the Pythagorean theorem to "solve the triangles." Solving a triangle means to find certain triangle elements (like sides and angles) using known elements (at least one of them should be a side). Such problems may appear not only in math, physics, astronomy and other scientific subjects, but also in our daily life. Here we consider only right triangles, leaving arbitrary triangles to Chapters 6 and 7.

Since we will use the Pythagorean theorem many times, and it is the most important result in geometry, we present here one of its simplest proofs. This theorem asserts that in any right triangle with legs a and b, and hypotenuse c, the following relation holds:

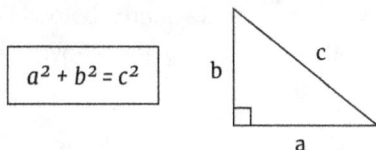

$$a^2 + b^2 = c^2$$

In words: the sum of the squares of two legs in a right triangle is equal to the square of the hypotenuse.

The fact that the hypotenuse depends on the legs is quite obvious, because if we construct two legs,

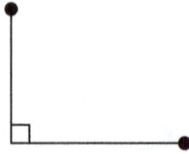

then we know both endpoints of the hypotenuse, and we can measure it. Therefore, two legs completely determine the hypotenuse. Thus, the real problem boils down to finding the **formula** for this dependency. This formula is the Pythagorean theorem described above. It has actually been discovered in many cultures even before Pythagoras (c. 570–c. 495 BC, Greek philosopher and mathematician). The Babylonians knew this result 1000 years earlier, and some informal proofs were found in China and India, also many years before Pythagoras. The formal proof has been found in the Pythagorean society. It is unknown whether this proof belongs to Pythagoras himself or one of his followers. This theorem has so many different proofs that it is even listed in the Guinness Book of Records as having the greatest number of proofs (there are more than 400).

Problem 2.1. Prove the Pythagorean Theorem.

Solution. Look at this triangle

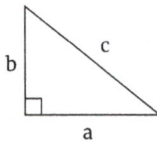

and at the figures of two large equal squares below. Both squares contain four white triangles that are equal to the above triangle, and colored squares with areas a^2, b^2, and c^2.

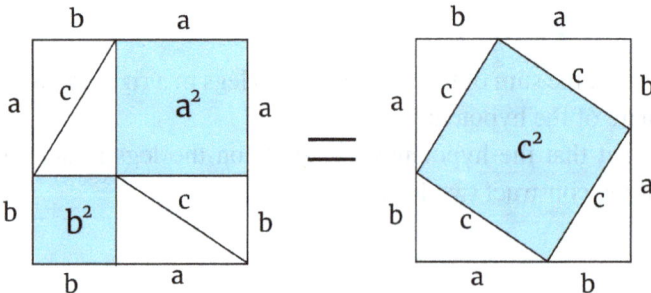

If we remove four white triangles from the square on the left, we will be left with $a^2 + b^2$. If we remove the same four triangles from the square on the right, c^2 will remain. Therefore, $a^2 + b^2 = c^2$, which is the Pythagorean theorem. ■

Note that the Pythagorean theorem does not state that if the sides a, b, and c of the triangle satisfy the equation $a^2 + b^2 = c^2$ then this triangle is a right one. It claims the opposite statement: if it is a right triangle, then $a^2 + b^2 = c^2$. However, the converse of the Pythagorean theorem is also true.

Problem 2.2. Let sides a, b, and c in a triangle satisfy the condition: $a^2 + b^2 = c^2$. Prove that this is a right triangle.

Solution. We will use the following result from geometry: if three sides of one triangle are equal to the corresponding three sides of another triangle, then the two triangles are equal (more formally, congruent). Let's denote the given triangle as T. Now, we construct a right triangle R with legs a and b. By the Pythagorean theorem, the hypotenuse of triangle R is $\sqrt{a^2 + b^2}$ which is the same as the third side of triangle T. Therefore, both triangles are equal, and T is a right triangle. ■

In Chapter 7, when we develop more theory, we'll show how to recognize by three given sides what type of triangle it is: an acute, right or obtuse triangle.

Among the right-angled triangles, one of the most famous is the so-called Egyptian triangle. The ratio of its sides is elegantly formed by whole numbers 3:4:5. It is believed that its name was given by Pythagoras when visiting Egypt in 535 BC, where he saw the pyramids. He was especially impressed by the pyramids of Khufu and Khafre because of their ideal proportions 3:4:5[1]:

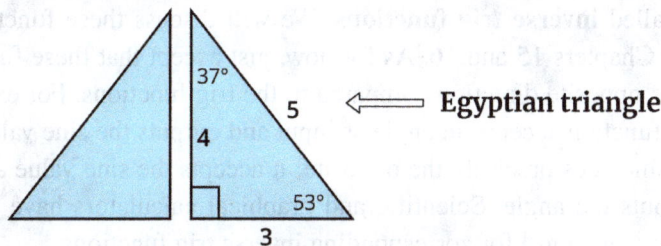

37°
5 ⟸ **Egyptian triangle**
4
53°
3

---

[1] However, there is no convincing evidence that Egyptians used the 3-4-5 triangle in building the pyramids.

Perhaps Pythagoras tried to generalize the relation $3^2 + 4^2 = 5^2$ to any right triangle, which led to the proof of his renowned theorem.

The pyramid of Khufu is also known as the pyramid of Cheops (in Greek) or the Great Pyramid of Giza. It is the oldest and largest of the three pyramids Khufu, Khafre, and Menkaure. All three are located on a plateau on the west bank of the Nile River, bordering modern-day Giza in Greater Cairo, Egypt. These pyramids were built as royal tombs for three Egyptian pharaohs Khufu, his son Khafre, and Khafre's son Menkaure.

The Great Pyramid of Giza is listed as the oldest structure in the famous "The Seven Wonders of the Ancient World" — a list of the most remarkable constructions in the world. It is the only one from the list that still exists. The pyramid has a base of about 230 m and the original height of about 147 m. It was built around 2532 B.C. and remained the tallest man-made structure in the world until the 14th century (for about 3,800 years). It took over 20 years to build and required the labor of about 20,000 people.

It is widely believed that slaves built the pyramids, but archaeological evidence suggests that the workers were probably native Egyptian agricultural laborers. They cut approximately 2.3 million blocks of stone, each weighing an average of 2.5 tons. Then they transported the stones and assembled the pyramid. The total weight of the pyramid is about 5.75-million-tons. It was a masterpiece of technical skill and engineering ability. No one knows with certainty how they did it.

Also, there is no exact answer to the question of why the ancient Egyptians chose the pyramid as a monument. One hypothesis is that it reflects the rays of the sun. It is interesting that the pyramid of Cheops is perfectly aligned with the constellation Orion.

Below we will calculate angles in the Egyptian triangle. To find angles in a triangle from the known value of the trig functions, we can use the so-called **inverse trig functions**. We will discuss these functions in detail in Chapters 15 and 16. As for now, just accept that these functions act in the opposite direction compared to the trig functions. For example, the sine function accepts an angle as input and outputs the sine value. The inverse sine does precisely the opposite: it accepts the sine value as input and outputs the angle. Scientific and graphical calculators have buttons $\sin^{-1}$, $\cos^{-1}$, and $\tan^{-1}$ for corresponding inverse trig functions.

Problem 2.3. Let *ABC* be an Egyptian triangle.

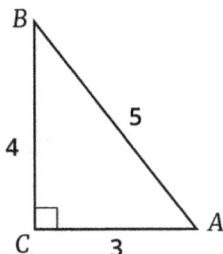

Calculate angles *A* and *B*. Round the answers to the nearest degrees.

Solution. We can use any of the basic trig functions: sine, cosine or tangent, since for all of them the inverse trig functions are available on the calculator. Let's use the sine function to find angle *A*. We have

$\sin A = \dfrac{4}{5} \Rightarrow A = \sin^{-1}\left(\dfrac{4}{5}\right) \approx 53°$. This is the angle that pleases our eyes,

when looking at the tops of Cheops and Khafre pyramids. Angle *B* is a complement to angle *A*, so $B = 90° - 53° = 37°$. ∎

    In the Appendix at the end of the book, we will look at some other remarkable triangles.

    Before we move on to the applications of the trig functions, let's explicitly define the angles of elevation and depression.

    The angle from the horizontal line going up is called the angle of **elevation**. For example, if you are looking up to the top of a building you are looking at the elevation angle. Similarly, the angle from the horizontal line going down is called the angle of **depression**. For example, looking down from the top of a building to the ground, you are looking at the depression angle:

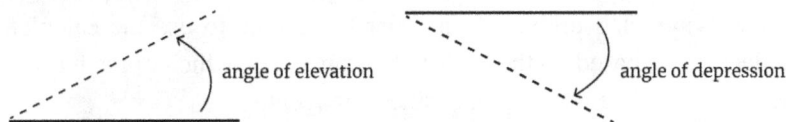

    In the figure below angle θ_1 is the angle of depression from point *A* and angle θ_2 is the angle of elevation for the point *B*:

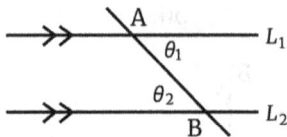

If lines L_1 and L_2 are parallel, then angles θ_1 and θ_2 are equal (as they are alternate interior angles). Therefore, we can always replace the angle of depression with the angle of elevation.

Now consider, as an example of using the Pythagorean theorem, the following typical "household" problem.

Problem 2.4. A ladder rests against a vertical wall and should stand 1 m away from the wall. The length of the ladder is 3 m. How high can the upper end of the ladder reach?

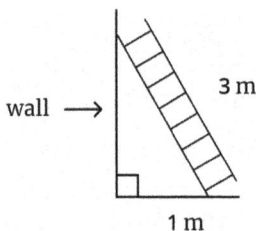

Solution. In terms of geometry, we have a right triangle with a horizontal leg $a = 1$ m and a hypotenuse $c = 3$ m. We need to find the vertical leg b. The problem is easily solved using the Pythagorean theorem: $c^2 = a^2 + b^2$. From here

$$b^2 = c^2 - a^2 \Rightarrow b = \sqrt{c^2 - a^2} = \sqrt{3^2 - 1^2} = \sqrt{8} \approx 2.83(\text{m}). \blacksquare$$

Now we consider problems that require the use of trigonometry. When solving such problems, in order to be able to use the calculator, first determine, based on the information provided, which of the basic trig functions should be used: sine, cosine or tangent.

Problem 2.5. A 3-m ladder leans against a vertical wall and, for safety reasons, should have steepness (angle of elevation) of 50° from the floor. How far from the wall should we set the lower end of the ladder to fulfill this requirement?

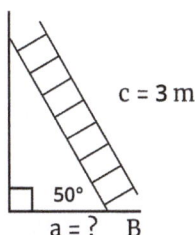

$$c = 3\text{ m}$$

Solution. We have a right triangle here. The hypotenuse denoted as c is 3 m, and the angle adjacent to the horizontal leg a denoted as B is 50°. The problem is to find the horizontal leg a. A suitable trig function is cosine of angle B (ratio of the adjacent side to the hypotenuse). Using the definition of $\cos B$, we have

$$\cos B = a/c \implies \cos 50° = a/3 \implies a = 3 \cdot \cos 50° = 3 \cdot 0.643 = 1.93 \text{ (m)}. \blacksquare$$

Problem 2.6. A giraffe's shadow is 8 m. How tall is the giraffe if the sun is 28° to the horizon?

Solution. Consider this figure:

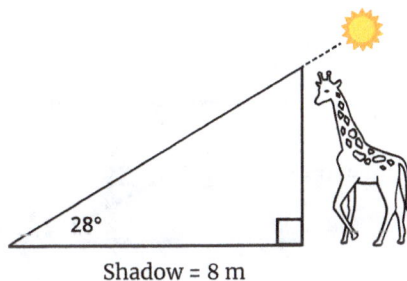

Shadow = 8 m

For the 28° angle, giraffe itself is the opposite side, and the giraffe's shadow is the adjacent side. A suitable trig function is tangent (ratio of the opposite side to the adjacent). Let's denote the giraffe's shadow as s and the giraffe's height as g. We have $\tan 28° = \dfrac{g}{s}$. From here,

$$g = s \tan 28° = 8 \cdot 0.5317 = 4.25 \text{ (m)}. \blacksquare$$

Problem 2.7. Nick launched a kite on a 120-foot thread. The angle of elevation of the thread is 40°. At what altitude is the kite flying?

Solution. Look at this figure:

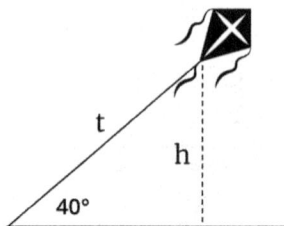

Let's denote the length of the thread as t. It is the hypotenuse and $t = 120$. The problem is to find the height h which is opposite to the 40° angle. A suitable trig function is sine (ratio of the opposite side to the hypotenuse). We have $\sin 40° = \dfrac{h}{t}$. From here,

$$h = t \sin 40° = 120 \cdot 0.6428 = 77.13 \text{ (ft)}. \blacksquare$$

Problem 2.8. Lilian is swimming in the sea and notices a coral reef at the sea bottom. The angle of depression from Lilian to the reef is 37° and the depth of the sea here is 50 feet. How far is she from the reef?

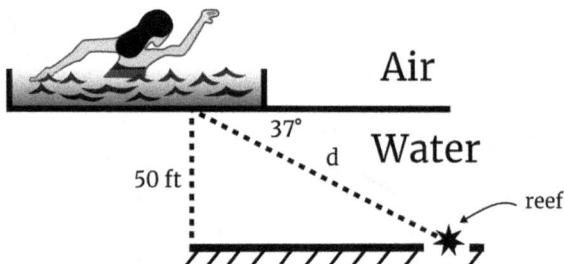

Solution. The angle of depression from Lilian is equal to the angle of elevation from the reef, which is 37°. In the above figure you can see a right triangle with a vertical leg of 50 feet and the opposite angle of 37°. The problem is to find the hypotenuse which we denote as d. A suitable trig function is sine (ratio of the opposite side to the hypotenuse). We have

$$\sin 37° = \frac{50}{d} \Rightarrow d = \frac{50}{\sin 37°} = 83.1 \text{ (ft)}. \blacksquare$$

Problem 2.9. A ship is 160 m away from the center of a horizontal barrier that measures 200 m from end to end. What is the minimum angle that the ship must turn to avoid hitting the barrier?

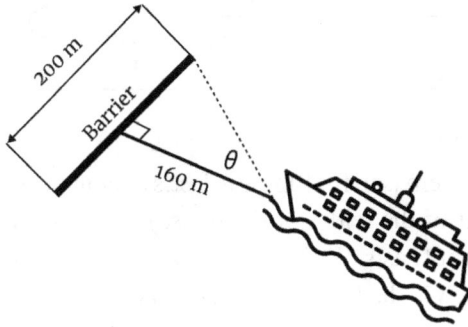

Solution. The problem is to find angle θ. Let's denote half of the length of the barrier as a. Then $a = 200/2 = 100$ m. In the figure, you can see a right triangle in which side a is opposite to angle θ. Another side is the distance from the ship to the barrier. We denote it as d. It is given that $d = 160$ m. This side is adjacent to angle θ. A suitable trig function here is tangent (ratio of the opposite side to the adjacent).

$$\tan \theta = \frac{a}{d} = \frac{100}{160} = 0.625.$$

To find angle θ from the tangent, we use the $\tan^{-1}$ button on the calculator:

$$\theta = \tan^{-1}(0.625) = 32°. \blacksquare$$

Problem 2.10. Michelle wants to shingle her roof. The roofer asked her for the angle of elevation of the roof to make sure he can climb it safely. Calculate the angle according to this figure:

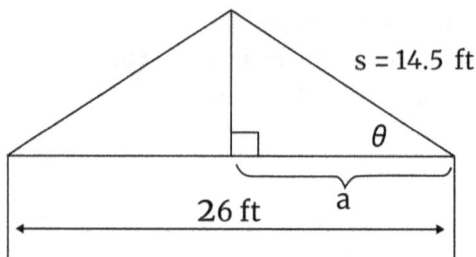

Solution. The problem is to find angle θ. Let's denote the marked horizontal line segment (half of the width of the house) as a, and slant line segment (the width of the roof) as s. We have $a = 26/2 = 13$ feet and $s = 14.5$ feet. In the figure, the side s is the hypotenuse and a is the side adjacent to angle θ. A suitable trig function here is cosine (ratio of the adjacent side to the hypotenuse).

$$\cos\theta = \frac{a}{s} = \frac{13}{14.5} = 0.9.$$

To find angle θ from the cosine, we use the $\cos^{-1}$ button on the calculator:

$$\theta = \cos^{-1}(0.9) = 26°. \blacksquare$$

Exercises for Solving on Your Own

Exercise 2.1. Let ABC be a right triangle with hypotenuse c and legs a and b:

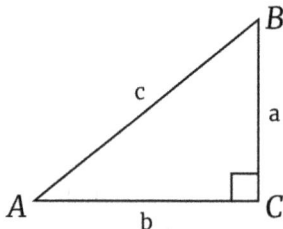

Solve the following problems (round the answer to the nearest hundredth):

(a) Find angle A, if $a = 4$, $b = 9$.

(b) Find angle B, if $b = 7$, $c = 8$.

(c) Find angle A, if $b = 6$, $c = 11$.

Exercise 2.2. Three integer numbers are called the **Pythagorean triple** if they can be the lengths of the sides in a right triangle. Let $a = m^2 - n^2$, $b = 2mn$, $c = m^2 + n^2$, where m and n are any integers such that $m > n > 0$. Prove that the triple (a, b, c) is a Pythagorean triple.

Exercise 2.3. In a right triangle, let h be the altitude to the hypotenuse, and m and n be altitudes to the legs. Prove the following "Reciprocal Pythagorean Theorem":

$$\frac{1}{m^2} + \frac{1}{n^2} = \frac{1}{h^2}.$$

Exercise 2.4. A triangle (not necessarily a right one) with integer sides, has a length of one side equal to 5 and the other to 1. What is the length of the third side?

Exercise 2.5. The angle of elevation from the airport to the airplane is $17°$. The distance from the plane to the airport is 25 miles. How high is the plane flying?

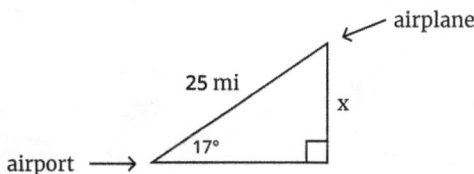

Exercise 2.6. A 20-foot ladder rests against a wall. Its angle of elevation from the ground is $55°$. How far from the wall is the base of the ladder?

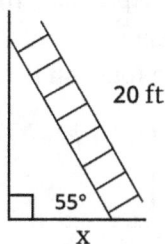

Exercise 2.7. Allison is looking at the top of a tall building. Her eyes are 5 feet above the ground. The angle of elevation is 75° and she is 15 feet from the building. How tall is the building?

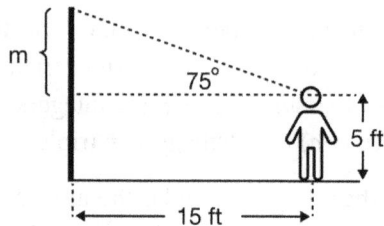

Exercise 2.8. A ladder is leaning against the wall such that the angle of depression from the top of the ladder is 56°. What is the length of the ladder if the distance from its lower end to the wall is 2 m?

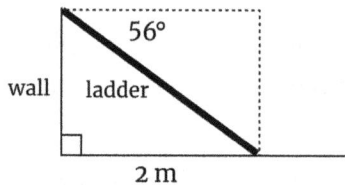

Exercise 2.9. A radar station detected an airplane at an altitude of 5.3 km. The angle of elevation from the radar station to the airplane is 35°. How far is the airplane from the radar station?

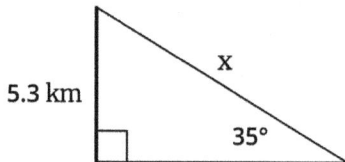

Exercise 2.10. The angle of depression from the top of an apartment building to the base of a fountain in a nearby park is 70°. If the building is 80 feet tall, how far away is the fountain from the building?

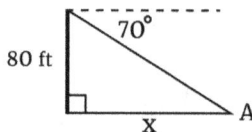

Exercise 2.11. A vertical pole stands on the ground and has a support wire that runs from its top to the ground. The support is 50 feet long and anchored 22 feet from the base of the pole. Find the angle of elevation from the anchor to the top of the pole.

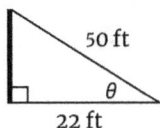

Exercise 2.12. An airplane is flying at an altitude of 2.5 miles and is preparing for landing. It is 8.6 miles from the runway. Find the angle of depression θ that the airplane must make to reach airport.

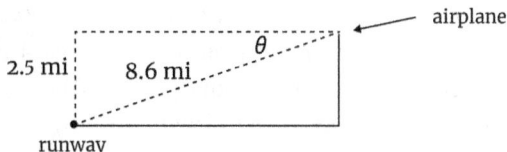

Exercise 2.13. At what angle should a telescope be directed to the top of a 40 m high tower from a distance of 15 m?

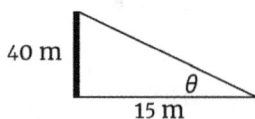

Exercise 2.14. Esther and Nick stand at points E and N, 2 m apart in a dark room with a large mirror. Esther stands 2 m from the mirror, and Nick stands 1 m. At what angle EBA should Esther shine a flashlight at the mirror so that the reflected light directly strikes Nick?

Note. According to the law of reflection, $\angle EBA = \angle NBC$.

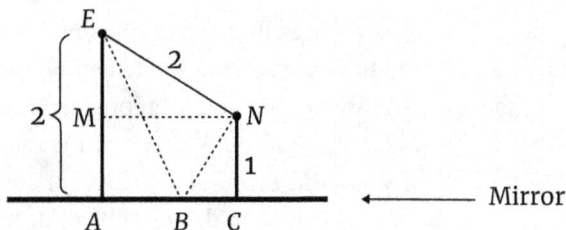

Exercise 2.15. Tom was solving a problem that required finding the area of a right triangle. After making a few steps in the solution, he calculated that the hypotenuse is 10 cm, and the altitude to it is 6 cm. Then Tom multiplied 10 by 6 and divided by 2. He got that the area of the triangle is 30 cm². Checking the answer in the textbook, he found that his answer was incorrect. What is wrong with his solution?

Exercise 2.16. A rectangular-shaped island is surrounded by a channel filled with water. Professor Smartman is standing on the opposite side of the channel and needs to get to the island. The channel width is the same everywhere and is equal to 2 m (see a fragment of the channel in the figure below). The professor has two metal boards which cannot float, each 2 m long. Using these boards, he is trying to get to the island by standing on them. The problem, however, is that in order for the boards to be held at the edges of the channel, they must protrude at least 4 cm beyond the edges. There is nothing to tie or bind them together (there are no tools for that). Also, the professor cannot enter the water or jump over the channel. How can he get to the island using these boards?

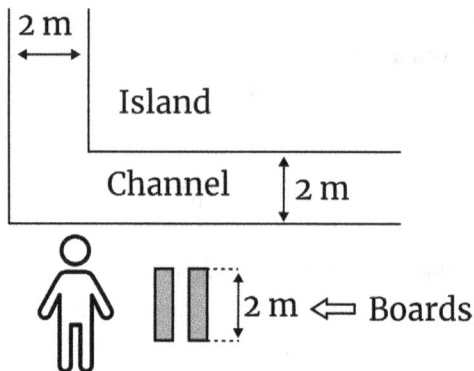

Exercise 2.17. Professor Smartman regularly rides a bus with his 7.5-foot fishing rod. Once the bus driver did not allow him to board due to a new rule: you may not carry objects longer than 6 feet (plus/minus 2 inch). The frustrated professor returned home. And then it dawned on him! The next day he, along with the same fishing rod, again approached the same bus (with the same driver) and was allowed on the bus without any objections (rules remained the same). What did Professor Smartman come up with? Note that the fishing rod may not be bent, disassembled, or hidden under clothing.

Entertainment Problems

Problem E2.1. Professor Smartman rode on this bus. Which direction (left or right) does the bus move?

Problem E2.2. Consider this two-player game: We have a table in the shape of a square. Taking turns, players place 25 cent coins anywhere on the table. Coins must be placed such that they lie entirely on the table (i.e., do not hang down). They may touch but should not overlap each other. Also, coins that have already been placed may not be moved. The player who no longer has a free spot to set his/her coin will lose the game. Is there an optimal strategy to play this game? If so, who will win — the player who goes first or second?

Chapter 3

Trigonometric Properties in Right Triangles

In this chapter, we will develop some formulas for trig functions in a right triangle and establish relationships between them. As a result, we'll express some functions through others. Also, we'll find the values of trig functions for some special angles.

Basic Formulas

Problem 3.1. Express functions cotangent, secant, and cosecant through the basic functions sine, cosine, and tangent.

Solution. In Chapter 1, we defined trig functions as ratios of sides in a right triangle:

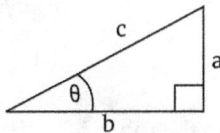

Figure 1.

Recall these ratios:

$$\sin \theta = a/c, \ \cos \theta = b/c, \ \tan \theta = a/b,$$

$$\csc \theta = c/a, \ \sec \theta = c/b, \ \cot \theta = b/a.$$

33

Note that the last three ratios are reciprocals to the first ones. Therefore, the last three functions are reciprocals to the basic functions: cotangent is reciprocal to tangent, secant is reciprocal to cosine, and cosecant is reciprocal to sine:

$$\cot \theta = \frac{1}{\tan \theta}, \quad \sec \theta = \frac{1}{\cos \theta}, \quad \csc \theta = \frac{1}{\sin \theta}. \quad \blacksquare$$

Problem 3.2. Prove that $\tan \theta = \sin \theta / \cos \theta$.

Solution. $\sin \theta / \cos \theta = (a/c)/(b/c) = a/b = \tan \theta. \quad \blacksquare$

This formula allows us to calculate tangent via sine and cosine. Solve problems 3.3 and 3.4 below on your own.

Problem 3.3. Prove that $\cot \theta = \cos \theta / \sin \theta. \quad \blacksquare$

Problem 3.4. Let hypotenuse c be equal to 1 (see Figure 1). Prove that $\sin \theta$ is equal to leg a, and $\cos \theta$ is equal to leg b. $\quad \blacksquare$

We can treat this statement as a geometrical interpretation of sine and cosine: they are the lengths of legs in a right triangle with a hypotenuse equal to 1.

Problem 3.5. Prove the following identity:

$$\boxed{\sin^2 \theta + \cos^2 \theta = 1}$$

Solution.

$$\sin^2 \theta + \cos^2 \theta = \frac{a^2}{c^2} + \frac{b^2}{c^2} = \frac{a^2 + b^2}{c^2} = \frac{c^2}{c^2} = 1. \quad \blacksquare$$

We will call this formula the **main identity** (or Pythagorean identity). It connects sine and cosine: knowing sine, we can find cosine, and vice versa. Namely,

$$\sin \theta = \sqrt{1 - \cos^2 \theta}, \quad \cos \theta = \sqrt{1 - \sin^2 \theta}.$$

Using the main identity, we can calculate tangent if we know, for instance, only sine.

Problem 3.6. Express tangent through sine.

Solution. Since $\tan \theta = \sin \theta / \cos \theta$ and $\cos \theta = \sqrt{1 - \sin^2 \theta}$, then

$$\tan \theta = \sin \theta / \sqrt{1 - \sin^2 \theta}. \blacksquare$$

Solve the following problem on your own.

Problem 3.7. Express tangent through cosine. $\blacksquare$

Two Special Right Triangles and Five Special Angles

Now, we are going to define the most commonly used so-called **special angles**.

When posing a question about what special angles could be in right triangles, the answer may depend on the opinion of different people. For example, we can say that angle 10° is a special angle, because we like the number 10. However, such an answer is subjective and artificial. Let's try to use more objective considerations that are related to mathematics. For example, it seems natural to consider the 90° angle as a special one, because it is a right angle and is often used in science as well as in everyday life.

To find out other special angles, let's turn our attention to such nice figures as regular polygons. These are polygons in which all angles are equal, and all sides are equal. The simplest polygons are an equilateral triangle (a regular polygon with three sides) and a square (a regular polygon with four sides). To extract right triangles from these figures, let's cut each of them in half according to the figures

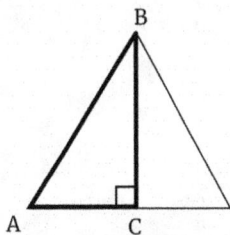

Δ ABC is a half
of an equilateral triangle

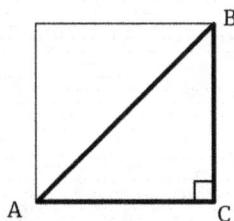

Δ ABC is a half
of a square

In the triangle *ABC* on the left, the acute angles are 30° and 60°. We call such a triangle a 30° – 60° triangle. In the triangle *ABC* on the right, both acute angles are 45°. We call such a triangle a 45° – 45° triangle. Both triangles, 30° – 60° and 45° – 45°, are called **special right triangles**, and angles 30°, 45° and 60° are called **special angles**. Let's calculate trig functions for these angles. We also consider the angles of 0° and 90°. We restrict ourselves only to the basic trig functions sine, cosine, and tangent, since the other three functions are easily expressed through them.

30° – 60° Triangle

Let's draw this triangle like this

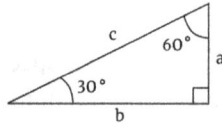

Recall that the hypotenuse *c* is the side of the equilateral triangle drawn above, so the leg *a* is half of the hypotenuse *c*: $a = \dfrac{c}{2}$ or $c = 2a$.

In any 30° – 60° triangle, the leg opposite to 30° is half of the hypotenuse (or the hypotenuse is twice as long as this leg).

From here we can get the expression for the side *b* (which is opposite to the 60° angle) through *a*. By the Pythagorean theorem,

$$b = \sqrt{c^2 - a^2} = \sqrt{(2a)^2 - a^2} = \sqrt{4a^2 - a^2} = \sqrt{3a^2} = a\sqrt{3}.$$

We get the following figure

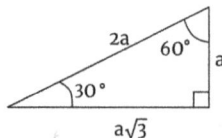

Solve two problems below on your own. In both, a 30° – 60° triangle is given. Denote by *c* the hypotenuse, and by *b* the leg opposite to 60°.

Problem 3.8. Prove that $b = c\sqrt{3}/2$. ∎

Problem 3.9. Prove that $c = 2b\sqrt{3}/3$. ∎

Based on the above figure, we can calculate basic trig functions sine, cosine, and tangent for 30° and 60° angles. Since trig functions do not depend on the size of a triangle, for simplicity we can set $a = 1$. Then the above figure becomes

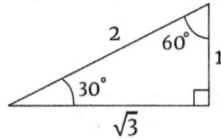

Now we use the definition of basic trig functions.

For the 30° angle: the opposite side is 1, the adjacent side is $\sqrt{3}$, and the hypotenuse is 2. Therefore,

$$\sin 30° = \frac{1}{2}, \quad \cos 30° = \frac{\sqrt{3}}{2}, \quad \tan 30° = \frac{1}{\sqrt{3}} = \frac{\sqrt{3}}{3}.$$

For the 60° angle: the opposite side is $\sqrt{3}$, the adjacent side is 1, and the hypotenuse is 2. Therefore,

$$\sin 60° = \frac{\sqrt{3}}{2}, \quad \cos 60° = \frac{1}{2}, \quad \tan 60° = \sqrt{3}.$$

45° – 45° Triangle

The triangle looks like this

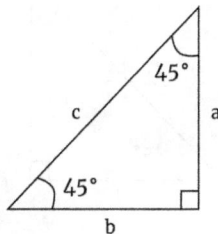

Both legs a and b are sides of the square above from which the triangle originated. Therefore, the legs are equal: $a = b$.

> In any $45° - 45°$ triangle both legs are equal.

Note. In any triangle with two equal angles the opposite sides are equal.

From here we can express the hypotenuse c through legs a and b. By the Pythagorean theorem,

$$c = \sqrt{a^2 + b^2} = \sqrt{a^2 + a^2} = \sqrt{2a^2} = a\sqrt{2}.$$

We get the following figure

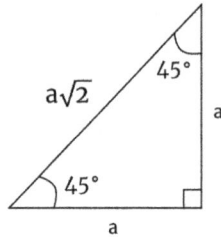

Solve the problem below on your own. In it, a $45° - 45°$ triangle is given. Denote by c the hypotenuse, and by a and b the two legs.

Problem 3.10. Prove that $a = b = c\sqrt{2}/2$. ■

To calculate basic trig functions sine, cosine, and tangent for $45°$, we can proceed similarly to the $30° - 60°$ triangle: set $a = b = 1$. We get the following figure:

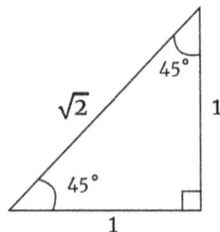

We have,

$$\sin 45° = \frac{1}{\sqrt{2}} = \frac{\sqrt{2}}{2}, \cos 45° = \frac{1}{\sqrt{2}} = \frac{\sqrt{2}}{2}, \tan 45° = 1.$$

0° Angle

This angle may look rather weird, since there are no real triangles with such an angle. However, we can imagine this angle as belonging to the so-called degenerate triangle that looks like a line segment. So, it makes sense to add 0° angle to the list of special angles.

Problem 3.11. How do you calculate the basic trig functions for angle 0°?

Solution. Certainly we cannot directly define trig functions for this angle as we did for acute angles. However, here is an idea of what we can do. Let's consider a right triangle with a small angle θ, and see what happens to sin θ if we decrease θ, bringing it closer and closer to zero:

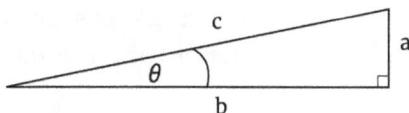

In this figure, we assume that angle θ is small, and the hypotenuse c does not change in length when θ is changing. As θ approaches zero (which means that we make angle θ smaller and smaller), then leg a will also approach zero. Since sin $\theta = a/c$ and a becomes zero, it is natural to accept by definition that sin 0° = 0/c = 0. To calculate cos θ, we can also use the same method. Namely, as θ approaches zero, the leg b approaches the hypotenuse c. Since cos $\theta = b/c$, we can define cos 0° = c/c = 1. Another way to define cos 0° is to use the main identity to express cosine through sine: $\cos\theta = \sqrt{1 - \sin^2\theta}$. From here we get: $\cos 0° = \sqrt{1 - \sin^2 0°} = \sqrt{1 - 0} = 1$. Tangent of 0° can be calculated by the formula: tan 0° = sin 0°/cos 0° = 0/1 = 0. Finally, we get the following values of basic trig functions for angle 0°:

$$\sin 0° = 0, \cos 0° = 1, \tan 0° = 0. \blacksquare$$

90° Angle

Problem 3.12. What are the values of basic trig functions for angle 90°?

Solution. To calculate sin 90°, we can use the formal definition of sine in a right triangle: this is the ratio of the opposite side to the hypotenuse. Since the opposite side for the 90° angle is the hypotenuse itself, then this ratio is equal to 1 (hypotenuse/hypotenuse = 1). So, sin 90° = 1. To calculate cos 90°, we express cosine through sine, using the main identity:

$$\cos 90° = \sqrt{1 - \sin^2 90°} = \sqrt{1 - 1} = 0.$$

So, cos 90° = 0. Now let's try to calculate tangent of 90° by the formula tan θ = sin θ/cos θ:

$$\tan 90° = \sin 90°/\cos 90° = 1/0 \ ?!$$

We've come to a situation where a number is being divided by zero. Since it is not possible, we must conclude that tan 90° does not exist. Finally, we get the following values of basic trig functions for angle of 90°:

sin 90° = 1, cos 90° = 0, tan 90° doesn't exist (undefined).

At this point, we identified five special angles:

0°, 30°, 45°, 60°, and 90°

and calculated basic trig functions for them.

Let's observe and compare the values of trig functions for special angles. We can notice that:

sin 0° = cos 90° = 0, sin 30° = cos 60° = 1/2, sin 45° = cos 45° = $\sqrt{2}$/2,
sin 60° = cos 30° = $\sqrt{3}$/2, sin 90° = cos 0° = 1.

This is not a coincidence. A common pattern here is that in each equality the angles are complementary to each other, which means that their sum is 90°:

$$0° + 90° = 90°, \ 30° + 60° = 90°, \ 45° + 45° = 90°.$$

Such a property is generalized in the following problem.

Problem 3.13. Let angles α and β be complementary to each other: $\alpha + \beta = 90°$. Prove that

$$\sin \alpha = \cos \beta \text{ and } \cos \alpha = \sin \beta.$$

Solution. Since $\alpha + \beta = 90°$, these angles can be interpreted as acute angles in a right triangle:

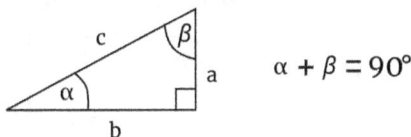

$$\alpha + \beta = 90°$$

From the definition of sine and cosine, we have $\sin \alpha = a/c$ and $\cos \beta = a/c$. Therefore, $\sin \alpha = \cos \beta$. Similarly, $\cos \alpha = \sin \beta$. ∎

Solve the problem below on your own.

Problem 3.14. Prove the following identities:

$$\sin(90° - \theta) = \cos \theta, \ \cos(90° - \theta) = \sin \theta. \ ∎$$

These identities give the reason for the name "cosine." Indeed, angles θ and $(90° - \theta)$ are complementary, so the sine of an angle is the cosine of its complementary angle. These identities are called **reduction formulas**. They allow us to reduce the value of any angle to 45° or less when calculating trig functions. The following problem demonstrates that.

Problem 3.15. We need to find sine and cosine for angle 73°, knowing the values of sine and cosine for angles less than 45°.

Solution. Using the reduction formulas, we have

$$\sin 73° = \sin (90° − 17°) = \cos 17°,$$

$$\cos 73° = \cos (90° − 17°) = \sin 17°. \blacksquare$$

Note. While calculators can be used for any angles, it is sometimes more convenient to work with smaller angles.

To finish this chapter, let's summarize our results.

Main Identity

$$\sin^2 \theta + \cos^2 \theta = 1$$

Simplest Relations between Trig Functions

$$\tan \theta = \sin \theta / \cos \theta$$
$$\cot \theta = \cos \theta / \sin \theta = 1/\tan \theta$$
$$\sec \theta = 1/\cos \theta$$
$$\csc \theta = 1/\sin \theta$$

Reduction Formulas

$$\sin (90° − \theta) = \cos \theta$$
$$\cos (90° − \theta) = \sin \theta$$

Basic Trig Functions of Special Angles

Angle θ	0°	30°	45°	60°	90°
$\sin \theta$	0	1/2	$\sqrt{2}/2$	$\sqrt{3}/2$	1
$\cos \theta$	1	$\sqrt{3}/2$	$\sqrt{2}/2$	1/2	0
$\tan \theta$	0	$\sqrt{3}/3$	1	$\sqrt{3}$	Undefined

Here's a practical way to help you memorize the above values of sine and cosine. Assign each finger of your hand the following angle measurements:

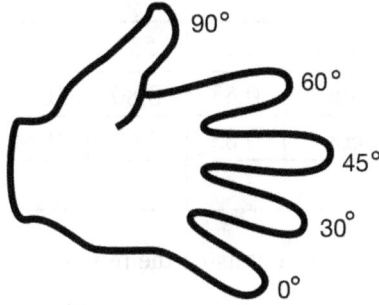

To find the sine for the desired angle, look at the finger that corresponds to that angle and use this formula:

$$\sin \theta = \frac{\sqrt{n}}{2},$$

where n is the number of fingers below the desired one.

This method is illustrated below for angles 60° and 45° degrees.

$$\sin = \frac{\sqrt{\text{Fingers below}}}{2}.$$

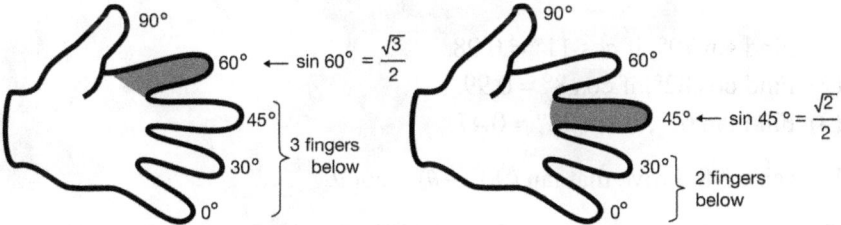

For cosine, a similar formula is used:

$$\cos = \frac{\sqrt{\text{Fingers above}}}{2}.$$

Exercises for Solving on Your Own

Exercise 3.1. Nick accidentally erased the fourth row in the following table:

Angle	32°	53°	70°
sin	0.53	0.80	0.94
cos	0.85	0.60	0.34
tan			

Can you help him restore it using the remaining rows?

Exercise 3.2. Calculate $\sin^2 37°42' + \cos^2 37°42'$ without using calculators or tables.

Exercise 3.3. Calculate cotangent for special angles.

Exercise 3.4. Express cotangent through sine.

Exercise 3.5. Prove that sine is expressed through tangent by the formula

$$\sin \theta = \frac{\tan \theta}{\sqrt{1 + \tan^2 \theta}} \ .$$

Exercise 3.6. Let θ be an acute angle. Which is greater: $\sin \theta$ or $\tan \theta$?

Exercise 3.7. Solve the following problems:

(1) Find $\sin 79°$, if $\cos 11° = 0.98$.
(2) Find $\cos 82°$, if $\cos 8° = 0.99$.
(3) Find $\tan 62°$, if $\sin 28° = 0.47$.

Exercise 3.8. Prove that $\tan (90° - \theta) = \cot \theta$.

Exercise 3.9. What is the value of $f(30°)$ for the function $f(x) = \sin x + \cos 2x$?

Exercise 3.10. Sofia tried to solve a trig problem concerning the angle α. She calculated that $\tan \alpha = 1/2$ and $\cot \alpha = 2/3$. Her answer is incorrect. Why?

Exercise 3.11. Let A, B, and C be angles in a $30° - 60°$ triangle: $\angle A = 30°$, $\angle C = 90°$. Sides a, b, and c are opposite to corresponding angles. Solve the following problems:

(a) $a = 6$. Find b and c; (b) $b = 3$. Find a and c; (c) $c = 8$. Find a and b.

Exercise 3.12. Let A, B and C be angles in a $45° - 45°$ triangle: $\angle A = 45°$, $\angle C = 90°$. Sides a, b, and c are opposite to corresponding angles. Solve the following problems:

(a) $a = 4$. Find b and c; (b) $b = 7$. Find a and c; (c) $c = 10$. Find a and b.

Entertainment Problem

Problem E3.1. A mint employs 30 people. Each day a worker is given a kilogram of gold and should produce 100 coins, weighing 10 g. It turned out that one worker makes counterfeit coins that weigh 1 g less. How can we identify this swindler using just one weighing?

Chapter 4

General Definitions of Trigonometric Functions

In the previous chapters, we have shown the use of trig functions for solving right triangles only. In many cases, however, we also need to solve arbitrary (oblique) triangles that may contain obtuse angles. For this, it is necessary to expand the definition of trig functions from acute angles to any angle. In this chapter, we will do this.

To understand how to come up with it, let's first discuss how we treat angles in geometry and trigonometry. We treat them somewhat differently.

Geometric and Trigonometric (Oriented) Angles

In geometry, we treat an angle as a figure that is formed by two rays coming from a common point, called the vertex of the angle. Here are the examples of angles:

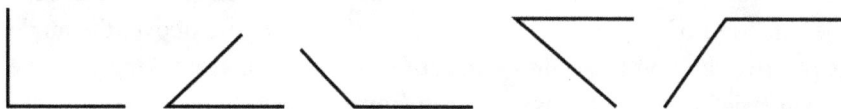

"Geometric" angles are mostly used in "static" (stationary) situations (some examples were given in Chapter 2), and these angles usually do not exceed 180° (actually, in concave polygons, angles can exceed 180°, but they are less than 360°). However, there are many practical cases where we need to measure angles in dynamic processes such as turns or rotations. Unlike angles in geometry, angles in such processes can take any large values. Moreover, the angles can even have negative values depending on the "direction of rotation." Adding the concept of direction of rotation is a new feature that we use in trigonometry.

Let's clarify what we mean by the direction of rotation. Consider a "geometric" angle

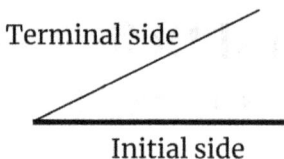

Terminal side

Initial side

We assume that one side, called the **initial side**, stands still, while the other side, called the **terminal side**, is rotating around the vertex. The terminal side starts rotating from the position of the initial side and then moves to its current position. To rotate, we have two directions: clockwise and counterclockwise. We can mark these two directions of rotation by arrows:

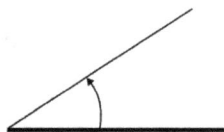

Counterclockwise rotation:
angle is positive

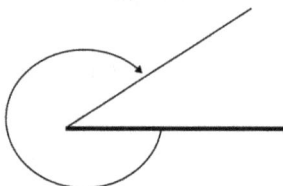

Clockwise rotation:
angle is negative

As a matter of convention, a **positive** measure of an angle is assigned if the direction of rotation is counterclockwise, and a **negative** measure if the direction of rotation is clockwise. On the left figure above, the angle is positive, and on the right — negative. As you can see, taking just one "geometric" angle (two rays, coming from the same point), we can consider two "trigonometric" (or oriented) angles: one is positive, and another is negative depending on the direction in which we rotate the terminal side. Even more, we can assign to a given "geometric" angle infinitely many "trigonometric" angles making multiple full rotations of the terminal side in either direction. All such "trigonometric" angles have the same position of the terminal side, and they are called **coterminal angles**. In the figures above, the two angles are coterminal.

If the value of an angle is specified, the arrows are not needed, since the sign of the angle indicates the direction of rotation.

Problem 4.1. Depict angles of turns of a ship's steering wheel (helm) according to the following commands:

(1) Turn the steering wheel left (counterclockwise) by 43°
(2) Turn the steering wheel right (clockwise) by 43°
(3) Turn the steering wheel left by 200°
(4) Turn the steering wheel right by 160°
(5) Make two full left rotations of the steering wheel plus 23°

Solution.

(1)   43°

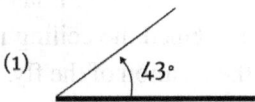

Counterclockwise turn to 43°

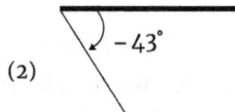

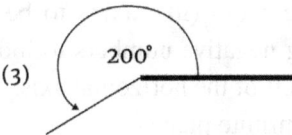

(2)   – 43°

Clockwise turn to 43°

(3)   200°

Counterclockwise turn to 200°

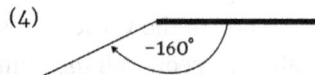

(4)   –160°

Clockwise turn to 160°

(5) One full rotation consists of 360°. To make two full rotations, we need to use angle $2 \cdot 360° = 720°$. After adding 23°, we get an angle of 743°. We can depict this angle like this

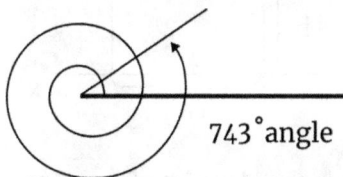

743° angle

Note that angles in parts 1 and 2 as geometrical angles have the same measure of 43°, so they are congruent. Angles in parts 3 and 4 as geometrical shapes are also congruent. The angle in part 5 is coterminal to angle of 23°. ■

Solve the problem below on your own.

Problem 4.2. Determine the value of the angle when making three full rotations clockwise and another 10° in the opposite direction. ■

General Definition of Trigonometric Functions

Our next step is to come up with the definitions of trig functions for arbitrary (trigonometric) angles. For this purpose, we use the so-called Cartesian system of coordinates. This system was invented by a French mathematician Rene Descartes (1596–1650), his Latin name is Cartesius.

Legend has it that Descartes, who liked to stay in bed until late, was watching a fly on the ceiling from his bed. He imagined the ceiling in the form of a rectangle, in which he could indicate the location of the fly, measuring how far it is located in the horizontal and in the vertical directions. These two numbers being the coordinates of the fly. Descartes expanded this idea by allowing the two sides of the room (like axes) to become infinitely long in both directions and using negative numbers to indicate the bottom part of the vertical axis and the left of the horizontal axis. Thus, he was able to specify all the points on an infinite plane.

Mathematically the Cartesian system on the plane is presented by two perpendicular number axes: x and y. Each point A on the plane is uniquely defined by two coordinates (a, b):

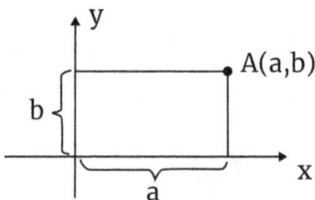

Axes x and y divide the plane into four parts, called the **quadrants**. They are numerated in the following order:

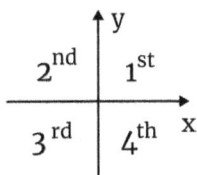

The Cartesian system will help us to define sine and cosine for any angle. For acute angles, we've already defined trig functions as the ratios of the sides in right triangles. We can simplify the calculation of these ratios by using triangles with hypotenuse equal to 1. In this case, sine and cosine will just be the lengths of legs in such triangles (we have seen this in Problem 3.4, Chapter 3):

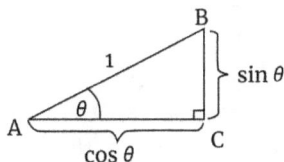

When angle θ changes, the hypotenuse AB (as terminal side) turns about point A. Since we keep the length of the hypotenuse equal to 1, point B moves along a circle of radius 1 centered at A. Such a circle is called a **unit circle**. In order to define $\sin \theta$ and $\cos \theta$ for different values of angle θ, it is convenient to position the triangle ABC together with the mentioned unit circle into the Cartesian system of coordinates:

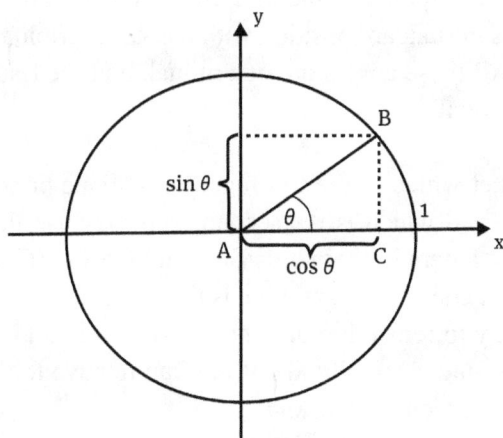

Here, we placed the vertex A of angle θ at the origin of the coordinate system and placed the initial side AC along the right (positive) part of the x-axis. We say that angle θ is in **standard position**.

The above figure shows a very important fact: **sine and cosine are the coordinates of point B** on the unit circle. Namely, sine is the second coordinate (y-coordinate), and cosine is the first coordinate (x-coordinate). So, we get a reformulation (i.e., a new definition) of sine and cosine for acute angles: they are coordinates of points on the unit circle. The main idea here is to use this reformulation as a general definition of sine and cosine for arbitrary angles. Thus, we get the following definition:

A general definition of sine and cosine

Let θ be an arbitrary angle in standard position: its vertex is located at the origin of the coordinate system, and the initial side goes along the positive (right) part of the x-axis. If B is the point of intersection of the terminal side with the unit circle, then, by definition, $\sin \theta$ is the y-coordinate of point B, and $\cos \theta$ is the x-coordinate:
$$\sin \theta = y, \cos \theta = x$$

Notes:
(1) Angles in the 1st quadrant of the coordinate system increase with counterclockwise rotation. This is one of the reasons why the angles in such a rotation were chosen as positive.
(2) When we pick a point on the unit circle, we can consider infinitely many angles in standard position with the same terminal side going to this point. All these angles are coterminal, and their sines as well as cosines are equal.

To remember which of the trig functions — sine or cosine — is the first coordinate, and which is the second, you may use the alphabetical order of the first letters in the words **Sine** and **Cosine** (C is before S, so Cosine is the first coordinate, and Sine is the second).

Another way to remember that the sine is a vertical coordinate is to imagine that the sine is the altitude of the Sun relative to the Earth (note that word *sine* is consonant with *sun*):

Movement of the Sun is a positive direction

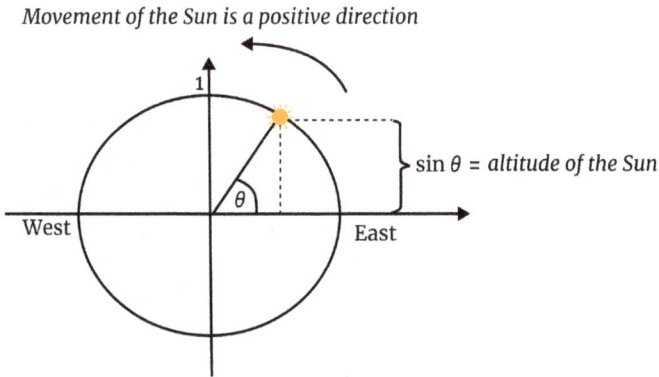

$\sin \theta = altitude\ of\ the\ Sun$

West

East

This figure also shows the reason for choosing the direction for positive angles: the sun moves from East (sunrise) to West (sunset) across the sky.

After we have defined sine and cosine for arbitrary angles, we can easily get the general definition of other trig functions. In Problem 3.2, Chapter 3, we stated that tan θ = sin θ/cos θ. We can use this formula as the general definition of tangent. We can also get the **geometric definition** of the tangent in the following way. On the right-hand side of the unit circle, draw a vertical line and extend the terminal side of the angle to meet this line. Then the tangent is the vertical coordinate of the point of intersection. Here is the figure of the tangent for the acute angle θ which is located in the 1st quadrant:

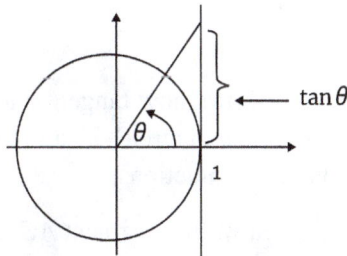

$\tan \theta$

In the 2nd quadrant, the tangent looks like this:

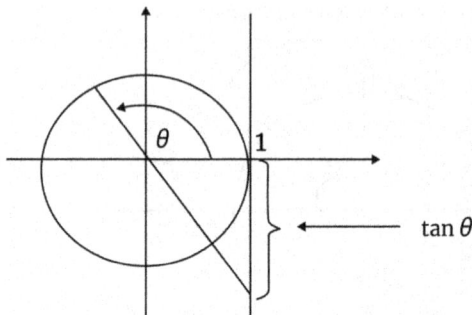

Now let's look at the cotangent. It can be defined for arbitrary angles in the same way as for acute angles: it is the reciprocal of tangent: $\cot \theta = 1/\tan \theta$. Since $\tan \theta = \sin \theta/\cos \theta$, we get $\cot \theta = \cos \theta/\sin \theta$. Also, there is a geometric definition of cotangent.

Solve the following problem on your own.

Problem 4.3. Verify that the figure below depicts $\cot \theta$:

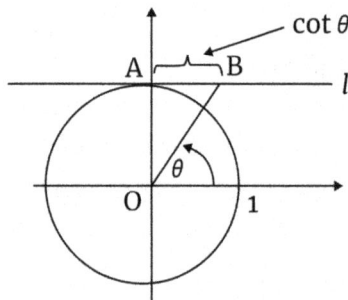

The following problem shows how tangent and cotangent functions are related with the tangent line to a circle. That is the reason for the term *tangent* in the name of these trig functions.

Problem 4.4. In the figure below, line segment AC touches the unit circle at point B (so, AC goes along a **tangent line** to the circle). Prove that $BC = \tan \theta$ and $AB = \cot \theta$.

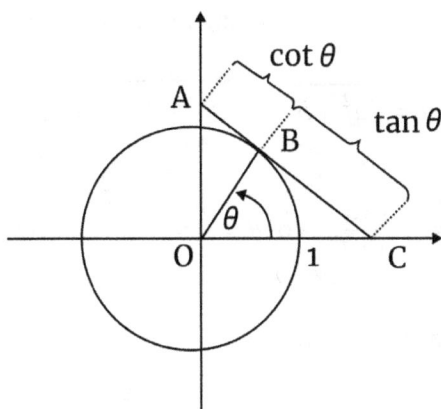

Solution. It is known from geometry that the tangent line AC is perpendicular to the radius OB, so $\triangle OBC$ is a right triangle. Also, $\angle OAB = \theta$ (as angles with mutually perpendicular sides). From $\triangle OBC$, $\tan \theta = BC/OB = BC/1 = BC$. From $\triangle OAB$, $\cot \angle OAB = AB/OB = AB/1 = AB$, which is $\cot \theta$. ∎

The other two trig functions — cosecant and secant — can be defined for any angle as reciprocals to the basic sine and cosine (see Chapter 3):

$$\csc \theta = \frac{1}{\sin \theta}, \quad \sec \theta = \frac{1}{\cos \theta}$$

It is possible to define trig functions using a circle with an arbitrary radius r (not only a unit circle with $r = 1$). Let $A(a, b)$ be any point on the terminal side of the angle θ (in a standard position). If r is the distance from point $A(a, b)$ to the origin, then sine, cosine and tangent of angle θ are defined by the formulas:

$$\sin \theta = \frac{b}{r}, \quad \cos \theta = \frac{a}{r}, \quad \tan \theta = \frac{b}{a}, \quad r = \sqrt{a^2 + b^2}.$$

Problem 4.5. Find the value of the six trig functions of angle θ if a point with coordinates $(2, -3)$ lies on the terminal side of angle θ, and θ is in standard position.

Solution. We have $a = 2$, $b = -3$. Using the above formulas,

$$r = \sqrt{a^2 + b^2} = \sqrt{2^2 + (-3)^2} = \sqrt{13},$$

$$\sin \theta = \frac{b}{r} = \frac{-3}{\sqrt{13}} = -\frac{3\sqrt{13}}{13}, \quad \cos \theta = \frac{a}{r} = \frac{2\sqrt{13}}{13}, \quad \tan \theta = \frac{b}{a} = \frac{-3}{2} = -\frac{3}{2}.$$

The other three trig function are

$$\csc \theta = \frac{1}{\sin \theta} = -\frac{\sqrt{13}}{3}, \quad \sec \theta = \frac{1}{\cos \theta} = \frac{\sqrt{13}}{2}, \quad \cot \theta = \frac{1}{\tan \theta} = -\frac{2}{3}. \blacksquare$$

Because sine and cosine are coordinates, trig functions may have both positive and negative values depending on the quadrant in which angle θ lies. The figures below show the signs of basic trig functions.

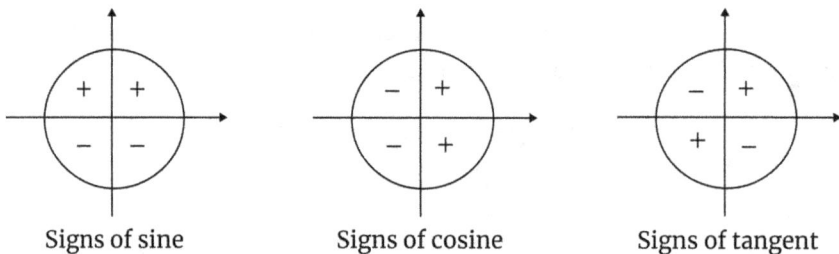

Signs of sine   Signs of cosine   Signs of tangent

Figure 1.

Note. The following phrase may help you to remember which of these functions (sine, cosine or tangent) is positive in each quadrant: "**A**ll **S**tudents **T**ake **C**alculus." This phrase hints that in the first quadrant all three are positive, in the second only **s**ine, in the third only **t**angent, and in the fourth only **c**osine.

Problem 4.6. Calculate the basic trig functions for the "quadrantal" angles $0°$, $90°$, $180°$, $270°$, and $360°$.

Solution. Here is the figure for quadrantal angles:

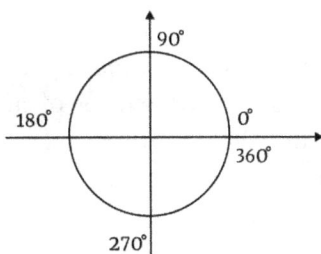

(1) For 0° and 360° angles, the corresponding point on the unit circle has coordinates $(1, 0)$. Therefore,

$$\sin 0° = \sin 360° = 0, \cos 0° = \cos 360° = 1, \tan 0° = \tan 360° = 0.$$

(2) For 90° angle, the corresponding point has coordinates $(0, 1)$. Therefore, $\sin 90° = 1$, $\cos 90° = 0$. By definition, $\tan \theta = \dfrac{\sin \theta}{\cos \theta}$. Because $\cos 90° = 0$, $\tan 90°$ is undefined (we cannot divide by zero).

(3) For 180° angle, the corresponding point has coordinates $(-1, 0)$. Therefore,

$$\sin 180° = 0, \cos 180° = -1, \tan 180° = 0.$$

(4) For 270° angle, the corresponding point has coordinates $(0, -1)$. Therefore,

$$\sin 270° = -1, \cos 270° = 0, \tan 270° \text{ is undefined. } \blacksquare$$

We summarize the results of the above problem in the following table:

Angle θ	0°	90°	180°	270°	360°
$\sin \theta$	0	1	0	−1	0
$\cos \theta$	1	0	−1	0	1
$\tan \theta$	0	Undefined	0	Undefined	0

We see that the maximum and minimum values of sine and cosine of the quadrantal angles are 1 and -1, respectively. For all other angles, sine and cosine are between -1 and 1, so the range of sine and cosine functions (i.e., all possible values that these functions may produce) is the interval $[-1, 1]$. There is no restriction for values of tangent. We present these properties in the following box:

Ranges of basic functions sine, cosine and tangent

$$-1 \leq \sin \theta \leq 1, -1 \leq \cos \theta \leq 1$$

or

$$|\sin \theta| \leq 1, |\cos \theta| \leq 1$$

$$-\infty < \tan \theta < \infty$$

Here, the notation $|N|$ means the absolute value of the number N. Informally speaking, it is an unsigned number. For example, $|-3| = |3| = 3$. A formal definition of the absolute value can be given using the so-called piecewise function:

$$|x| = \begin{cases} x, & x \geq 0 \\ -x, & x < 0. \end{cases}$$

We can also get the ranges of three other trig functions: cosecant, secant and cotangent. Cosecant and secant are reciprocals for sine and cosine, respectively. Since the later are located in the interval $[-1, 1]$, the cosecant and secant are located outside this interval. Cotangent is the reciprocal of tangent. The range for tangent is the entire number line, and the same is true for cotangent. We can represent these results in the following interval form:

Ranges of cosecant, secant and cotangent

cosecant and secant: $(-\infty, -1] \cup [1, \infty)$

cotangent: $(-\infty, \infty)$

Here the symbol $\cup$ means the union of two intervals.

Exercises for Solving on Your Own

Exercise 4.1. Let angle α equal 35°:

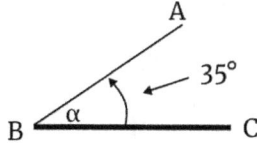

What are the values of the following angles?

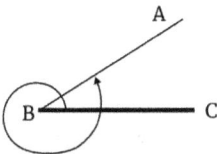

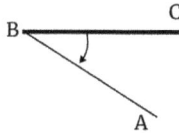

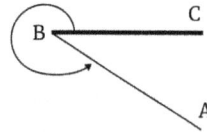

Angle 1   Angle 2   Angle 3

Exercise 4.2. In which quadrant does the terminal side of the 500° angle lie?

Exercise 4.3. Draw angle $\alpha = -430°$.

Exercise 4.4. Which function (sine, cosine or tangent) of the angle α represents the length of the segment PQ in the unit circle?

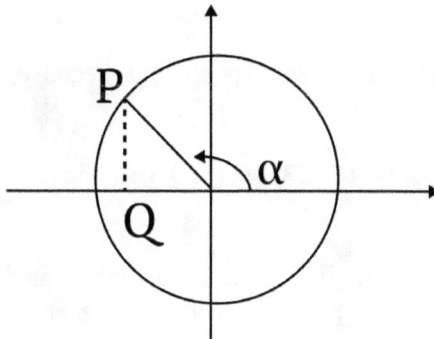

Exercise 4.5. Find these values:

(1) $\sin(-180°)$; (2) $\cos(-180°)$; (3) $\tan(-180°)$;

(4) $\sin(-90°)$; (5) $\cos(-90°)$; (6) $\tan(-90°)$;

(7) $\cot 0°$; (8) $\cot 180°$; (9) $\cot(-90°)$; (10) $\cot(-270°)$.

Exercise 4.6. For which values of x from the interval $[0, 360°]$ is $\tan(x + 30°)$ undefined?

Exercise 4.7. For which angles is the tangent undefined?

(a) $180°$; (b) $90°$; (c) $60°$; (d) $270°$.

Exercise 4.8. Point A lies on a circle of radius 2.3 in the 2nd quadrant. Its x-coordinate is equal to -1.2. What are the sine, cosine, and tangent of the corresponding angle?

Exercise 4.9. Point A lies on the unit circle in the 2nd quadrant. The tangent of the corresponding angle is -3.3. What are the coordinates of point A?

Exercise 4.10. Point $(2, -3)$ lies on the terminal side of an angle θ. Find $\sin \theta$, $\cos \theta$ and $\tan \theta$.

Exercise 4.11. Each statement is incorrect. Explain why:

(1) $\sin \alpha = 0.4$ and α is in the 3rd quadrant.
(2) $\cos \alpha = -0.2$ and α is in the 4th quadrant.

Exercise 4.12. What are the signs (positive or negative) for the following values?

(1) $\sin(-50°)$; (2) $\cos(-50°)$; (3) $\tan(-50°)$; (4) $\sin(200°)$; (5) $\cos(200°)$; (6) $\tan(200°)$.

Exercise 4.13. Prove that if θ is not a quadrantal angle, then $\sin \theta$ and $\tan(\theta/2)$ have the same sign (both positive or both negative).

Exercise 4.14. Determine the signs of cotangent in each quadrant.

Exercise 4.15. In which quadrant does angle θ lie, if:

(1) $\sin \theta < 0$ and $\cos \theta > 0$; (2) $\sin \theta > 0$ and $\cos \theta < 0$;
(3) $\sin \theta > 0$ and $\tan \theta < 0$; (4) $\sin \theta < 0$ and $\tan \theta > 0$;
(5) $\cos \theta < 0$ and $\tan \theta > 0$; (6) $\cos \theta > 0$ and $\tan \theta < 0$.

Exercise 4.16. What are the coordinates of the points on the unit circle at which the sine and cosine of the corresponding angles are equal? In which quadrants are these points located?

Entertainment Problem

Problem E4.1. Traveling Professor Smartman is on the shore of a round lake with a diameter of 200 m. In the middle of the lake, there is a small island on which a tree grows. There is also a tree on the shore near the water. The professor needs to get from the shore to the island, but he cannot swim. However, he has a rope, which is slightly over 200 m long. How can he get to the island with this rope?

Chapter 5

Basic Properties of General Trigonometric Functions. Methods for Reducing to Acute Angles

In the previous chapter, we introduced the definitions of trigonometric functions for arbitrary angles. Here we consider some of their properties. We also develop methods of reducing the calculation of trig functions for arbitrary angles to the acute ones.

Basic Properties

In Chapter 3, we considered the main identity for acute angles. Let's transfer it to arbitrary angles. Solve the problem below on your own.

Problem 5.1. Prove that the following equality is satisfied for any angle θ:

Main Identity

$$\boxed{\sin^2 \theta + \cos^2 \theta = 1} \quad \blacksquare$$

From the main identity, we can derive two more identities: one that relates tangent with secant and the other that relates cotangent with cosecant. To derive the first one, divide both sides of the main identity by $\cos^2 \theta$:

$$\frac{\sin^2 \theta}{\cos^2 \theta} + \frac{\cos^2 \theta}{\cos^2 \theta} = \frac{1}{\cos^2 \theta} \Rightarrow \tan^2 \theta + 1 = \sec^2 \theta.$$

Solve the following problem on your own.

Problem 5.2. Prove the identity: $\cot^2 \theta + 1 = \csc^2 \theta.$ ■

Next, we consider properties related to the so-called even and odd functions. A function $f(x)$ is called **even** if its value does not change when the argument x changes to the opposite value: $f(-x) = f(x)$. A function is called **odd** if its value changes to the opposite when the argument changes to the opposite: $f(-x) = -f(x)$. Note that a function may be neither even nor odd (a simple example is the function $f(x) = x + 1$). Let's see whether sine and cosine have "even-odd" properties.

Problem 5.3. Find out whether sine and cosine functions are even, odd, or neither.

Solution. We consider the case when angle θ lies in the 1st or 4th quadrants. Look at the figure below:

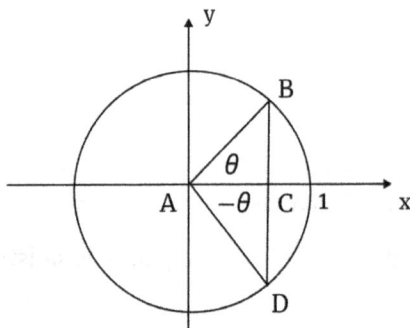

Here, angle θ lies in the 1st quadrant (it is angle *BAC*), and angle $-\theta$ lies in the 4th quadrant (it is angle *CAD*). The sine of the angle θ is the y-coordinate of the point *B*, and it is a positive number: sin $\theta = BC$. The sine of the angle $-\theta$ is the y-coordinate of the point *D*, and it is a negative number: sin $(-\theta) = -CD$. Since the lengths of *BC* and *CD* are equal, sin $(-\theta) = -\sin \theta$. Therefore, **sine is an odd function**. Cosine of both angles, θ and $-\theta$, is the same: it is the x-coordinates of points *B* and *D*, which is *AC*: cos $(-\theta) = \cos \theta = AC$. Therefore, **cosine is an even function**. ■

Solve the problems below on your own.

Problem 5.4. Prove that sine is an odd function, and cosine is an even function in the case when angles lie in 2nd or 3rd quadrants. ■

Problem 5.5. Prove that tangent is an odd function: $\tan \theta = -\tan(-\theta)$. ■

Another property that we consider here is a **periodic** property. Let's describe it in a general way. Informally, a function $f(x)$ is called periodic if it keeps repeating its values after a constant interval of length T. More formally, this means that there exists a number $T > 0$, such that the following equation is true:

$$f(x + T) = f(x) \text{ for all } x \text{ in the domain of the function } f[1].$$

Problem 5.6. Let function $f(x)$ satisfy the above equation and for any integer n, the number $x + nT$ belongs to the domain of $f(x)$. Prove that $f(x + nT) = f(x)$.

Solution. $f(x + 2T) = f[(x + T) + T] = f(x + T) = f(x)$,

$$f(x + 3T) = f[(x + 2T) + T] = f(x + 2T) = f(x), \text{ and so on.}[2] ■$$

This result shows that the function $f(x)$ repeats its values after an infinite number of intervals with lengths $T, 2T, 3T, \ldots$. A number $T > 0$ is called a **period** of a function $f(x)$ if it is the smallest of all other positive numbers U with the same property $f(x + U) = f(x)$.

Problem 5.7. Prove that sine is a periodic function with a period of $360°$.

Solution. First of all, note that $\sin \theta = \sin(\theta + 360°)$. Indeed, angles θ and $\theta + 360°$ correspond to the same point on the unit circle, and the sine is the vertical coordinate of this point. Therefore, sine for angles θ and $\theta + 360°$ is the same. What's left to prove is that $360°$ is the smallest "period" number. More formally, it means that there is no positive number T less than $360°$ that satisfies the equation $\sin \theta = \sin(\theta + T)$ for all angles θ. We prove this statement by contradiction. Suppose that the equation $\sin \theta = \sin(\theta + T)$

---

[1] Recall that the domain of a function $f(x)$ is the set of all allowable values for the argument x — (input numbers).

[2] Readers familiar with the mathematical induction, may provide a more formal proof.

is true for all θ, and T is positive and less than 360°. We are free to substitute for θ any number. Let's substitute $\theta = 0°$. Then $\sin 0° = \sin (T) = 0$. Inside the interval (0°, 360°), the equation $\sin (T) = 0$ has only one solution $T = 180°$ (you can see this on the unit circle). Now substitute the values $\theta = 90°$ and $T = 180°$ into the equation $\sin \theta = \sin (\theta + T)$. We get $\sin(90°) = \sin (90° + 180°) = \sin (270°)$. But $\sin (90°) = 1$, and $\sin (270°) = -1$. We've got a contradiction. Therefore, our assumption that the period T is less than 360° is wrong. ■

Below in the Reduction Formulas section we show that cosine is also a periodic function with the same period of 360°.

Now, let's look at tangent. Since $\tan \theta = \dfrac{\sin \theta}{\cos \theta}$ and both sine and cosine have the same period of 360°, it may seem that tangent also has the same period. However, this is not true.

Problem 5.8. Prove that tangent is a periodic function with a period of 180°.

Solution. We consider the case when angle θ lies in the 1ˢᵗ quadrant. Let's recall the geometric interpretation of tangent (see Chapter 4):

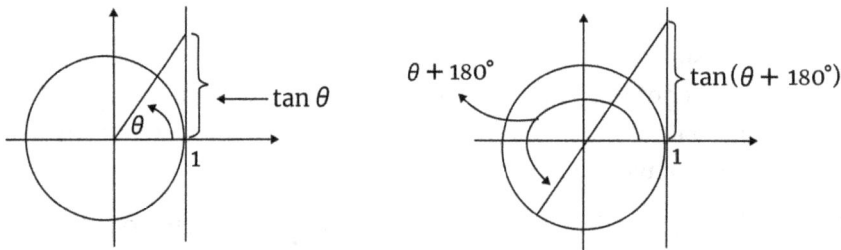

You can see from these figures that $\tan \theta = \tan (\theta + 180°)$. What's left to prove is that for positive T, the value $T = 180°$ is the smallest possible, such that $\tan \theta = \tan (\theta + T)$ for any angle θ. Again, we prove this statement by contradiction. Assume that this is not true and $T < 180°$. Plug in 0° for θ: $\tan 0° = \tan (T) = 0$. However, inside the interval (0°, 180°), the equation $\tan (T) = 0$ does not have a solution (tangent is equal to zero at the same points as sine, only at the endpoints 0° and 180°). Therefore, the period of tangent cannot be less than 180°. ■

To conclude this section, let's summarize the basic properties of trig functions that we have discussed.

Main identity

$$\sin^2 \theta + \cos^2 \theta = 1$$

Even-Odd Properties

$\sin (-\theta) = -\sin \theta$ (odd function)

$\cos (-\theta) = \cos \theta$ (even function)

$\tan (-\theta) = -\tan \theta$ (odd function)

Periodic properties

$\sin (\theta + 360°) = \sin \theta$ (period is 360°)

$\cos (\theta + 360°) = \cos \theta$ (period is 360°)

$\tan (\theta + 180°) = \tan \theta$ (period is 180°)

Now we consider two methods (or tools) to reduce the calculation of trig functions for arbitrary angles to acute angles. These are reduction formulas and reference angles.

Reduction Formulas ("Forehead Rule")

Recall that in Problem 3.14, Chapter 3, we proved that for acute complement angles (angles which add up to 90°), the sine of one of them is equal to the cosine of the other (and vice versa). We called these properties **reduction formulas**. They are:

$$\sin (90° - \theta) = \cos \theta, \cos (90° - \theta) = \sin \theta.$$

We will see that the same property is true for any angle (not just for an acute).

Problem 5.9. Prove the above reduction formulas for angle θ located in the 2nd quadrant.

Solution. In the figure below, angle θ lies in the 2nd quadrant. Angle $\alpha = \theta - 90°$.

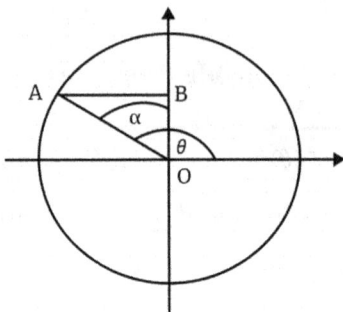

Point A has coordinates $x = -AB$ and $y = OB$. These coordinates are cosine and sine of angle θ: $\cos \theta = -AB$ and $\sin \theta = OB$.

Now take a look at angle α. It is an acute angle in the right triangle AOB with the hypotenuse $AO = 1$. By definition,

$$\sin \alpha = \frac{AB}{AO} = \frac{AB}{1} = AB. \text{ Similarly } \cos \alpha = \frac{OB}{AO} = OB.$$

Since $\alpha = \theta - 90°$ then, using the odd property of sine (see above), we have

$$\sin (90° - \theta) = -\sin (\theta - 90°) = -\sin \alpha = -AB = \cos \theta.$$

So, $\sin (90° - \theta) = \cos \theta$. Similarly, using the even property of cosine, we have

$$\cos (90° - \theta) = \cos (\theta - 90°) = \cos \alpha = OB = \sin \theta.$$

Therefore, $\cos (90° - \theta) = \sin \theta$. ■

From the reduction formula $\sin (90° - \theta) = \cos \theta$ we conclude that the cosine is a periodic function with the same period of 360° as the sine.

In the same way we can check that the above reduction formulas are true for angles in other quadrants.

We can extend the reduction formulas for the quadrantal angles 90°, 180°, 270°, and 360°. Here is a complete list of all of them.

Reduction formulas

$$\sin{(90^\circ - \theta)} = \cos{\theta} \qquad \cos{(90^\circ - \theta)} = \sin{\theta}$$

$$\sin{(90^\circ + \theta)} = \cos{\theta} \qquad \cos{(90^\circ + \theta)} = -\sin{\theta}$$

$$\sin{(180^\circ - \theta)} = \sin{\theta} \qquad \cos{(180^\circ - \theta)} = -\cos{\theta}$$

$$\sin{(180^\circ + \theta)} = -\sin{\theta} \quad \cos{(180^\circ + \theta)} = -\cos{\theta}$$

$$\sin{(270^\circ - \theta)} = -\cos{\theta} \quad \cos{(270^\circ - \theta)} = -\sin{\theta}$$

$$\sin{(270^\circ + \theta)} = -\cos{\theta} \quad \cos{(270^\circ + \theta)} = \sin{\theta}$$

$$\sin{(360^\circ - \theta)} = -\sin{\theta} \quad \cos{(360^\circ - \theta)} = \cos{\theta}$$

$$\sin{(360^\circ + \theta)} = \sin{\theta} \qquad \cos{(360^\circ + \theta)} = \cos{\theta}$$

As you see, these formulas look very similar, and you need to have very good memory to memorize them all. Fortunately, there is an excellent method to do this. We call this method the **Forehead Rule.** As you observe reduction formulas, you may notice that some formulas contain the same function on the left and right sides, while others contain different functions. Also, some formulas have a negative sign on the right side, and some do not. So, to get the reduction formulas, we need to answer two questions:

(1) Should we write a negative sign on the right side of the formula?
(2) Should we change the function (i.e., change sine to cosine or vice versa)?

The Forehead Rule suggests answers to these questions. It works like this. Always assume that angle θ is acute.

(1) To answer the first question about the negative sign, determine in which quadrant the angle in the left side of the formula is located. Then, based on the quadrant, determine the sign of the given trig function. You can review Chapter 4 for that (see Figure 1). If you get a negative sign, then write a negative sign on the right side. Otherwise, do not.

(2) To answer the second question about changing the function, ask yourself this: "Should I change the function?" Then move your head along the axis on which the quadrantal angle lies. You will automatically get the answer "yes" or "no" (your head without the participation of your brain will give the answer).

Let's see how the Forehead Rule works in the problem below.

Problem 5.10. Apply the Forehead Rule to get reduction formulas for

$$\sin (180° - \theta) \text{ and } \cos (270° - \theta).$$

Solution.
For $\sin (180° - \theta)$:

(1) Angle $180° - \theta$ is located in the 2nd quadrant. The sine is positive here. So, do not write a negative sign on the right side.
(2) The quadrantal angle $180°$ points to the horizontal axis. Move your head along it (left–right). Your head says "No," so do not change the function. Therefore, the answer is

$$\sin (180° - \theta) = \sin \theta.$$

For $\cos (270° - \theta)$:

(1) Angle $270° - \theta$ is located in the 3rd quadrant. The cosine is negative here. So, write a negative sign on the right side.
(2) The quadrantal angle $270°$ points to the vertical axis. Move your head along it (up-down). Your head says "Yes," so change cosine to sine. The answer is

$$\cos (270° - \theta) = -\sin \theta. \blacksquare$$

Solve the following problem on your own.

Problem 5.11. Apply the Forehead Rule to get the reduction formulas for $\sin (90° + \theta)$ and $\cos (180° + \theta)$. $\blacksquare$

Note that the reduction formulas in the last row of the table above represent periodic properties of sine and cosine: both functions are periodic with a period of 360°. The reduction formulas, together with periodic properties, allow us to reduce any angle in trig functions to an angle less than (or equal to) 45°.

Problem 5.12. Reduce the angles in sin 1250° and cos 590° to angles less than 45°.

Solution.

sin 1250° = sin (3·360° + 170°) = sin170° = sin (180° − 10°) = sin 10°.

cos 590° = cos (360° + 230°) = cos 230° = cos (270° − 40°) = −sin 40°. ∎

Solve the following problem on your own.

Problem 5.13. Reduce the angles in sin 925° and cos 430° to be less than 45°. ∎

Reference Angle

Here, we describe another method for reducing the calculation of trig functions for arbitrary angles to acute ones. These acute angles are called reference angles and are defined as follows.

Definition of the reference angle

Let θ be an arbitrary angle in standard position. Angle θ_r is called the **reference angle** to θ, if it satisfies three conditions:
(1) The terminal side of θ_r coincides with the terminal side of θ.
(2) The initial side of θ_r is horizontal (it coincides with either positive or negative parts of the x-axis).
(3) Angle θ_r is an acute angle.

Let's see how the reference angle θ_r looks depending on the quadrant in which the original angle θ is located. We assume that angle $\theta > 0$. If we need to get a reference angle for a negative angle, we can always change the sign of the angle according to even-odd properties: sin $(-\theta) = -\sin \theta$, cos $(-\theta) = \cos \theta$, tan $(-\theta) = -\tan \theta$.

(1) Angle θ is located in the 1st quadrant. Then θ_r coincides with θ: $\theta_r = \theta$.

(2) Angle θ is located in the 2nd quadrant. Then $\theta_r = 180° - \theta$:

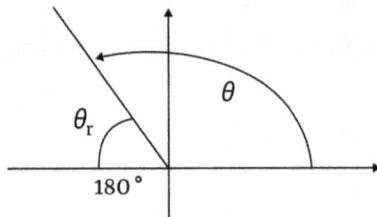

(3) Angle θ is located in the 3rd quadrant. Then $\theta_r = \theta - 180°$:

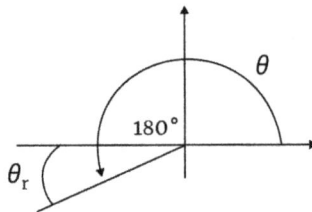

(4) Angle θ is located in the fourth quadrant. Then $\theta_r = 360° - \theta$:

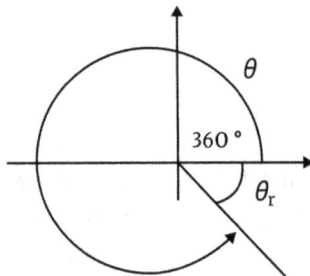

A reference angle is useful, because the values of any trig function of θ coincide with the values of the same trig function for the reference angle θ_r (if we are ignoring the +/– sign). You can check this using the reduction formulas described above.

Main Property of Reference Angle

The absolute value of any trig function of any angle is equal to the value of the same trig function of the reference angle.

Hence, to calculate the value of a trig function, it is enough to find the sign of the function and the value of this trig function of the reference angle.

To find the sign of the function and the reference angle, determine the quadrant where the original angle is located. Keeping in mind that one of the sides of the reference angle is always horizontal (goes along the x-axis), you can find the reference angle even without remembering the above formulas.

Problem 5.14. Calculate $\cos 120°$.

Solution. Angle $120°$ is located in the 2nd quadrant, so $\cos 120° < 0$. This is case 2 in the figures above. Reference angle $\theta_r = 180° - 120° = 60°$. We have $\cos 60° = \dfrac{1}{2}$. Therefore,

$$\cos 120° = -\frac{1}{2}. \blacksquare$$

Problem 5.15. Calculate $\sin 225°$.

Solution. Angle $225°$ is located in the 3rd quadrant, so $\sin 225° < 0$. This is case 3 above. Reference angle $\theta_r = 225° - 180° = 45°$. We have $\cos 45° = \dfrac{\sqrt{2}}{2}$. Therefore,

$$\sin 225° = -\frac{\sqrt{2}}{2}. \blacksquare$$

Problem 5.16. Calculate $\tan 330°$.

Solution. Angle $330°$ is located in the 4th quadrant, so $\tan 330° < 0$. This is case 4 above. Reference angle $\theta_r = 360° - 330° = 30°$. We have $\tan 30° = \dfrac{\sqrt{3}}{3}$. Therefore,

$$\tan 330° = -\frac{\sqrt{3}}{3}. \blacksquare$$

Problem 5.17. Find the values of the other five trig functions, if $\cos \theta = -\dfrac{5}{6}$ and $\tan \theta > 0$.

Solution. For reference angle θ_r, $\cos\theta_r = |\cos\theta| = \dfrac{5}{6}$, and angle θ_r is acute. Let's draw a right triangle, using the definition of $\cos\theta_r$ as a ratio of the adjacent side to the hypotenuse:

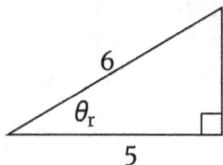

By the Pythagorean theorem, the vertical leg of this triangle is $\sqrt{6^2 - 5^2} = \sqrt{11}$. From here, $\sin\theta_r = \dfrac{\sqrt{11}}{6}$ and $\tan\theta_r = \dfrac{\sin\theta_r}{\cos\theta_r} = \dfrac{\sqrt{11}}{5}$. Since $\cos\theta < 0$ and $\tan\theta > 0$, angle θ lies in the 3rd quadrant. Therefore, $\sin\theta = -\dfrac{\sqrt{11}}{6}$ and $\tan\theta = \dfrac{\sqrt{11}}{5}$. The other three trig functions are:

$$\cot\theta = \frac{1}{\tan\theta} = \frac{5}{\sqrt{11}} = \frac{5\sqrt{11}}{11}, \quad \sec\theta = \frac{1}{\cos\theta} = -\frac{6}{5},$$

$$\csc\theta = \frac{1}{\sin\theta} = -\frac{6}{\sqrt{11}} = -\frac{6\sqrt{11}}{11}. \quad \blacksquare$$

Exercises for Solving on Your Own

Exercise 5.1. In solving a trig problem, Sofia calculated that $\sin\theta = 0.4$ and $\cos\theta = 0.6$. What is wrong with her calculation?

Exercise 5.2. Prove that $\cot\theta = -\tan(\theta + 90°)$.

Exercise 5.3. Find the reference angle to the given angles.
(a) 130°, (b) 320°, (c) 250°, (d) 760°, (e) −210°.

Exercise 5.4. Using the reduction formulas and the Forehead Rule, calculate the exact values without a calculator.
(a) $\sin 210°$, (b) $\cos 330°$, (c) $\tan 135°$.

Exercise 5.5. Using the reference angle, calculate the exact values without a calculator.

(a) sin 315°, (b) cos 150°, (c) tan 240°.

Exercise 5.6. Find the values of the other five trig functions, if

(a) $\sin \theta = -2/3$ and $\tan \theta < 0$; (b) $\cos \theta = -2/5$ and $\sin \theta > 0$; (c) $\tan \theta = 3/5$ and $\cos \theta < 0$.

Exercise 5.7. Let $f(x) = ax^2 + bx + c$. Assume that f is a periodic function with the period P. Prove that $a = b = 0$ (so, f is a constant function).[3]

In problems below, we ignore the condition that a period has a minimum value (this minimum value is also called the fundamental period).

Exercise 5.8. Let the numbers 3 and 5 be periods of the function $f(x)$. Prove that the number 1 is also a period.

Exercise 5.9. Let function f satisfy the equation $f(x + P) = -f(x)$ for any x and some constant $P \neq 0$. Prove that f is a periodic function with a period of $2P$. Give an example of such a function.

Entertainment Problems

Problem E5.1. You have two candles. Each of them can burn for 1 h. Candles do not burn evenly and can be lit from either end. How would you measure out 45 min?

Problem E5.2. Brothers Eli and Ben live on the opposite banks of a river. A ferry runs between the banks. Through the ferry captain, Eli needs to pass Ben a valuable item safeguarded in a special container. There is a large chain around the container which can be securely locked with several padlocks. However, Ben doesn't have the keys to Eli's padlocks. How will Ben be able to retrieve the item? It is unsafe to send an unlocked container.

---

[3] Readers familiar with mathematical induction or the fundamental theorem of algebra, may try to prove a general statement: a periodic polynomial is a constant.

Chapter 6

Solving Oblique Triangles: Law of Sines

In Chapter 2, we dealt with solving right triangles, and the only tools we used were the Pythagorean theorem and the definition of trig functions. In many cases, however, the triangle in question is not a right one, and we need to get additional knowledge in trigonometry to solve arbitrary (oblique) triangles. Recall that by solving a triangle we mean to find its basic elements — sides and angles, given some of them.

First of all, let's see what elements must be given. Obviously, if only angles are given and no sides, this information is not enough to determine sides, since triangles with the same angles are similar and may have different sizes. So, at least one side must be given. Let's consider all possible cases when one, two or all three sides are given as well as some number of angles. More precisely, the following four cases are possible in solving triangles:

(1) One side and two angles are given.
(2) Two sides and the angle opposite one of them are given.
(3) Two sides and the angle between them are given.
(4) Three sides are given.

The main tools for solving these problems are two theorems in trigonometry. These theorems are so important that they are even given the status of laws: Law of Sines and Law of Cosines. In this chapter we explore the Law of Sines and the first two cases. In the next chapter, the Law of Cosines and the last two cases will be considered.

Case 1: One Side and Two Angles are Given

Problem 6.1. A vertical post is standing on the ground and is supported by two wires (one on each side, going in opposite directions). The ends of the wires on the ground are 7 feet apart. The angle of elevation of one of the wires is 65°, and of the other wire is 80°. Find the length of each wire.

Solution. The following figure illustrates the problem:

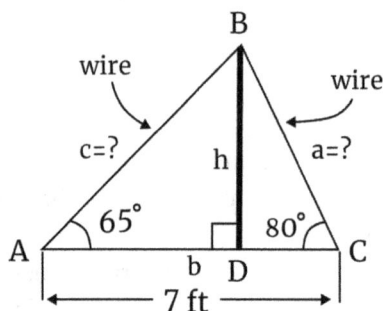

Let's consider this problem in the general form: in triangle ABC, side b and two angles A and C are given. Based on these data, we need to find sides a and c. For simplicity, we'll consider only an acute triangle (reasoning for an obtuse triangle is similar, but slightly different).

The height h of the triangle ABC splits it into two right triangles: ABD and BCD. Look at the sines of angles A and C:

From triangle ABD, $\sin A = \dfrac{h}{c}$. Solve for h: $h = c \sin A$.

From triangle BCD, $\sin C = \dfrac{h}{a}$. Solve for h: $h = a \sin C$.

Equate the above two expressions for h: $c \sin A = a \sin C$. Divide both sides of this equation by $\sin A \cdot \sin C$ and get

$$\frac{a}{\sin A} = \frac{c}{\sin C}.$$

A similar ratio is true for side b and angle B. We've got the following important result (theorem) which is called the **Law of Sines:**

In any triangle the ratio of any side to the sine of the opposite angle remains the same for all three sides:

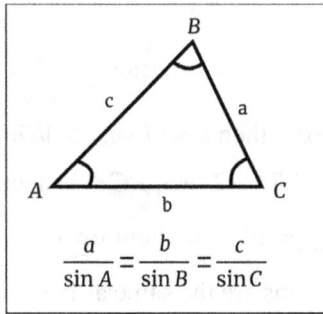

Law of Sines

Here is **another proof of the Law of Sines** based on the area of a triangle.

In Problem 1.5, Chapter 1, we stated that the area S of a triangle ABC can be calculated by the formula $S = \frac{1}{2}ab \cdot \sin C$. Let's write the formula for S, using sides b and $c : S = \frac{1}{2}bc \cdot \sin A$.

Now equate the two expressions for S:

$$\frac{1}{2}ab \cdot \sin C = \frac{1}{2}bc \cdot \sin A \implies a \cdot \sin C = c \cdot \sin A \implies \frac{a}{\sin A} = \frac{c}{\sin C}.$$

Note. In any triangle, the larger the side, the larger the opposite angle. However, sides are **not** proportional to the opposite angles. For example, in the right triangle $30° - 60°$, if we denote the side opposite to $30°$ angle as a, then the side opposite to $60°$ angle is $\sqrt{3}a$, which is not $2a$. Instead, the sides are proportional to the **sines** of opposite angles.

Let's apply the Law of Sines to our particular Problem 6.1. First, find angle B:

$B = 180° - A - C = 180° - 65° - 80° = 35°$. Next, according to the Law of Sines,

$$\frac{b}{\sin B} = \frac{c}{\sin C}.$$

Solving this proportion for side c, we get $c = \dfrac{b \sin C}{\sin B}$. It is given that $b = 7$, $C = 80°$, and we calculated $B = 35°$. Therefore, $c = \dfrac{7 \sin 80°}{\sin 35°} = 12$ feet.

To find side a, apply this proportion: $\dfrac{a}{\sin A} = \dfrac{b}{\sin B}$. From here, $a = \dfrac{b \sin A}{\sin B}$. Since $A = 65°$, then $a = (7 \sin 65°)/\sin 35° = 11$ feet.

Here is the final answer: $AB = 12$ feet, $BC = 11$ feet. ■

Solve the following problem on your own.

Problem 6.2. The conditions are the same as in the previous problem. The goal is to find the height of the post. ■

Ratios in the Law of Sines have an interesting geometric interpretation which is presented in the next problem.

Problem 6.3. Prove that the ratios in the Law of Sines are equal to the diameter of a circumscribed circle (a circle that passes through all three vertices of the given triangle).

Solution. We will use the following result from geometry: If two inscribed angles of a circle have the same intercepted arc, then they are equal, and their measure is equal to half the measure of the intercepted arc:

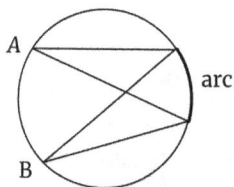

$$\angle A = \angle B = \frac{1}{2} \, arc$$

In particular, if the intercepted arc is a semicircle, then the inscribed angle is a right one.

Now, let a triangle *ABC* be inscribed in a circle of diameter *d*:

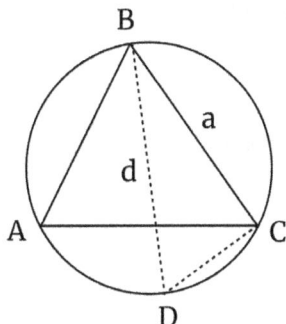

Angles *A* and *D* are inscribed angles with the same intercepted arc *BC*, so they are equal. Also, since *BD* is a diameter, angle *BCD* is a right one. From the right triangle *BCD*, we have

$$\sin D = \frac{a}{d} \;\Rightarrow\; d = \frac{a}{\sin D} \;\Rightarrow\; d = \frac{a}{\sin A} \;. \;\blacksquare$$

This result can be reformulated as follows. A side of a triangle is equal to the product of the sine of the opposite angle and the diameter of the circumscribed circle: $a = \sin A \cdot d$.

Case 2: Two Sides and the Angle Opposite One of Them are Given

Problem 6.4. Three friends, Alice, Bob and Carol, are camping in their own tents on a flat meadow in the woodland. Alice and Bob are 17 m apart, and Bob and Carol are 25 m apart. The angle going from Alice to Bob and Carol is 50°. What are the angles going from Bob to Alice and Carol, and from Carol to Alice and Bob? Also, how far apart are Alice and Carol?

Solution. Here is the figure:

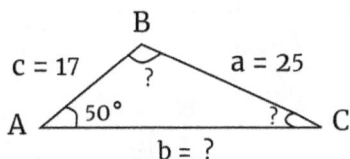

The problem is to find angles B and C, as well as side b. First, find angle C using the Law of Sines:

$\dfrac{a}{\sin A} = \dfrac{c}{\sin C}$. Solving this proportion for $\sin C$, we get

$$\sin C = \frac{c \sin A}{a} = \frac{17 \sin 50°}{25} = 0.5209.$$

From here,

$C = \sin^{-1}(0.5209) = 31.4°$ and
$B = 180° - A - C = 180° - 50° - 31.4° = 98.6°$.

To find the side b, we use the Law of Sines again:

$$\frac{b}{\sin B} = \frac{a}{\sin A} \Rightarrow b = \frac{a \sin B}{\sin A} = \frac{25 \sin 98.6°}{\sin 50°} = 32.3 \text{ m}.$$

Final answer: $B = 98.6°$, $C = 31.4°$, $b = 32.3$ *m.* ∎

Warning. When solving a problem similar to the one above, we must be very careful in using the Law of Sines in order to avoid a possible mistake. **The solution to the problem below is wrong!**

Problem 6.5. One of Sofia's friends suggested to her the following problem:

Eli and Ben came to the forest to pick some wildflowers. That day was rather foggy, and at some point, the guys decided to return to the main road, where they left their car. Eli went along a straight trail that forms a 55° angle with the main road. He walked 120 m to get to the main road. Ben took another straight trail and walked 100 m to reach the main road. Because of the fog, the guys could see each other at a distance of no more than 80 m. Is it possible that they will see each other when they both get to the main road (assuming that they reach the main road at the same time)?

Wrong solution. Sofia drew the following figure, in which Eli and Ben began to move from the top point A:

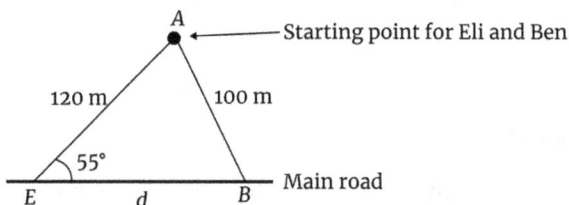

Sofia realized that the problem is to calculate the distance d between points E and B on the main road. If $d \leq 80$ m, then the guys would see each other, otherwise not. First, she found sine of angle B, using the Law of Sines:

$$\frac{120}{\sin B} = \frac{100}{\sin 55°} \Rightarrow \sin B = \frac{120 \sin 55°}{100} = 0.983.$$

Then she pressed $\sin^{-1}$ button on the calculator to get the angle from the sine:

$$B = \sin^{-1}(0.983) = 79.4°.$$

Next, she calculated angle A:

$$A = 180° - E - B = 180° - 55° - 79.4° = 45.6°.$$

Finally, using the Law of Sines again, she found the distance d:

$$\frac{d}{\sin A} = \frac{100}{\sin 55°} \Rightarrow \frac{d}{\sin 45.6°} = \frac{100}{\sin 55°} \Rightarrow d = \frac{100 \sin 45.6°}{\sin 55°} = 87.2 \text{ m}.$$

Since $d > 80$ m, Sofia concluded that it would not be possible for the guys to see each other. However, her conclusion is incorrect. Can you guess why? ∎

The answer to the question "What's wrong with Sofia's solution?" relates to the risk of getting the "ambiguous case" in using the Law of Sines. Let's discuss this case in a more general form, and then return to Sofia's problem.

Ambiguous Case

Consider in a general form what may happen in solving a triangle when two sides and the angle opposite to one of them are given. It turns out that a triangle is not always uniquely defined by these data. In some cases, it is possible to have two triangles. These are the ambiguous cases that we discuss here. We will assume that the following data are given in a triangle: two sides a and b, and angle A opposite to the side a (so, angle A is adjacent to side b). We separately consider cases when angle A is obtuse and acute.

Case: Angle A is obtuse (no ambiguity)

This is a simple case, since only two options are possible: a triangle does not exist or a triangle is unique (so, there is no ambiguity in this case). To understand why, let's draw the angle A and mark side b on its slant side:

To get a triangle, we need to draw side a from the top point to meet the horizontal side of angle A. Obviously, if side a is too short, it will not meet the horizontal side, and the triangle does not exist:

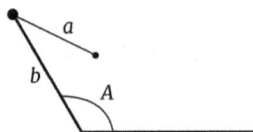

For a triangle to exist, side a must be greater than b. Then the triangle is defined uniquely. We come up to the following

Proposition 6.1. Let two sides a and b, and an **obtuse** angle A opposite to side a be given. Then

(1) If $a \leq b$, the triangle does not exist.
(2) If $a > b$, the triangle exists and is unique. ∎

Note. Conclusion in part 1 is also clear for the following reason: if $a \leq b$, then $\angle A \leq \angle B$. Since angle A is obtuse, then angle B must also be obtuse. But a triangle cannot have two obtuse angles.

Case: Angle A is acute (ambiguity is possible)

Similar to an obtuse angle, let's draw angle A and mark side b on its slant side:

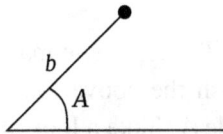

To create a triangle, we draw side a from the top point. Four cases are possible here:

(1) Side a is too short to meet with the horizontal side:

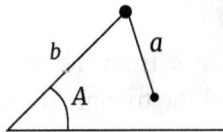

The triangle does not exist.

(2) Side a touches the horizontal side at exactly one point:

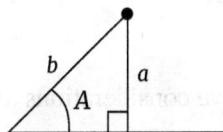

The triangle is a right one and unique.

(3) Side *a* intersects the horizontal side at two points:

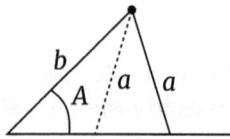

There are two triangles with sides *a*, *b* and angle *A* (this is the ambiguous case).

(4) Side *a* is long enough and intersects horizontal side only at one point:

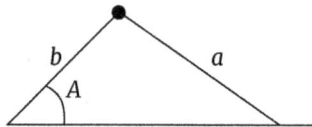

The triangle is unique. The top angle may be acute, right, or obtuse.

How can we distinguish the above four cases using the numerical values of sides *a*, *b* and angle *A*? Take a look at this figure

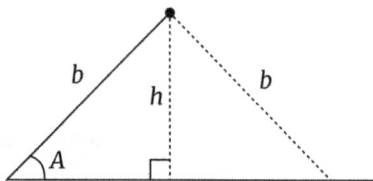

In your mind, draw side *a* from the top point. You can see that if *a* < *h*, side *a* is too short, and the triangle does not exist. If *a* = *h*, we can draw only one right triangle. If *a* is between *h* and *b*: *h* < *a* < *b*, then we can draw side *a* on both sides (left and right) of height *h*, and we have two triangles. Finally, if *a* ≥ *b*, we can draw only one triangle. Note that

$$\sin A = \frac{h}{b}, \text{ so } h = b \sin A.$$

We summarize the above considerations as follows:

Proposition 6.2. Two sides a and b, and an **acute** angle A opposite side a are given. Also, we denote $h = b \sin A$ (this is the height of the triangle). Four cases are possible:

(1) If $a < h$, then the triangle does not exist.
(2) If $a = h$, then the triangle is unique. It is a right triangle.
(3) If $a \geq b$, then the triangle is unique. This triangle may be acute, right, or obtuse (related to the top angle).
(4) If $h < a < b$, then there are two triangles. One of them (whose side a is to the left of height h) is always obtuse, and the other (whose side a is to the right of height h) can be an acute, obtuse or a right triangle (see Problem 6.7). ∎

Now let's return to Problem 6.5. Sofia's mistake is that she a priori assumed that Ben's path forms an acute angle ABE with the main road. But this angle may be obtuse, and Ben could walk 100 m from point A according to this figure:

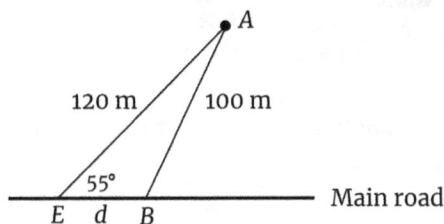

The angle B calculated by Sofia (which is 79.4°) is supplementary to the angle B from this figure. Therefore, in the above obtuse triangle, $B = 180° - 79.4° = 100.6°$. From here,

$$A = 180° - E - B = 180° - 55° - 100.6° = 24.4°.$$

Using the Law of Sines, we calculate the distance d:

$$\frac{d}{\sin 24.4°} = \frac{100}{\sin 55°} \Rightarrow d = \frac{100 \sin 24.4°}{\sin 55°} = 50.4 \text{ m}.$$

Now $d = 50.4$ m < 80 m, so in this case the guys could see each other.

As you can see, in Problem 6.5 we have an ambiguous case, and there are two triangles that satisfy the given data. That means that the information provided in Problem 6.5 is not sufficient to conclude whether or not the guys would see each other. We can only say that it is possible for them to see each other if Ben will go along the path AB shown in the above figure. ∎

In defense of Sofia, we can say that she received an incorrectly stated problem that does not have a unique solution.

In Proposition 6.2, we considered four possible cases of solving triangles with an acute angle A when two sides a and b, and **acute** angle A opposite to side a are given. Based on this proposition, we can describe the following.

Procedure to Recognize and Resolve an Ambiguous Case

Apply Law of Sines $\dfrac{a}{\sin A} = \dfrac{b}{\sin B}$ and solve this equation for $\sin B : \sin B = \dfrac{b \sin A}{a}$. Three cases are possible here:

(1) $\sin B > 1$. The triangle does not exist (sine of any angle cannot be greater than 1).
(2) $\sin B = 1$. Angle $B = 90°$. The triangle is unique. It is a right triangle.
(3) $\sin B < 1$. At least one triangle exists with the angle $B_1 = \sin^{-1}(\sin B)$. To see whether the second triangle exists, calculate the supplementary angle $B_2 = 180° - B_1$. If $B_2 + A < 180°$, then the second triangle exists with the angle B_2, otherwise, not.

Problem 6.6. Return to Problems 6.4 and 6.5. Using the above procedure, once again check the number of their solutions.

Solution.

(a) In Problem 6.4: $a = 25$, $c = 17$, $A = 50°$.

By the Law of Sines, $\dfrac{a}{\sin A} = \dfrac{c}{\sin C}$. From here

$$\sin C = \frac{c \sin A}{a} = \frac{17 \sin 50°}{25} = 0.521 \;\Rightarrow\; C_1 = \sin^{-1}(0.521) = 31.4°.$$

Calculate the supplementary angle C_2 :

$$C_2 = 180° - C_1 = 180° - 31.4° = 148.6°.$$

Since $C_2 + A = 148.6° + 50° = 198.6° > 180°$, the second triangle does not exist, and the triangle is unique. The only angle opposite to the side c is $C = 31.4°$. Therefore, Problem 6.4 has one solution.

(b) In Problem 6.5: $b = 120$, $e = 100$, $E = 55°$.

By the Law of Sines, $\dfrac{b}{\sin B} = \dfrac{e}{\sin E}$. From here

$$\sin B = \frac{b \sin E}{e} = \frac{120 \sin 55°}{100} = 0.983 \Rightarrow B_1 = \sin^{-1}(0.983) = 79.4°.$$

Calculate the supplementary angle B_2:

$$B_2 = 180° - B_1 = 180° - 79.4° = 100.6°.$$

Since $B_2 + E = 100.6° + 55° = 155.6° < 180°$, there are two triangles. Two angles opposite to side b are $B_1 = 79.4°$ and $B_2 = 100.6°$. So, Problem 6.5. has two solutions. ∎

The following problem provides a more detailed analysis of the possible shape of triangles in an ambiguous case.

Problem 6.7. Two sides a and b, and an **acute** angle A opposite side a in a triangle ABC are given. Denote $h = b \sin A$ (this is the height to the side c). Let side c be drawn horizontally, so the height h is drawn vertically (see the figures below in the Solution). Assume that $h < a < b$. As we showed in Proposition 6.2, there are two triangles with the given sides a, b and angle A. One them, whose side a is to the left of height h, is always obtuse. Prove the following statements about the second triangle whose a is to the right of the height h:

(1) If $a = b \tan A$, the triangle is right.
(2) If $a < b \tan A$, the triangle is acute.
(3) If $a > b \tan A$, the triangle is obtuse.

Solution. You can see the proof just looking at the following figures:

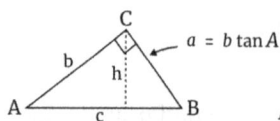

| Triangle *ABC* is right | Triangle *ABC* is acute | Triangle *ABC* is obtuse ∎ |

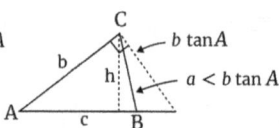

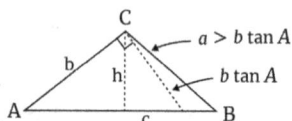

Exercises for Solving on Your Own

- **Three friends go on a camping trip.**

Exercise 6.1. Three friends, Alice, Bob and Carol like to hike together. While camping, everyone stays in their own tent. Here is the information about the location of tents for each trip.

Trip 1: Alice and Carol are 25 m apart. The angle going from Alice to Bob and Carol is 20°, and the angle going from Bob to Alice and Carol is 110°. How far apart are Bob and Carol?

Trip 2: Alice and Carol are 12 m apart. The angle going from Alice to Bob and Carol is 50°, and the angle going from Carol to Alice and Bob is 110°. How far apart are Alice and Bob?

Trip 3: Alice and Bob are 23 m apart, and Bob and Carol are 18 m apart. The angle going from Carol to Alice and Bob is 55°. What is the angle going from Alice to Bob and Carol?

Trip 4: Alice and Bob are 17 m apart, and Bob and Carol are 25 m apart. The angle going from Alice to Bob and Carol is 50°. What is the angle going from Bob to Alice and Carol?

Trip 5: Alice and Carol are 12 m apart, and Alice and Bob are 19 m apart. The angle going from Carol to Alice and Bob is 110°. How far apart are Bob and Carol?

- **Nina tries to measure a parcel of land.**

Exercise 6.2. Nina intends to purchase a parcel of land in the shape of a triangle *ABC* (with sides *a*, *b*, and *c*). She hired a few assessors to measure the parcel. Unfortunately, each of the assessors submitted an incorrect report, and Nina had to fire them. Try to figure out, what was wrong in each of the reports.

Report 1: $a = 370$ feet, $A = 91°$, $b = 400$ feet.
Report 2: $a = 370$ feet, $A = 71°$, $B = 115°$.
Report 3: $a = 370$ feet, $A = 71°$, $B = 65°$, $C = 40°$.
Report 4: $a = 370$ feet, $A = 71°$, $b = 500$ feet, $C = 42°$.
Report 5: $a = 370$ feet, $b = 500$ feet, $c = 120$ feet.
Report 6: $a = 370$ feet, $A = 71°$, $b = 400$ feet.
Report 7: $a = 370$ feet, $A = 71°$, $b = 350$ feet, $C = 54°$.
Report 8: $a = 370$ feet, $A = 41°$, $b = 400$ feet.

Exercise 6.3. Nina made another attempt and hired one more assessor. The assessor submitted the following report: $a = 370$ feet, $A = 61°$, $b = 400$ feet. This time Nina decided not to fire the assessor but to request additional information about this parcel of land. What additional (non-numeric) information is needed?

- **Flagpole problem**

Exercise 6.4. A 2.5 m flagpole is NOT standing up straight on the ground. It is supported by two wires (one on each side going in opposite directions), each 3 m long. Both wires make a 55° angle with the ground. How far apart is each wire from the flagpole?

Entertainment Problem

Problem E 6.1. There is a windowless room with three light bulbs. Outside the room, there are three light switches. You need to figure out which switch controls which light bulb. To do this, you are allowed to manipulate the switches (with the door closed) and then enter the room only once and make your determination. In the initial position, all switches are off.

Chapter 7

Solving Oblique Triangles:
Law of Cosines

In the previous chapter we investigated two out of four cases on solving oblique triangles using the Law of Sines:

(1) One side and two angles are given.
(2) Two sides and an angle opposite one of these sides are given.

Here we look at two other types of cases that cannot be solved directly using the Law of Sines. Namely, we consider the following cases:

(3) Two sides and the angle between them are given.
(4) Three sides are given.

In both cases, the triangle is unique, so we do not have any ambiguity. In case 3, a triangle always exists no matter what sides and angle (acute, right or obtuse) are given. In case 4, however, a triangle does not exist if one of the sides is greater than or equal to the sum of the other two sides. We will assume that this will not happen in our cases.

Case 3: Two sides and the angle between them are given.
Find the third side and the other angles.

Problem 7.1. Eli and Ben went on a forest hike to pick some mushrooms. They started from the same point and each walked in a straight line at a 40° angle relative to each other. Every few minutes, or so, they would call out to each other to avoid being lost. A sound in this forest can be heard up to 60 m away. After 5 minutes Eli walked 80 m and Ben walked 70 m. Would they hear each other at that time?

Solution. Depict this problem graphically:

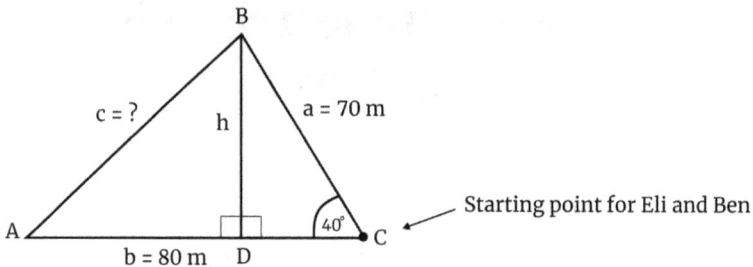

This problem is reduced to finding side c, through sides a and b, and angle C. Since these data completely define side c, we can guess that there should be a formula expressing a side in terms of the other two sides and the angle between them. To derive this formula, draw height h to the side b. Height h splits the triangle ABC into two right triangles: ABD and BCD.

From $\triangle ABD$, $c^2 = AD^2 + h^2 = (AC - DC)^2 + h^2 = (b - DC)^2 + h^2$.
From $\triangle BCD$, $DC = a \cos C$, $h = a \sin C$.
Substituting expressions for DC and h into the above formula for c^2, we get

$$c^2 = (b - a \cos C)^2 + a^2 \sin^2 C = b^2 - 2ab \cos C + a^2 \cos^2 C + a^2 \sin^2 C$$
$$= b^2 - 2ab \cos C + a^2 \left(\cos^2 C + \sin^2 C\right) = b^2 - 2ab \cos C + a^2.$$

Here we used the main identity for trig functions: $\sin^2 C + \cos^2 C = 1$. Thus, we arrive at the following important result which is called the **Law of Cosines**:

In any triangle, the third side is expressed through two others and the angle between them by the formula:

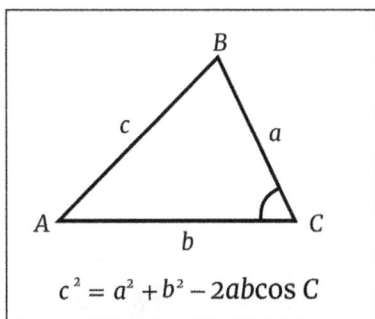

$$c^2 = a^2 + b^2 - 2ab\cos C$$

Law of Cosines

Notes:

(1) Consider a special case when $C = 90°$ (the case of a right triangle). Then $\cos C = \cos 90° = 0$ and the above formula becomes $c^2 = a^2 + b^2$ which is exactly the Pythagorean theorem. Therefore, the Law of Cosines can be considered as a generalization of the Pythagorean theorem to oblique triangles.

(2) We proved the Law of Cosines for an acute angle C. For an obtuse angle, the proof is similar, but slightly different.

To finish solving Problem 7.1, we just need to plug in the given values of a, b and C into the Law of Cosines. We have $a = 70$ m, $b = 80$ m, $C = 40°$. Therefore,

$$c^2 = 70^2 + 80^2 - 2 \cdot 70 \cdot 80 \cdot \cos 40° \approx 2{,}720.3 \Rightarrow c = \sqrt{2{,}720.3} \approx 52.2 \text{ m.}$$

So, after 5 minutes, Eli and Ben will be 52.2 m apart from each other. Since $52.2 < 60$, the guys will be able to hear each other. ∎

In deriving the Law of Cosines, we have expressed side c through sides a, b and the angle C, which is between them. Since all three sides play the same role, no one side has any privileges over the others. Therefore, we can write similar expressions for sides a and b:

$$a^2 = b^2 + c^2 - 2bc \cos A \quad \text{and} \quad b^2 = a^2 + c^2 - 2ac \cos B.$$

The Law of Cosines also allows us to express the cosine of any angle through three sides. To do this, just solve the above equations for cosine:

$$\cos A = \frac{b^2 + c^2 - a^2}{2bc}, \cos B = \frac{a^2 + c^2 - b^2}{2ac}, \cos C = \frac{a^2 + b^2 - c^2}{2ab}. \quad (1)$$

From these formulas, to find the value of the angle from cosine, we can use the inverse cosine function on a calculator (via $\cos^{-1}$ button). We will discuss this function in detail in Chapter 16. We have

$$A = \cos^{-1}\left(\frac{b^2 + c^2 - a^2}{2bc}\right), B = \cos^{-1}\left(\frac{a^2 + c^2 - b^2}{2ac}\right),$$

$$C = \cos^{-1}\left(\frac{a^2 + b^2 - c^2}{2ab}\right).$$

In the next problem, the Law of Cosines is used 3 times.

Problem 7.2. The circle inscribed in triangle ABC touches sides AB, BC, and AC at points D, E and F, respectively. Find DE if $AF = 3$, $FC = 2$ and $\angle A = 60°$:

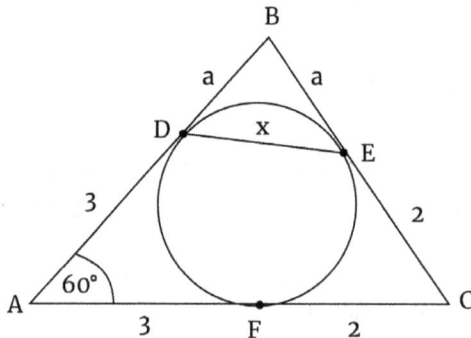

Solution.

(1) Denote $a = DB = BE$. First, we will find a, using the Law of Cosines for $\triangle ABC$:

$$BC^2 = AB^2 + AC^2 - 2AB \cdot AC \cdot \cos A.$$

From here, obtain the equation for *a*:

$$(2+a)^2 = (3+a)^2 + (3+2)^2 - 2(3+a)(3+2)\cos 60°.$$

Or

$$4+4a+a^2 = 9+6a+a^2 +25-2\cdot(3+a)5\cdot\frac{1}{2} \Rightarrow 3a=15 \Rightarrow a=5.$$

(2) Next, we will find cos *B*, using the Law of Cosines for $\triangle ABC$ one more time:

$$AC^2 = AB^2 + BC^2 - 2AB \cdot BC \cdot \cos B.$$

From here

$$(3+2)^2 = (3+5)^2 + (2+5)^2 - 2(3+5)(2+5)\cos B \Rightarrow \cos B = \frac{11}{14}.$$

(3) Finally, using the Law of Cosines for $\triangle DBE$, we get

$$DE^2 = DB^2 + BE^2 - 2DB \cdot BE \cdot \cos B.$$

Or

$$DE^2 = 2a^2 - 2a^2\frac{11}{14} = 50 - 50\cdot\frac{11}{14} = \frac{25\cdot 3}{7} \Rightarrow$$

$$DE = \sqrt{\frac{25\cdot 3}{7}} = 5\sqrt{\frac{3}{7}} = \frac{5\sqrt{21}}{7}. \quad \blacksquare$$

When solving problems to find an unknown angle in a triangle, we should be cautious in certain cases. Look at this problem.

Problem 7.3. Ann and Dan tried to solve the following problem:

Three friends, Alice, Bob and Carol, are camping in their own tents on a flat meadow in the woodland. Alice and Carol are 15 m apart, and Carol and Bob are 50 m apart. The angle going from Carol to Alice and Bob is 55°. What is the angle going from Alice to Bob and Carol?

Solution. Both Ann and Dan realized that geometrically they have a triangle ABC in which two sides are given ($a = 50$ and $b = 15$), as well as the angle in between ($C = 55°$). The problem is to find angle A. The friends decided to solve this problem in two steps: first, find side c, and then angle A.

- **Step 1:** Ann and Dan started with the Law of Cosines:

$$c^2 = a^2 + b^2 - 2ab \cos C = 50^2 + 15^2 - 2 \cdot 50 \cdot 15 \cdot \cos 55°$$

$$= 2500 + 225 - 1500 \cdot 0.5736 = 1,864.6$$

$$\Rightarrow c = \sqrt{1,864.6} = 43.18.$$

- **Step 2:** At this point, Ann and Dan know all three sides of the triangle and angle C. To find angle A, they see two approaches: to continue with the Law of Cosines or use the Law of Sines. Ann decided to continue with the Law of Cosines, while Dan chose to go with the Law of Sines. See what happened.

  Ann's solution. Using the Law of Cosines, she calculated

$$\cos A = \frac{b^2 + c^2 - a^2}{2bc} = \frac{15^2 + 1,864.63 - 50^2}{2 \cdot 15 \cdot 43.18} = -0.3168.$$

Pressing $\cos^{-1}$ on a calculator, she got $A = \cos^{-1}(-0.3168) = 108.5°$.
  Dan's solution. Using the Law of Sines, he calculated

$$\frac{a}{\sin A} = \frac{c}{\sin C} \Rightarrow \sin A = \frac{a \sin C}{c} = \frac{50 \sin 55°}{43.18} = 0.9485.$$

Pressing $\sin^{-1}$ button on a calculator, he got $A = \sin^{-1}(0.9485) = 71.5°$. The friends got two different answers: $A = 108.5°$ by using the Laws of Cosines and $A = 71.5°$ by using the Law of Sines! Which answer is correct? And why did they get different results? Can you guess?

Well, the inverse sine function $\sin^{-1}$ does not return values greater than 90°. It only returns **acute** angles (we will study inverse sine in detail in Chapter 15). However, the actual angle in a triangle may be obtuse. So, this problem has an issue. Let's check the acute angle $A = 71.5°$ in the following way. Using it and the given angle $C = 55°$, calculate angle B: $B = 180° - A - C = 180° - 71.5° - 55° = 53.5°$. Now check the Law of Sines: $\dfrac{a}{\sin A} = \dfrac{b}{\sin B}$. The left side is $\dfrac{a}{\sin A} = \dfrac{50}{\sin 71.5°} \approx 52.7$, but the right side is $\dfrac{b}{\sin B} = \dfrac{15}{\sin 53.5°} \approx 18.7$. We got a violation of the Law of Sines, and we must conclude that the answer $A = 71.5°$ is wrong. Therefore, Dan's way of using the Law of Sines is incorrect. Ann got the correct answer $A = 108.5°$ using the Laws of Cosines. ■

We have learned from this example that we need to be careful in using the Law of Sines. One of the recommendations is this: if possible, use the Law of Cosines instead of Law of Sines because cosine takes positive values for acute angles and negative values for obtuse ones. Therefore, inverse cosine can distinguish acute and obtuse angles. On the contrary, sine takes only positive values for both acute and obtuse angles, so the inverse sine cannot distinguish them. If the correct angle is obtuse, and we use $\sin^{-1}$ to find it, we will get the acute angle, which is wrong.

Another recommendation in using the Law of Sines is to NOT start your calculation with the angle opposite the longest side (this angle is the largest and might be obtuse). Start with another angle, which is definitely acute.

In Problem 7.3, if Dan wants to use the Law of Sines in the second step, he should not start with angle A which is opposite the longest side $a = 50$. Instead, he should first calculate angle B, which is guaranteed to be acute since it is opposite the shortest side $b = 15$:

$$\frac{b}{\sin B} = \frac{c}{\sin C} \Rightarrow \sin B = \frac{b \sin C}{c} = \frac{15 \sin 55°}{43.18}$$

$$\approx 0.2846 \Rightarrow B = \sin^{-1}(0.2846) = 16.5°.$$

After that, Dan could calculate the required angle A by subtracting angles B and C from 180°:

$$A = 180° - B - C = 180° - 16.5° - 55° = 108.5°.$$

This is the correct answer.

Case 4: Three sides are given. Find the three angles.

Using the Law of Cosines, we can start with any side. We recommend starting from the longest side and finding the opposite angle. In doing so, we guarantee that the other two angles are acute, and to find them we can use either the Law of Cosines again, or the Law of Sines (there will not be such a mistake as Dan experienced in Problem 7.3). Here are our general recommendations:

> (1) When using the Law of Sines, start with the shortest side.
> (2) When using the Law of Cosines, start with the longest side.

Problem 7.4. On the next trip, our three friends, Alice, Bob, and Carol, set up tents such that Alice and Bob are 21 m apart, Alice and Carol are 25 m apart, and Bob and Carol are 7 m apart. Find the angles going from Alice to Bob and Carol, from Bob to Alice and Carol, and from Carol to Alice and Bob.

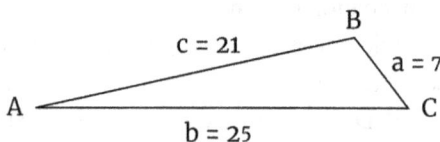

Solution. The problem is to find ∡A, ∡B, and ∡C.

(1) According to the above recommendation, we use the Law of Cosines starting with the longest side $b = 25$ m and calculate $\cos B$:

$$\cos B = \frac{a^2 + c^2 - b^2}{2ac} = \frac{7^2 + 21^2 - 25^2}{2 \cdot 7 \cdot 21} = -0.4592.$$

From here, $B = \cos^{-1}(-0.4592) = 117.3°$.

(2) Next, we use the Law of Sines to find another angle. At this stage, it is safe (in the sense of avoiding a mistake) to calculate either angle A or C, since both are acute. We calculate angle C:

$$\frac{b}{\sin B} = \frac{c}{\sin C} \Rightarrow \sin C = \frac{c \sin B}{b} = \frac{21 \sin 117.3°}{25} \approx 0.7464,$$

$$C = \sin^{-1}(0.7464) = 48.3°.$$

(3) What remains is to calculate angle A: $A = 180° - B - C = 180° - 117.3° - 48.3° = 14.4°$.

Final answer: $A = 14.4°$, $B = 117.3°$, $C = 48.2°$. ∎

Note. If you do not follow the above recommendation, and in step 1 you do not start from the longest side b, but from sides a or c, and then in step 2 use the Law of Sines to calculate angle B, the result will be wrong: angle B will be acute.

In conclusion of this chapter, we describe a method on how to determine, by three given sides, what type of triangle it is: acute, right, or obtuse.

Problem 7.5. Let a, b, and c be the three sides of a triangle, and let c be the longest side. Denote $E = a^2 + b^2 - c^2$. Prove that

(1) If $E > 0$, then the triangle is acute.
(2) If $E < 0$, then the triangle is obtuse.
(3) If $E = 0$, then the triangle is right.

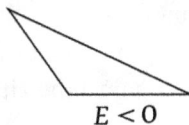

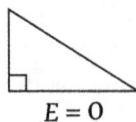

$$E > 0 \qquad E < 0 \qquad E = 0$$

Solution. Using the Law of Cosines, $\cos C = \dfrac{a^2 + b^2 - c^2}{2ab} = \dfrac{E}{2ab}$.

(1) If $E > 0$, then $\cos C > 0$ and $C < 90°$. Since c is the longest side, C is the largest angle. Therefore, two other angles are also less than $90°$ and the triangle is acute.

(2) If $E < 0$, then cos $C < 0$ and $C > 90°$. The triangle is obtuse.

(3) If $E = 0$, then cos $C = 0$ and $C = 90°$. The triangle is right. ∎

Exercises for Solving on Your Own

Exercise 7.1. Eli and Ben are in two separate boats moving away from each other at an angle of 70°. Eli traveled 750 m and Ben, 830 m. What is the distance between the boats?

Exercise 7.2. Lillian needs to measure the distance between two trees A and B, which are on opposite sides of a small pond.

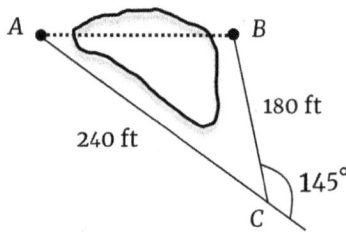

She started at tree A and walked 240 feet in a straight line along the pond. Then she made a 145° turn toward the second tree B and walked 180 feet until she reached the second tree. What is the distance between the trees?

Exercise 7.3. The distance from a small airplane and the airport is 60 miles. The airplane is going down for landing. However, its navigation device malfunctions and incorrectly shows the distance to be 65 miles. The dispatcher notices this mistake and figures out that if the airplane continues on its current course, it will end up 16 miles from the airport. By how many degrees should the dispatcher adjust the airplane's heading?

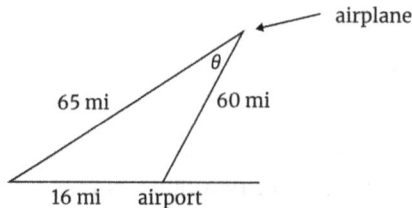

Exercise 7.4. Eli (*E*), Ben (*B*) and their father (*F*) are located at the following distances from each other: $EF = 500$ m, $BF = 800$ m, and $BE = 700$ m. The father has a small telescope that allows him to observe objects at a maximum angle of 40°. Could the father see both children at the same time through the telescope?

Exercise 7.5. Straight roads connect three cities *A*, *B*, and *C*. The roads from city *A* to cities *B* and *C* diverge at a 30° angle. It is known that the distance between cities *A* and *B* is 9 km, and between *A* and *C* is 4 km. At what angle do the roads from city *C* diverge?

Exercise 7.6. After Hurricane Sandy, a 10-foot-long tree started to lean. To keep it from falling, a 19-foot strap was tied to the top of the tree. The strap was attached to the ground 11 feet from the base of the tree. At what angle did the tree lean?

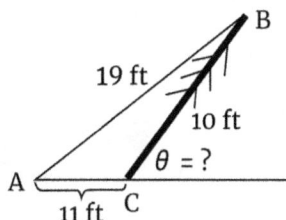

Exercise 7.7. Sides of a triangle are 5, 13, and 9. What kind of triangle is it: acute, right, or obtuse?

Exercise 7.8. Let the hour and minute hands on a clock have the lengths of 4 and 6 cm, respectively. Find the distance between their ends at 1:25 pm.

Entertainment problem

Problem E7.1. Professor Smartman was so busy with his research that he forgot to wind his cuckoo wall clock, and it stopped. In addition, his wristwatch also broke. After finishing his work, the professor went to visit his friend who lives within walking distance. Having spent the evening with his friend, the professor returned home and was able to set his wall clock correctly. How was he able to do it, if the travel time was not known in advance?

Chapter 8

Addition and Double Angle Formulas

In some previous chapters, we considered problems that require finding the distance between different objects. Here, we investigate more "distance" problems. We start with a relatively simple geometric problem. Then we look at much more complex problems that will lead us to important new results in trigonometry.

Problem 8.1. Two tourists, A and B, became separated and lost in the mountains. They were located by a plane, and their coordinates were established as A (6.2, 4.8) and B (4.5, 3.4) (for simplicity, we use Cartesian coordinates, not real coordinates on Earth). All measurements are made in miles. We need to determine how far they are from each other.

Solution. This problem can easily be solved in the general case. In the coordinate system, consider two points A and B with coordinates $A(x_1, y_1)$ and $B(x_2, y_2)$:

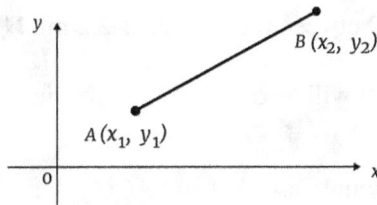

Mark the coordinates of the points A and B on the coordinate axes:

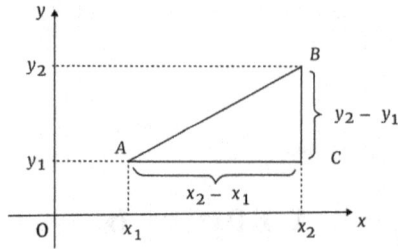

By the Pythagorean Theorem, $AB = \sqrt{AC^2 + BC^2}$. You can see that $AC = x_2 - x_1$ and $BC = y_2 - y_1$. From here, $AB = \sqrt{(x_2 - x_1)^2 + (y_2 - y_1)^2}$. We have obtained the following result.

The distance AB between points $A(x_1, y_1)$ and $B(x_2, y_2)$ is calculated by the formula:

Distance Formula

$$AB = \sqrt{(x_2 - x_1)^2 + (y_2 - y_1)^2}$$

You can verify that this formula is valid regardless of the quadrants in which points A and B are located. Note that this formula is symmetric with respect to points A and B: it remains the same if we exchange x_1 and x_2, and/or y_1 and y_2.

Applying this formula to the given points A (6.2, 4.8) and B (4.5, 3.4), we get:

$$AB = \sqrt{(6.2 - 4.5)^2 + (4.8 - 3.4)^2} = \sqrt{1.7^2 + 1.4^2}$$
$$= \sqrt{2.89 + 1.96} = \sqrt{4.85} \approx 2.2.$$

So, the distance between the tourists is 2.2 mi. ■

The next problem will lead us to one of the most important results in trigonometry.

Problem 8.2. Two mountains ABD and ACD stand next to each other. It was possible to measure the following distances: $AB = a$, $BD = e$, $AD = c$, $AC = d$, and $CD = b$. Based on these data, we need to calculate the distance between the peaks B and C of the mountains.

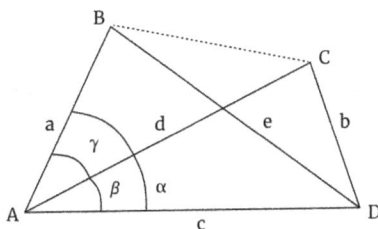

Figure 1.

Solution. Mathematically, we have two triangles *ABD* and *ACD* with a common side *AD*. All sides of both triangles are given. The problem is to find the distance between vertices *B* and *C*. In the figure above, we also marked angles α, β, and γ such that $\alpha = \beta + \gamma$.

Another interpretation of this problem is this: in a convex quadrilateral *ABCD* (a polygon with four sides), three sides and both diagonals are given. The problem is to find the fourth side.

To find *BC*, we use the Law of Cosines for $\triangle ABC$: $BC^2 = a^2 + d^2 - 2ad \cos \gamma$. Since $\gamma = \alpha - \beta$, we get

$$BC = \sqrt{a^2 + d^2 - 2ad \cos(\alpha - \beta)}.$$

In this expression, what's left is to find $\cos(\alpha - \beta)$. Using the Law of Cosines again, we can find separately $\cos \alpha$ and $\cos \beta$. Indeed, from $\triangle ABD$, $\cos \alpha = (a^2 + c^2 - e^2)/(2ac)$ and from $\triangle ACD$, $\cos \beta = (c^2 + d^2 - b^2)/(2cd)$. If we could express $\cos(\alpha - \beta)$ through $\cos \alpha$ and $\cos \beta$, then our problem would be completely solved. Fortunately, such a formula exists! It is called the **cosine difference formula,** and it is as follows:

$$\cos(\alpha - \beta) = \cos \alpha \cos \beta + \sin \alpha \sin \beta.$$

We'll prove this formula below. As for now, we'll just use it. Note that this formula contains not only $\cos \alpha$ and $\cos \beta$ but also $\sin \alpha$ and $\sin \beta$. To complete the solution, we'll need expressions for each of them. We can use the main identity that we discussed in Chapter 5.

Using it, we have $\sin \alpha = \sqrt{1 - \cos^2 \alpha}$, $\sin \beta = \sqrt{1 - \cos^2 \beta}$.

If we denote

$$M = \frac{a^2 + c^2 - e^2}{2ac}, \quad N = \frac{c^2 + d^2 - b^2}{2cd}.$$

then

$$\cos \alpha = M, \cos \beta = N, \sin \alpha = \sqrt{1 - M^2}, \sin \beta = \sqrt{1 - N^2}.$$

Substitute these expressions into the cosine difference formula:

$$\cos(\alpha - \beta) = MN + \sqrt{(1 - M^2)(1 - N^2)}.$$

Finally, we get an expression for the distance BC through all known data:

$$BC = \sqrt{a^2 + d^2 - 2ad \left[MN + \sqrt{(1 - M^2)(1 - N^2)} \right]}. \quad \blacksquare$$

Note. A Great Swiss mathematician, Leonhard Euler (1707–1783), discovered a wonderful formula, now called the **Euler's Quadrilateral Theorem,** which describes the relationship between the sides and diagonals of a convex quadrilateral. In this theorem, additional data are used: a line segment that connects the midpoints of the diagonals. If we denote the length of this line segment as g, denote sides of the quadrilateral as a, b, c, f, and diagonals as d and e (as in Figure 1), then Euler's Quadrilateral Theorem asserts that

$$a^2 + b^2 + c^2 + f^2 = d^2 + e^2 + 4g^2.$$

The above formula for BC (which, of course, does not look as nice as Euler's formula) relates only six line segments: four sides and two diagonals, whereas Euler's formula adds to these data the seventh (unnecessary) line segment.

Now we present a proof of the cosine difference formula.

The Cosine Difference Formula

$$\cos(\alpha - \beta) = \cos \alpha \cos \beta + \sin \alpha \sin \beta$$

To prove it, we draw angles α and β in the unit circle within a coordinate system. To show the figure more clearly, we consider the case when the angle α is in the 2nd quadrant, and angle β is in the 1st. Here is the figure:

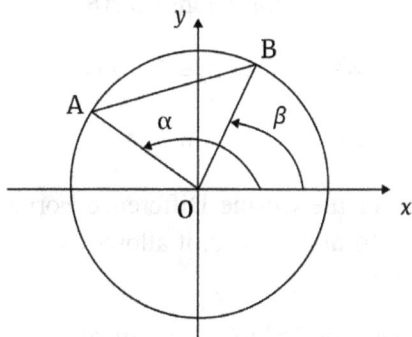

You can see that the angle $AOB = \alpha - \beta$. Point A corresponds to the angle α on the unit circle. Therefore, the coordinates of this point are $(\cos \alpha, \sin \alpha)$. Similarly, the coordinates of point B are $(\cos \beta, \sin \beta)$. To get the formula for $\cos(\alpha - \beta)$, we calculate the length of the chord AB in two ways: by the distance formula (see Problem 8.1), and by the Law of Cosines. Then we equate the two expressions. For further convenience, we calculate the squares of these distances.

Using the distance formula,

$$\begin{aligned}
AB^2 &= (\cos\alpha - \cos\beta)^2 + (\sin\alpha - \sin\beta)^2 \\
&= \cos^2\alpha - 2\cos\alpha \cdot \cos\beta + \cos^2\beta + \sin^2\alpha - 2\sin\alpha \cdot \sin\beta + \sin^2\beta \\
&= (\cos^2\alpha + \sin^2\alpha) + (\cos^2\beta + \sin^2\beta) - (2\cos\alpha \cdot \cos\beta + 2\sin\alpha \cdot \sin\beta) \\
&= 2 - 2(\cos\alpha \cdot \cos\beta + \sin\alpha \cdot \sin\beta).
\end{aligned}$$

Here we also used the main identity (see Chapter 5). Now, we calculate the same AB^2 using the Law of Cosines for the triangle AOB:

$$AB^2 = AO^2 + BO^2 - 2AO \cdot BO \cdot \cos(\alpha - \beta).$$

Since AO and BO are radii of the unit circle, $AO = BO = 1$. We get

$$AB^2 = 1 + 1 - 2\cos(\alpha - \beta) = 2 - 2\cos(\alpha - \beta).$$

Now we equate the two expressions for AB^2:

$2 - 2(\cos \alpha \cdot \cos \beta + \sin \alpha \cdot \sin \beta) = 2 - 2\cos(\alpha - \beta)$. From here,

$\cos(\alpha - \beta) = \cos \alpha \cos \beta + \sin \alpha \sin \beta.$ ■

The importance of the Cosine Difference Formula is that it is the source of many other formulas. Also, it allows us to make some calculations in exact radical form.

Problem 8.3. Calculate $\cos 15°$ in exact radical form, without the use of a calculator.

Solution. We have $\cos 15° = \cos(45° - 30°)$

$= \cos 45° \cdot \cos 30° + \sin 45° \cdot \sin 30° = \dfrac{\sqrt{2}}{2} \cdot \dfrac{\sqrt{3}}{2} + \dfrac{\sqrt{2}}{2} \cdot \dfrac{1}{2} = \dfrac{\sqrt{6} + \sqrt{2}}{4}.$ ■

The following problem leads to another, so-called the **cosine sum formula**.

Problem 8.4. Four friends A, B, C, and D live in different houses, as shown below. Everyone knows the distance from his/her house to the others, except for the distance between B and D. The problem is to find the distance BD.

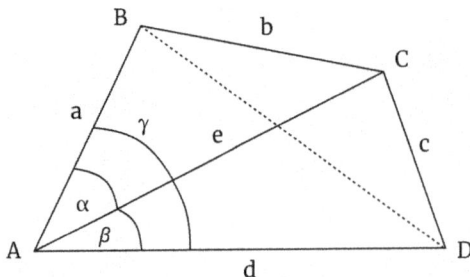

Solution. Mathematically, this problem is to find a diagonal in a convex quadrilateral, given all four sides and one of the diagonals. In the figure above, we also marked angles α, β and γ such that $\gamma = \alpha + \beta$. We will use the Law of Cosines. From $\triangle ABC$, $\triangle ACD$, and $\triangle ABD$ we get, respectively,

$$\cos \alpha = (a^2 + e^2 - b^2) / (2a \cdot e)$$
$$\cos \beta = (d^2 + e^2 - c^2) / (2d \cdot e)$$
$$BD^2 = a^2 + d^2 - 2a \cdot d \cdot \cos(\alpha + \beta).$$

Now we need to express $\cos(\alpha + \beta)$ in terms of $\cos \alpha$ and $\cos \beta$. Such a formula is easily derived from the formula for $\cos(\alpha - \beta)$. Indeed, if we denote $\theta = -\beta$ and use the even property of cosine and odd property of sine, then

$$\cos(\alpha + \beta) = \cos(\alpha - \theta) = \cos \alpha \cdot \cos \theta + \sin \alpha \cdot \cos \theta$$
$$= \cos \alpha \cdot \cos(-\beta) + \sin \alpha \cdot \sin(-\beta) = \cos \alpha \cdot \cos \beta - \sin \alpha \cdot \sin \beta.$$

So, we obtained the formula for cosine of a sum:

The Cosine Sum Formula

$$\boxed{\cos(\alpha + \beta) = \cos \alpha \cdot \cos \beta - \sin \alpha \cdot \sin \beta}$$

Combining this formula with the above expressions for $\cos \alpha$, $\cos \beta$ and BD^2, we can get a final result just as in Problem 8.2. Details are left to the reader. ■

Solve the problem below on your own.

Problem 8.5. Calculate $\cos 75°$ in exact radical form. ■

The next problem will lead us to another useful formula.

Problem 8.6. The figure below shows swamps, marked by shaded regions. Using special structural supports in the form of right triangles *AED* and *FBG*, it was possible to strengthen the ground and measure some distances.

Here are the results of all measurements (to generalize the problem, we use letters instead of specific numbers):

$$AC = a, DE = b, AE = c, FG = d, \text{ and } FB = e.$$

Based on these data, we need to determine the distance between points A and B.

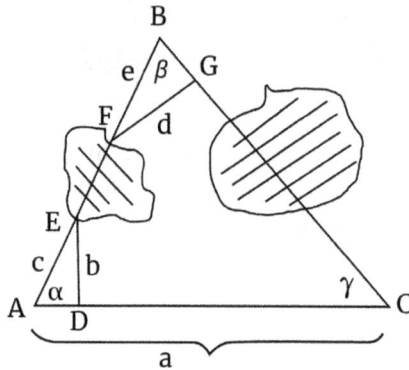

In the figure above, we also marked angles α, β and γ.

Solution. To find the distance AB, we will apply the Law of Sines to the triangle ABC:

$$\frac{AB}{\sin \gamma} = \frac{AC}{\sin \beta} \Rightarrow AB = \frac{AC \cdot \sin \gamma}{\sin \beta} = \frac{a \cdot \sin \gamma}{\sin \beta}.$$

Since $\gamma = 180° - \alpha - \beta$, we can use the reduction formula for $\sin \gamma$:

$$\sin \gamma = \sin(180° - \alpha - \beta) = \sin[180° - (\alpha + \beta)] = \sin(\alpha + \beta).$$

From here,

$$AB = \frac{a \cdot \sin \gamma}{\sin \beta} = \frac{a \cdot \sin(\alpha + \beta)}{\sin \beta}.$$

To complete the solution, we need to express $\sin(\alpha + \beta)$ and $\sin \beta$ through known data. To get an expression for $\sin(\alpha + \beta)$, we can use the expression for $\cos(\alpha - \beta)$ and the reduction formula:

$$\begin{aligned}
\sin(\alpha + \beta) &= \cos[90° - (\alpha + \beta)] = \cos[(90° - \alpha) - \beta] \\
&= \cos(90° - \alpha) \cdot \cos \beta + \sin(90° - \alpha) \cdot \sin \beta \\
&= \sin \alpha \cdot \cos \beta + \cos \alpha \cdot \sin \beta.
\end{aligned}$$

So, we get

The Sine Sum Formula

$$\sin(\alpha + \beta) = \sin \alpha \cdot \cos \beta + \cos \alpha \cdot \sin \beta$$

To find sine and cosine of the angles α and β, we consider right triangles *AED* and *FBG*. By the Pythagorean theorem we can find *AD* (from the triangle *AED*) and *BG* (from the triangle *FBG*):

$$AD = \sqrt{c^2 - b^2}, BG = \sqrt{e^2 - d^2}.$$

From here,

$$\sin \alpha = \frac{b}{c}, \cos \alpha = \frac{AD}{c} = \frac{\sqrt{c^2 - b^2}}{c}, \sin \beta = \frac{d}{e}, \cos \beta = \frac{BG}{e} = \frac{\sqrt{e^2 - d^2}}{e}.$$

Now we can get an expression for *AB* through the given data:

$$\begin{aligned}
AB &= \frac{a \cdot \sin(\alpha + \beta)}{\sin \beta} = \frac{ae}{d}\left[\frac{b}{c} \cdot \frac{\sqrt{e^2 - d^2}}{e} + \frac{\sqrt{c^2 - b^2}}{c} \cdot \frac{d}{e}\right] \\
&= \frac{a\left(b\sqrt{e^2 - d^2} + d\sqrt{c^2 - b^2}\right)}{cd}. \blacksquare
\end{aligned}$$

Solve the problem below on your own.

Problem 8.7. Prove the following:

The Sine Difference Formula

$$\sin(\alpha - \beta) = \sin \alpha \cdot \cos \beta - \cos \alpha \cdot \sin \beta \quad \blacksquare$$

As direct corollaries of the cosine and sine sum formulas, we can get formulas for double angles by simply setting $\beta = \alpha$.

Problem 8.8. Prove the following formula:

Double Angle Cosine Formula

$$\cos 2\alpha = \cos^2 \alpha - \sin^2 \alpha$$
$$= 2 \cos^2 \alpha - 1$$
$$= 1 - 2 \sin^2 \alpha$$

Solution. Take the cosine sum formula $\cos(\alpha + \beta) = \cos \alpha \cdot \cos \beta - \sin \alpha \cdot \sin \beta$ and set $\beta = \alpha$. We will immediately get the first expression $\cos^2 \alpha - \sin^2 \alpha$. To get the second expression, we use the main identity in the form $\sin^2 \alpha = 1 - \cos^2 \alpha$ and substitute it into the first expression:

$$\cos^2 \alpha - \sin^2 \alpha = \cos^2 \alpha - (1 - \cos^2 \alpha) = 2 \cos^2 \alpha - 1.$$

Similarly, to get the third expression, we replace in the first expression $\cos^2 \alpha$ with $1 - \sin^2 \alpha$. $\blacksquare$

Solve the problem below on your own.

Problem 8.9. Prove the following formula:

Double Angle Sine Formula

$$\sin 2\alpha = 2 \sin \alpha \cos \alpha \quad \blacksquare$$

We will apply double angle formulas in the next problem.

Problem 8.10. To renovate a round circus arena, it became necessary to measure its diameter. However, since repair work is being carried out at its center, it prevents direct measurements. Nevertheless, it was possible to measure the chords between points *A, B, C,* and *D* according to this figure:

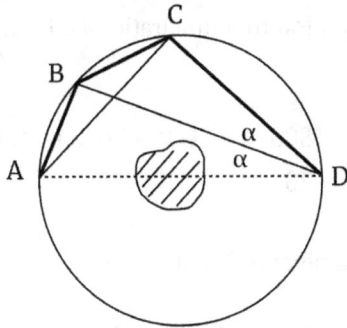

It was measured that $AB = BC = 30$ m, $CD = 70$ m. In the figure, we need to calculate the diameter *AD* which is marked with a dotted line.

Solution. Triangles *ABD* and *ACD* are inscribed in a circle with a common side *AD*, which is the diameter of the circle. Since *AD* is the diameter, triangles *ABD* and *ACD* are right triangles. Also, since $AB = BC$, angles *ADB* and *BDC* are equal: $\angle ADB = \angle BDC = \alpha$, so $\angle ADC = 2\alpha$. From the right triangle *ACD*,

$$\cos 2\alpha = \frac{CD}{AD} = \frac{70}{AD}.$$

From the right triangle *ABD*,

$$\sin \alpha = \frac{AB}{AD} = \frac{30}{AD}.$$

We can get an equation for *AD* if we express cos 2α through sin α. To do this, use the above double angle cosine formula $\cos 2\alpha = 1 - 2\sin^2 \alpha$. We have

$$\cos 2\alpha = \frac{70}{AD} = 1 - 2\sin^2 \alpha = 1 - 2 \cdot \frac{900}{AD^2},$$

or

$$\frac{70}{AD} = 1 - \frac{1800}{AD^2}.$$

We can reduce this equation to a quadratic one by multiplying both sides by AD^2. We get

$$70AD = AD^2 - 1800 \implies AD^2 - 70AD - 1800 = 0 \implies$$
$$(AD - 90)(AD + 20) = 0.$$

From here, the diameter $AD = 90$ m. ■

As we saw, using only the Cosine Difference Formula, we derived formulas of the sum and difference for sine. All these formulas are called **Addition Formulas**, and they express trigonometric functions of angles $(\alpha \pm \beta)$ in terms of angles α and β. Special cases of addition formulas are double angle formulas. In the next chapter, we will see other useful formulas that we will derive from the basic Cosine Difference Formula.

Let's summarize the results obtained.

Distance Formula

The length AB of the line segment between endpoints $A(x_1,y_1)$ and $B(x_2,y_2)$ is determined by the formula:

$$AB = \sqrt{(x_2 - x_1)^2 + (y_2 - y_1)^2}$$

Addition Formulas (Sum and Difference)

For any angles α and β, the following equalities are true:

$$\sin(\alpha + \beta) = \sin \alpha \cdot \cos \beta + \cos \alpha \cdot \sin \beta$$
$$\cos(\alpha + \beta) = \cos \alpha \cdot \cos \beta - \sin \alpha \cdot \sin \beta$$
$$\sin(\alpha - \beta) = \sin \alpha \cdot \cos \beta - \cos \alpha \cdot \sin \beta$$
$$\cos(\alpha - \beta) = \cos \alpha \cdot \cos \beta + \sin \alpha \cdot \sin \beta$$

Double Angle Formulas

$$\cos 2\alpha = \cos^2 \alpha - \sin^2 \alpha$$
$$= 2 \cos^2 \alpha - 1$$
$$= 1 - 2 \sin^2 \alpha$$
$$\sin 2\alpha = 2 \sin \alpha \cos \alpha$$

Exercises for Solving on Your Own

Exercise 8.1. Derive the equation of a circle with radius 3, centered at the origin of the coordinate system.

Exercise 8.2. Express $\cos (330° + \theta)$ through $\sin \theta$ and $\cos \theta$.

Exercise 8.3. Prove that $\cos \theta \cos 2\theta \cos 4\theta = \sin 8\theta/(8 \sin \theta)$.

Exercise 8.4. Prove the following Tangent of the Sum formula:

$$\tan(\alpha + \beta) = (\tan \alpha + \tan \beta)/(1 - \tan \alpha \cdot \tan \beta).$$

Exercise 8.5. Derive the Double Angle Tangent formula:

$$\tan 2\alpha = \frac{2 \tan \alpha}{1 - \tan^2 \alpha}.$$

Exercise 8.6. Calculate the exact values without using a calculator.

(a) $\sin 10° \cos 20° + \cos 10° \sin 20°$;
(b) $\sin 55° \sin 65° - \sin 25° \sin 35°$.

Exercise 8.7. Calculate the exact value of $\cos 36° \cos 72°$ without using a calculator.

Exercise 8.8. In an isosceles triangle ABC, the base AC is b, and the height to BC is h. Find angle B.

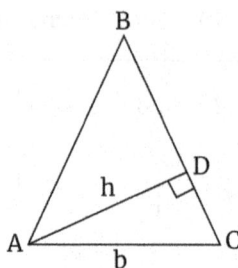

Exercise 8.9. Two radar stations are located 15 miles apart. They detect an aircraft between them. The angle of elevation measured by the first station is 20°, and the angle of elevation measured by the second station is 40°. Find the altitude of the aircraft.

Exercise 8.10. Returning to the previous exercise, solve it in a general form. Let d be the distance between the radar stations, and let A and B be the angles of elevation to the aircraft from the stations. As before, the aircraft is between the radar stations. Prove that the altitude h of the aircraft can be calculated by the formula $h = \dfrac{d \sin A \sin B}{\sin (A + B)}$. It is possible to input this formula into a computer program to calculate the altitude instantly and track it during the aircraft's flight between the radar stations.

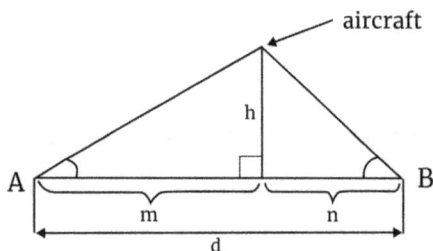

Exercise 8.11. Back to Exercise 8.9. This time, an aircraft is detected on the right side of both stations. The distance between the stations is 15 miles, and the angles of elevation to the aircraft are 20° and 70°. Find the altitude of the aircraft.

Exercise 8.12. Returning to the previous exercise, solve it in a general form. Let d be the distance between the radar stations, and let A and B be the angles of elevation to the aircraft. As before, the aircraft is on the same side (left or right) of both radar stations. Prove that the altitude h of the aircraft can be calculated by the formula $h = \dfrac{d \sin A \sin B}{\sin |A - B|}$.

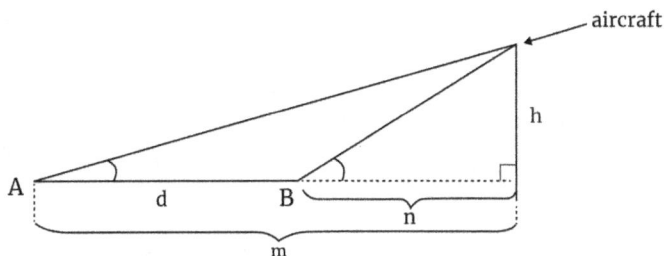

Entertainment Problems

Problem E8.1. How would you measure out 4 L of water using 3 L and 5 L measuring containers?

Problem E8.2. There are two glasses: one filled with coffee, and the other with milk. From the glass of milk, you take a spoonful of milk and transfer it to the glass of coffee and stir. Then take the same spoonful of the resulting mixture and transfer it back to the glass of milk. Question: looking at the two glasses, which quantity is greater, milk in coffee or coffee in milk? What is the answer if you make several transfers?

Chapter 9

Product, Sum, and Difference of Trigonometric Functions

In the previous chapter, we derived addition formulas from the main Cosine Difference Formula. In this chapter, we will derive many more useful formulas by employing the formulas we have already obtained. As one of the results, for instance, we'll be able to get exact values for trig functions of various "non-special" angles.

Problem 9.1. A parcel of land has a shape of a quadrilateral $ABCD$ and is divided into two triangular parts: ABD and BCD.

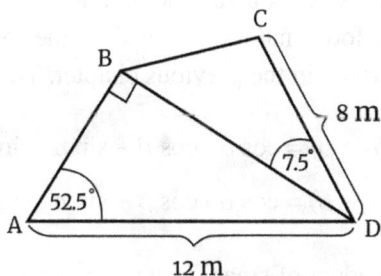

We need to plant flowers in area BCD (we don't need flowers in area ABD) with a density of 3 flower bushes/m^2. The following measurements were made:

$$AD = 12 \ m, CD = 8 \ m, \angle ABD = 90°, \angle BAD = 52.5°, \angle BDC = 7.5°.$$

How many flower bushes can we plant in the area *BCD*? Assume that only a simple calculator is available that cannot calculate trig functions.

Solution. The problem is reduced to calculating the area *S* of the triangle *BCD*. Using the result of Problem 1.5, Chapter 1, we have

$$S = \frac{1}{2} CD \cdot BD \cdot \sin \angle BDC = 4 \cdot BD \cdot \sin 7.5°.$$

From the right triangle *ABD*,

$$BD = AD \cdot \sin 52.5° = 12 \cdot \sin 52.5°.$$

Substitute this expression into the previous and get

$$S = 48 \cdot (\sin 52.5° \cdot \sin 7.5°).$$

To continue, we need to calculate the product

$$\sin 52.5° \cdot \sin 7.5°.$$

According to our assumption, a calculator is not available for that. Note, however, that the sum of the angles 52.5° and 7.5° is 60°, and the difference is 45°. These are two special angles, for which we can get values of trig functions without a calculator. To calculate the product $\sin 52.5° \cdot \sin 7.5°$, it looks natural to consider the formulas for sum and difference that we derived in the previous chapter. The most suitable are:

$$\cos(\alpha + \beta) = \cos \alpha \cdot \cos \beta - \sin \alpha \cdot \sin \beta,$$
$$\cos(\alpha - \beta) = \cos \alpha \cdot \cos \beta + \sin \alpha \cdot \sin \beta.$$

To extract the product of sines from them, we subtract the first identity from the second and divide by 2. We get

Sine Product Formula

$$\sin \alpha \cdot \sin \beta = \frac{\cos(\alpha - \beta) - \cos(\alpha + \beta)}{2}.$$

Applying this formula to angles $\alpha = 52.5°$ and $\beta = 7.5°$, we get:

$$\sin 52.5° \cdot \sin 7.5° = \frac{\cos(52.5° - 7.5°) - \cos(52.5° + 7.5°)}{2}$$

$$= \frac{\cos 45° - \cos 60°}{2} = \frac{\sqrt{2} - 1}{4}.$$

Now we can return to the area S of the triangle BCD:

$$S = 48 \cdot (\sin 52.5° \cdot \sin 7.5°) = 12 \cdot (\sqrt{2} - 1).$$

This is the exact value. To get a decimal approximation, we can use our simple calculator: $S \approx 5$ m². According to the given problem, we can plant 3 flower bushes/m². Therefore, we can plant in total $3 \cdot 5 = 15$ flower bushes. ∎

Solve the following problem on your own.

Problem 9.2. Prove the following identities:

Product Formulas

$$\cos \alpha \cdot \cos \beta = \frac{\cos(\alpha + \beta) + \cos(\alpha - \beta)}{2},$$

$$\sin \alpha \cdot \cos \beta = \frac{\sin(\alpha + \beta) + \sin(\alpha - \beta)}{2}.$$

∎

As direct corollaries of the product formulas, we can get the following power reduction formulas simply by setting $\beta = \alpha$. They reduce square degrees of trig functions to linear quantities.

Problem 9.3. Prove the following formula:

Power Reduction Formula for Sine

$$\sin^2 \alpha = \frac{1 - \cos 2\alpha}{2}.$$

Solution. In sine product formula above, set $\beta = \alpha$. We get

$$\sin^2 \alpha = \frac{\cos 0 - \cos 2\alpha}{2} \quad \Rightarrow \quad \sin^2 \alpha = \frac{1 - \cos 2\alpha}{2}. \quad \blacksquare$$

Solve the problem below on your own.

Problem 9.4. Prove the following:

Power Reduction Formula for Cosine

$$\cos^2 \alpha = \frac{1 + \cos 2\alpha}{2}. \quad \blacksquare$$

Making slight modification of the power reduction formulas, we can obtain the following half-angle formulas. Just replace α with $\alpha/2$:

Half-Angle Formulas

$$\sin^2 \frac{\alpha}{2} = \frac{1 - \cos \alpha}{2},$$

$$\cos^2 \frac{\alpha}{2} = \frac{1 + \cos \alpha}{2}.$$

Half-angle cosine formula works in the problem below.

Problem 9.5. In a right triangle ABC, hypotenuse c and leg b are given. Find the length of the bisector AD of angle A.

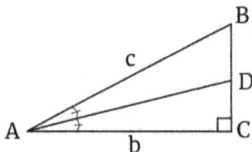

Solution. Denote $\angle BAC = \alpha$. Then $\angle DAC = \alpha/2$. From the triangle ADC, using the half-angle cosine formula, we get

$$AD = \frac{b}{\cos(\alpha/2)} = \frac{b}{\sqrt{(1+\cos\alpha)/2}} = \frac{b\sqrt{2}}{\sqrt{1+\cos\alpha}}.$$

From the triangle ABC, $\cos\alpha = b/c$. Therefore,

$$\sqrt{1+\cos\alpha} = \sqrt{1+\frac{b}{c}} = \sqrt{\frac{b+c}{c}} = \frac{\sqrt{b+c}}{\sqrt{c}}.$$

Finally,

$$AD = \frac{b\sqrt{2}}{\sqrt{1+\cos\alpha}} = b\sqrt{2} \div \sqrt{1+\cos\alpha} = b\sqrt{2} \div \frac{\sqrt{b+c}}{\sqrt{c}}$$

$$= b\sqrt{2} \cdot \frac{\sqrt{c}}{\sqrt{b+c}} = b\sqrt{\frac{2c}{b+c}}. \blacksquare$$

The problem below will lead us to another useful formula.

Problem 9.6. A piece of cloth was used to patch a ripped tent. The patch was in the form of a triangle ABC with sides $AB = 23$ cm and $AC = 34$ cm, and the angle between them was equal to 15°. For another tent, a second patch was needed. It was also in the shape of a triangle $A'B'C'$ with the same length of the sides, but the angle between them was equal to 75°. How much more fabric was needed for the second patch (in cm²)? As in the previous problem, we may use a calculator for arithmetic operations and square roots, but not for trig functions.

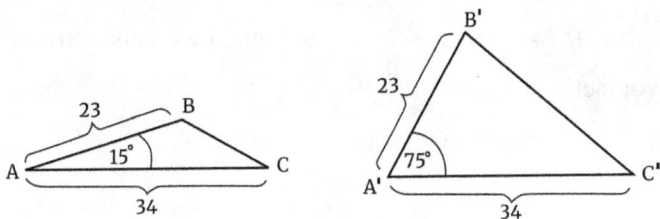

Solution. Mathematically, the problem is to find the difference between the areas of the two triangles. Denote by S and S' the areas of triangles ABC and $A'B'C'$, respectively. These areas are determined by the formulas:

$$S = \frac{1}{2} AB \cdot AC \cdot \sin 15°, \; S' = \frac{1}{2} A'B' \cdot A'C' \cdot \sin 75°.$$

Since $AB = A'B' = 23$ and $AC = A'C' = 34$, the difference of the areas, denoted as Δ, is

$$\Delta = S' - S = \frac{1}{2} \cdot 23 \cdot 34 \cdot (\sin 75° - \sin 15°) = 391 \cdot (\sin 75° - \sin 15°). \quad (1)$$

In order to continue, we need to calculate the difference of sines

$$\sin 75° - \sin 15°.$$

Note that $75° + 15° = 90°$, and $75° - 15° = 60°$. Recall that we know the exact values of trig functions for angles $90°$ and $60°$. If we look at the product formulas above, it seems that the formula for $\sin \alpha \cdot \cos \beta$ can be used if we read it from right to left. The only thing is that on the right side it contains the sum of sines, but we need their difference. We can get it by using the odd property of sine. Let's rewrite formulas for $\sin \alpha \cdot \cos \beta$ from right to left in the form

$$\sin (\alpha + \beta) + \sin (\alpha - \beta) = 2 \sin \alpha \cos \beta. \quad (2)$$

Denote $\alpha + \beta = u$ and $\alpha - \beta = -v$. Solving these equations for α and β (do this yourself), we get $\alpha = \dfrac{u - v}{2}$, $\beta = \dfrac{u + v}{2}$. Substitute these expressions into (2) and use the odd property of sine:

$$\sin u + \sin(-v) = \sin u - \sin v = 2 \sin \frac{u - v}{2} \cdot \cos \frac{u + v}{2}.$$

We reassign $u = \alpha$ and $v = \beta$. In this way we obtain an important formula for the difference of sines:

Difference of Sines Formula

$$\sin \alpha - \sin \beta = 2\sin\frac{\alpha - \beta}{2}\cos\frac{\alpha + \beta}{2}.$$

Now we can continue and find the difference Δ between areas from formula (1). We have

$$\Delta = 391 \cdot (\sin 75° - \sin 15°)$$
$$= 391 \cdot 2\sin\frac{75° - 15°}{2}\cos\frac{75° + 15°}{2} = 782\sin 30°\cos 45°$$
$$= 782 \cdot \frac{1}{2} \cdot \frac{\sqrt{2}}{2} \approx 276.5\,\text{cm}^2. \quad \blacksquare$$

Using the above difference of sines formula, we can get similar formulas involving sines and cosines with sums and differences. To do this, we can also use the reduction formulas, as well as odd and even properties of sine and cosine.

Problem 9.7. Express the sum of $\cos \alpha + \cos \beta$ through the product of trig functions.

Solution. $\cos \alpha + \cos \beta = \sin(90° - \alpha) + \sin(90° - \beta)$
$$= \sin(90° - \alpha) - \sin(\beta - 90°)$$
$$= 2\sin\frac{[(90° - \alpha) - (\beta - 90°)]}{2} \cdot \cos\frac{[(90° - \alpha) + (\beta - 90°)]}{2}$$
$$= 2\sin\frac{[180° - (\alpha + \beta)]}{2} \cdot \cos\frac{-\alpha + \beta}{2} = 2\sin\left(90° - \frac{\alpha + \beta}{2}\right) \cdot \cos\frac{\alpha - \beta}{2}$$
$$= 2\cos\frac{\alpha + \beta}{2} \cdot \cos\frac{\alpha - \beta}{2}.$$

We've got the following formula.

Sum of Cosines Formula

$$\cos \alpha + \cos \beta = 2 \cos \frac{\alpha + \beta}{2} \cdot \cos \frac{\alpha - \beta}{2}.$$

∎

Solve the problem below on your own.

Problem 9.8. Derive the formulas for

(a) $\sin \alpha + \sin \beta$, (b) $\cos \alpha - \cos \beta$.   ∎

In the problem below, be ready to see much more sophisticated calculations.

Problem 9.9. Traveling Professor Smartman needs to climb to peak B. At the edge of the peak there is a sturdy stump, which can be used to fasten his rope:

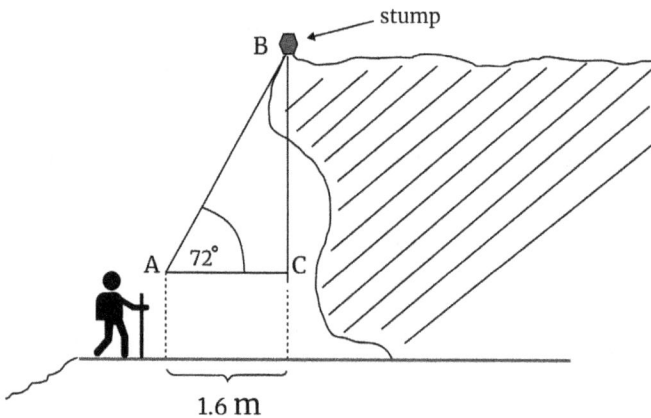

The professor has a nifty device that allows him to shoot a rope and attach it to a target with a special hook. The length of the rope is 5.5 m. To make sure that the rope is long enough, the professor needs to calculate the length of AB, since he has only one attempt to shoot it. He found that angle $A = 72°$. The professor also determined that length of $AC = 1.6$ m.

He understood that from the right triangle *ABC*, he can calculate *AB* by a simple formula:

$$AB = \frac{AC}{\cos A} = \frac{1.6}{\cos 72°}.$$

Unfortunately, the professor does not have a scientific calculator to calculate trig functions. He only has a simple calculator that can perform basic arithmetic operations and take square roots. He analyzed all possible alternative methods and came to the conclusion that the only approach to the solution of ensuring he can climb to the peak is to calculate cos 72°. How can he do that without the help of a scientific calculator? For him, it's a matter of life or death!

Solution. To save his life, Professor Smartman would need to bring out his skills in using the various trigonometric formulas to calculate cos 72°. Fortunately, he is an expert in trigonometry. Here is how he conquered this difficult problem. He started with this expression:

$$\cos 36° - \sin 18° = \cos(90° - 54°) - \sin 18° = \sin 54° - \sin 18°$$

$$= 2 \sin 18° \cos 36° = \frac{\sin 18° \cdot 2 \cos 36° \cdot \sin 36°}{\sin 36°}$$

$$= \frac{\sin 18° \cdot \sin 72°}{\sin 36°} = \frac{\sin 18° \cdot \sin(90° - 18°)}{\sin 36°}$$

$$= \frac{\sin 18° \cdot \cos 18°}{\sin 36°} = \frac{2 \sin 18° \cdot \cos 18°}{2 \sin 36°} = \frac{\sin 36°}{2 \sin 36°} = \frac{1}{2}.$$

So, he got the formula:

$$\cos 36° - \sin 18° = \frac{1}{2}.$$

At this point, the professor finished the more difficult part of the problem. After that, he transformed cos 36° to $1 - 2\sin^2 18°$ using the double angle cosine formula $\cos 2\alpha = 1 - 2\sin^2 \alpha$ for $\alpha = 18°$:

$$\cos 36° - \sin 18° = 1 - 2\sin^2 18° - \sin 18° = 1/2.$$

He got a quadratic equation for sin 18°. The professor set $x = \sin 18°$ and came to an equation in the form

$$1 - 2x^2 - x = \frac{1}{2} \implies 4x^2 + 2x - 1 = 0.$$

This equation has two roots: one positive, and another negative. Since sin 18° > 0, the negative root is rejected. Solving the quadratic equation, he got a positive root:

$$x = \sin 18° = \frac{\sqrt{5} - 1}{4}.$$ Since cos 72° = cos (90° − 18°) = sin 18°, then

$$\cos 72° = \frac{\sqrt{5} - 1}{4}.$$

And, finally, he found AB:

$$AB = \frac{1.6}{\cos 72°} = \frac{1.6 \cdot 4}{\sqrt{5} - 1} \approx 5.18 \text{ m}.$$

Therefore, the professor's 5.5 m rope is long enough to climb the peak. ∎

Let's summarize the results obtained.

Product Formulas

$$\sin \alpha \cdot \sin \beta = \frac{\cos(\alpha - \beta) - \cos(\alpha + \beta)}{2}$$

$$\cos \alpha \cdot \cos \beta = \frac{\cos(\alpha + \beta) + \cos(\alpha - \beta)}{2}$$

$$\sin \alpha \cdot \cos \beta = \frac{\sin(\alpha + \beta) + \sin(\alpha - \beta)}{2}$$

Power Reduction Formulas

$$\sin^2 \alpha = \frac{1 - \cos 2\alpha}{2}$$

$$\cos^2 \alpha = \frac{1 + \cos 2\alpha}{2}$$

Half-Angle Formulas

$$\sin^2 \frac{\alpha}{2} = \frac{1 - \cos \alpha}{2}$$

$$\cos^2 \frac{\alpha}{2} = \frac{1 + \cos \alpha}{2}$$

Sum and Difference Formulas

$$\sin \alpha + \sin \beta = 2 \sin \frac{\alpha + \beta}{2} \cos \frac{\alpha - \beta}{2}$$

$$\cos \alpha + \cos \beta = 2 \cos \frac{\alpha + \beta}{2} \cos \frac{\alpha - \beta}{2}$$

$$\sin \alpha - \sin \beta = 2 \sin \frac{\alpha - \beta}{2} \cos \frac{\alpha + \beta}{2}$$

$$\cos \alpha - \cos \beta = 2 \sin \frac{\alpha + \beta}{2} \sin \frac{\beta - \alpha}{2}$$

Exercises for Solving on Your Own

Exercise 9.1. Calculate exact values without using a calculator.
(a) $\cos 15° \cos 75°$; (b) $\sin 15° + \sin 75°$.

Exercise 9.2. Prove that
(a) $\sin 20° + \sin 40° = \sin 80°$; (b) $\cos 50° + \cos 70° = \cos 10°$.

Exercise 9.3. Prove that $\sin (60° + \theta) - \sin (60° - \theta) = \sin \theta$.

Exercise 9.4. Prove that $\sin (\alpha + \beta) \sin (\alpha - \beta) = \sin^2 \alpha - \sin^2 \beta$.

Exercise 9.5. Prove that $\tan\dfrac{\theta}{2} = \dfrac{\sin\theta}{1+\cos\theta} = \dfrac{1-\cos\theta}{\sin\theta} = \csc\theta - \cot\theta$.

Entertainment Problems

Problem E9.1. In Problem 9.9, Professor Smartman had to climb to the peak of a rock. This time he needs to come down. He stands on top of a mountain 100 m high. At the halfway point with an altitude of 50 m, there is a ledge where he can make an intermediate stop. The professor has a 75-m long rope and scissors. How does he come down from the mountain? [Do not worry about the slight amount of rope needed for tying. Assume he has enough for that as well.]

Problem E9.2. As Professor Smartman descended from the mountain top, he accidentally stepped on two poisonous snakes at once: a cobra and a viper and was bitten by both. Fortunately, he had antidotes with him. In total four pills: two against cobra poison and two against viper. All pills look identical and so the professor was careful and kept them in two separate packages.

Life-saving directions: The first pill should be taken immediately after the bite (one against cobra and one against viper), and the second pill the next day.

The professor placed one cobra pill (C) from the package into his palm. But as he was trying to get one viper pill (V), his hand trembled and both V pills dropped out of the package into his palm.

Now he has three identical-looking pills in his hand: one C and two Vs. This is problematic, as he needs to immediately take one C and one V, leaving the second pill for each of the antidotes for tomorrow. What should he do? There is no time to wait for help.

Chapter 10

Radian Measure

So far, we have measured angles in degrees. As we know, one degree is $1/360^{th}$ part of a circle. Therefore, to get one degree, we need to divide a circle into 360 equal parts. Then the central angle from one part (sector) will be equal to one degree.

The number 360 was introduced by astronomers in ancient Babylon (at least 3000 BC). No one knows for sure why they settled on this number. At that time, it was already known that the annual cycle consists of 365 days. It is reasonable to guess that they decided to treat one degree as one day. But instead of 365 they selected 360, since this number contains significantly more factors (it is divisible by every number from 2 to 10 except 7). In other words, the number 360 can be divided into whole parts much better than 365.

Another possible reason for choosing 360 is that the Babylonians used a sexagesimal number system based on the number 60 instead of 10. The advantage of using the number 60 was that this number has 10 different factors, more than any other smaller number.

Babylonians also divided the day (and night) into 12 h. They divided each hour by their favorite number 60 and got the minutes, and then divided each minute again by 60 and got the seconds.

In the modern-day, calculation of time is not based on astronomical events. The seconds are defined on the atomic level. Today the official definition of a second is based on the duration of the vibration of a cesium atom (cesium is a liquid metal).

In addition to the degree system, other systems are sometimes used, in which the angle unit is also obtained by dividing a circle by a certain number. Here are some such systems.

- **The rhumb:** A unit of angle measurement which is obtained by dividing a circle into 4 parts, and then each quarter is divided by 8. So, the rhumb is 1/32nd of the full circle. It was introduced in 1537 in Portugal and was used in sea navigation.
- **The gradian or grad:** A unit of angle measurement which is obtained by dividing the circle into 4 parts, and then each quarter is divided by 100. So, a grad is 1/400th of a full circle. It was introduced in 1793 in France at the time of the Revolution and was used for geodesic measures.
- **The thousandth:** A unit of angle measurement which is obtained by dividing a circle into 6000 parts. So, it is 1/6000th of the full circle. The thousandth are used in some countries in artillery (for example, in Russia).
- **The revolution:** A unit of angle measurement is a full circle. It represents the number of rotations of the initial side of the angle to return back to its initial position. It is used, for instance, to measure the angular speed[1] of rotating mechanisms, such as automobile engines (revolutions per minute or RPM).
- **The radian:** A unit of angle measurement which is obtained by dividing the circle into 2π parts. So, the radian is $1/2\pi$ part of the full circle. Recall that π is the ratio of a circle's circumference to its diameter: $\pi \approx 3.14$, so one radian is approximately 57° (since 360°/$(2\pi) \approx 57°$). The concept of the radian measure is credited to Roger Cotes (1682–1716), an English mathematician who worked closely with Isaac Newton. However, the term radian was introduced later by James Thomson (1822–1892), Ireland.

At first glance, it may seem that all of the above systems, except radians, are natural since each of them is obtained by dividing the full circle by some whole number. And only the Radian system divides by

---

[1] We will discuss angular speed in more detail in Chapter 12.

this "strange" 2π value, which is not even a fraction, but some irrational number. However, the actual situation is just the opposite.

In fact, only the Radian system is a natural one, and all others are artificial. Indeed, why is the division done by 1000, or 400? Such a division is provided in the decimal system, which is only used because we have 10 fingers on our hands. So, it is not based on the law of nature. In science, especially in computer science, it is often more convenient to use binary or octal systems, in which the bases are powers of two. In the Degree system, the division is made by the number 360, which is also artificial, although useful.

Let us turn to the Radian system. Its main idea is to measure angles through such a natural unit as a radius of the circle. More precisely, an angle of one radian is obtained if the endpoint of the radius is moved along the circumference by a distance equal to the radius. Hence, the term "radian" is used. Here is a formal definition:

Definition of Radian Measure

> Angle of 1 radian is an angle, for which the length of its arc is equal to the radius.

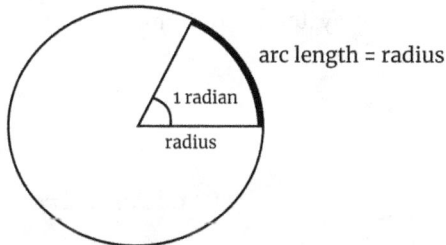

arc length = radius

1 radian

radius

We can also say that a one-radian angle is an angle in a "curvilinear" equilateral triangle (sector) in which two sides are radii, the third side is an arc, and all three sides are equal in length. From the definition of radians, we obtain that their number in the entire circle is equal to the number of radii on the circumference, which is 2π. Thus, a full circle contains 2π radians. In the Degree system, a full circle contains $360°$. Therefore,

> $360°$ corresponds to 2π radians

As we will see below, measuring angles in radians has many advantages over the Degree system.

Problem 10.1. Consider a central angle θ in a circle of radius r bounded by an arc of length s:

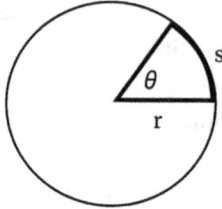

Express the arc length s through the angle θ and radius r. Do this in:

(1) Degree system.
(2) Radian system.

Solution. Since the angle is proportional to the arc's length, then, to solve the problem, we can set up a proportion. For angle θ, we will use the notation $\theta°$ in the Degree system and θ_r in the Radian system.

(1) In the Degree system, a full circle contains 360°. We can create this proportion:

$$\theta° \text{ relates to } s$$

$$\text{as}$$

$$360° \text{ relates to } 2\pi r.$$

Or, in algebraic notation,

$$\frac{\theta°}{s} = \frac{360°}{2\pi r} .$$

By cross-multiplying we get: $360° \cdot s = 2\pi r \cdot \theta°$. From here

$$s = \frac{2\pi}{360°}\theta°r \approx 0.017 \cdot \theta° \cdot r.$$

(2) In the Radian system, a full circle contains 2π radians, and we create this proportion:

$$\frac{\theta_r}{s} = \frac{2\pi}{2\pi r} \, .$$

From here,

$$\boxed{s = \theta_r \cdot r} \quad \blacksquare$$

The last equality $s = \theta_r \cdot r$ shows that the Radian system provides the simplest possible relationship between linear and angular measurement. Since for any measurement of angles, the arc length should be proportional to the central angle and the radius, the relationship between s, θ, and r for an arbitrary measuring system has the form

$$s = k \cdot \theta \cdot r,$$

where k is some numerical coefficient. This coefficient is a characteristic of a given system of angle measurement. As we see in part 1 from the problem above, $k = 0.017$ for the degree measure. Radian measure differs from all others by having the coefficient $k = 1$.

Solve the following problem on your own.

Problem 10.2. Find the coefficient k in the formula $s = k \cdot \theta \cdot r$, for the gradian measure described above. ■

In a **unit circle** (a circle with a radius of 1), the arc length is simply equal to the angle in radians:

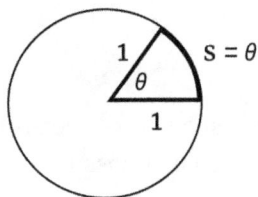

Angle θ is expressed in radians.

The equality $s = \theta$ literally sounds like "the arc is equal to the angle."

How do we understand this equality? Consider, for example, an angle of 1.3 radians in the circle with a radius of 1 m:

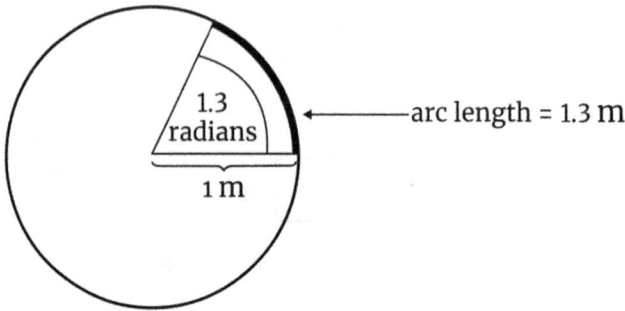

In this figure, the angle equals **1.3 radians**, and the length of arc equals **1.3 m**. Therefore, the equality $s = \theta$ means that arc s and angle θ have only the same **numerical** value equaling 1.3. But each of them has its own unit of measure: the angle θ is expressed in radians, and the length of arc s is expressed in meters (in our case).

Let's rewrite the formula $s = \theta \cdot r$, as $\theta = s/r$. Using it, we can give

Another Definition of Radian Measure

> The radian measure of a central angle θ is defined as a ratio of arc s to radius r

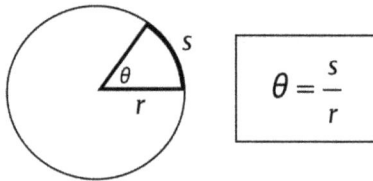

$$\theta = \frac{s}{r}$$

The advantage of the Radian system leads to simplifications in various scientific and technical calculations. The next problem demonstrates this.

Problem 10.3. A bus with wheels with a radius of 0.3 m moves along the road. Calculate its speed if:

(1) The wheels rotate at an angular speed of 2,900°/s.

(2) The wheels rotate at an angular speed of 50 radians/s.

Solution.

(1) **Calculation in degrees.** First, we determine the time t of one full rotation of the wheel according to the proportion:

$$\frac{2,900°}{1\,\text{s}} = \frac{360°}{t}.$$

From here, $t = 360/2,900 = 0.124$ s. Next, we calculate the length l of the circle: $l = 2\pi r = 2 \cdot 3.14 \cdot 0.3 = 1.885$ m. Finally, the speed of the bus is

$$\frac{l}{t} = \frac{1.885}{0.124} \approx 15 \text{ m}/\text{s}.$$

(2) **Calculation in radians.** One radian corresponds to the arc length equal to the radius (which is 0.3 m), and 50 radians correspond to the arc length 50 times longer: $0.3 \cdot 50 = 15$ m. This is the length that the wheels pass in one second, which is the speed of the bus: 15 m/s. ■

As we see, the solution in radians is much simpler: when calculating in degrees, we had to use a calculator several times. In radians, all calculations could easily be done in your head.

Solve the problem below on your own.

Problem 10.4. Let the central angle be 1/2 radians. What is the relation between its arc and the radius? ■

Problem 10.5. Are angles in 2π radians and 2 radians have the same value?

Solution. No, these angles are different. The numerical value of the angle in 2π radians is about $2 \cdot 3.14 = 6.28$ radians, and the second angle equals just 2 radians. ■

Problem 10.6. On a winch with a 20 cm diameter, the cable is wound by a handle, the initial position of which is at point A. There is a cart at the end of the cable, which must be moved 7 m by turning the handle. After moving the cart, the handle stops at some point B and must be locked.

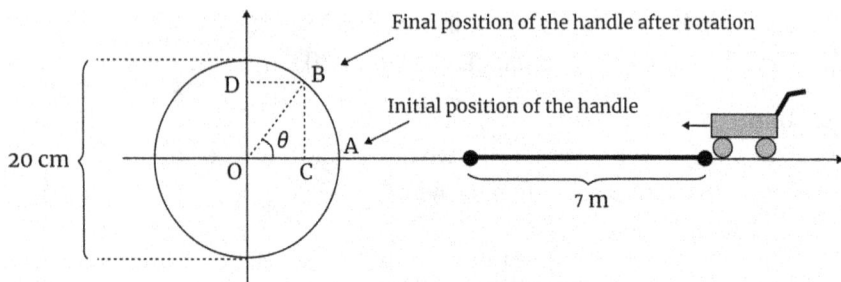

To do this, it is necessary to determine the coordinates of point B (in cm). The diameter of the cable can be neglected. The handle rotates counterclockwise.

Solution. Point A moves to point B along a circle with radius $r = 10$ cm for a distance $s = 7$ m $= 700$ cm. According to the definition of radian measure, distance s, angle θ, and radius r are related by the formula: $\theta = \dfrac{s}{r} = \dfrac{700}{10} = 70$ radians. From the right triangle OBC

$$OC = OB \cdot \cos \theta = r \cdot \cos \theta = 10 \cdot \cos 70,$$
$$BC = OB \cdot \sin \theta = r \cdot \sin \theta = 10 \cdot \sin 70.$$

Using a calculator in radian mode, we get: $\cos 70 \approx 0.633$ and $\sin 70 \approx 0.774$.

From here, $OC \approx 10 \cdot 0.633 = 6.33$ cm, $BC \approx 10 \cdot 0.774 = 7.74$ cm. Therefore, the coordinates of point B are $(6.33, 7.74)$. ∎

The following problems show more benefits of radians.

Problem 10.7. Derive and compare formulas for a sector area in degrees and radians.

Solution. Let S be the area of a sector subtended by angle θ in a circle with the radius r. Recall that the entire circle area is πr^2, and the full circle consists of $360°$ and 2π radians. From geometry, we know that the area of a sector is proportional to its angle. Therefore, for the angle in degrees, we can set up this proportion:

$$\frac{\theta°}{S} = \frac{360°}{\pi r^2}$$

From here,

$$S = \frac{\pi r^2 \theta^\circ}{360^\circ}.$$

Similarly, for the angle in radians, the proportion is: $\dfrac{\theta_r}{S} = \dfrac{2\pi}{\pi r^2}$. From here,

$$S = \frac{r^2 \theta_r}{2}.$$

As we see, the formula with radians is simpler than the formula with degrees. ■

Problem 10.8. Let θ be an acute angle. Prove that if angle θ is in radians, then

$$\sin \theta < \theta < \tan \theta.$$

Solution. Compare the shaded areas S_1, S_2 and S_3 in three unit circles with the angle θ:

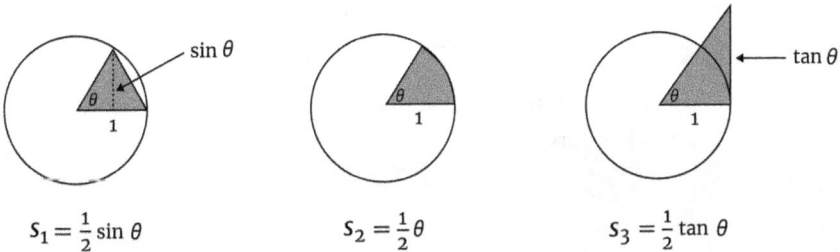

$$S_1 = \frac{1}{2}\sin \theta \qquad\qquad S_2 = \frac{1}{2}\theta \qquad\qquad S_3 = \frac{1}{2}\tan \theta$$

Note that area S_2 is calculated using the result of the previous problem by setting $r = 1$.

It is obvious that $S_1 < S_2 < S_3$. From here

$$\frac{1}{2}\sin \theta < \frac{1}{2}\theta < \frac{1}{2}\tan \theta \Rightarrow \sin \theta < \theta < \tan \theta. ■$$

If we divide the last inequality by $\sin \theta$, then we get:

$$1 < \theta/\sin \theta < 1/\cos \theta.$$

Replacing ratios in these inequalities by the reciprocals and reversing the signs, we obtain

$$\cos \theta < \frac{\sin \theta}{\theta} < 1.$$

If θ approaches 0, then $\cos \theta$ approaches 1, and, therefore, $\frac{\sin \theta}{\theta}$ will also approach 1 (by the so-called "Squeeze" theorem). Thus, the limit of the expression $\frac{\sin \theta}{\theta}$ is equal to 1 when θ approaches 0. This remarkable limit is widely used in calculus. Note that such a result occurs only when using radians.

Problem 10.9. Set up a relationship between the radians and degrees.

Solution. Since a full circle contains $360°$ and 2π radians, then $180°$ corresponds to π radians. For any angle, we denote its degree measure as $\theta°$, and the radian measure as θ_r. The relationship between $\theta°$ and θ_r can be established using the proportion:

$$\boxed{\frac{180°}{\pi} = \frac{\theta°}{\theta_r}}$$

We call this proportion the **main proportion**. From it we can express $\theta°$ through θ_r and vice versa:

$$\theta° = \frac{180}{\pi} \cdot \theta_r, \quad \theta_r = \frac{\pi}{180} \cdot \theta°. \ \blacksquare$$

Note. You do not need to memorize these formulas. Just remember that $180°$ corresponds to π radians:

$$\boxed{180° = \pi_{\text{rad}}}$$

and then use the main proportion.

Problem 10.10. Express angle of 1 radian in degrees.

Solution. For angle of 1 radian, the main proportion takes the form

$$\frac{180°}{\pi} = \frac{\theta°}{1_r}.$$

From here,

$$\theta° = \frac{180}{\pi} \approx 57.3°.$$

So, 1 radian approximately equals to 57°. ∎

Note. If an angle in radians is given in terms of π, there is no need to use a proportion to convert this angle into degrees: simply replace π with 180. Using this method, we can immediately say that

$$\frac{\pi}{2} \text{ is } 90°, \frac{3\pi}{2} \text{ is } 270°, 2\pi \text{ is } 360° \text{ and so on.}$$

Solve the problem below on your own.

Problem 10.11. Prove that the angle of $\frac{5\pi}{12}$ radians equals to 75°. ∎

Problem 10.12. Express angle of 1° in radians.

Solution. For a 1-degree angle, the main proportion takes the form

$$\frac{180°}{\pi} = \frac{1°}{\theta_r}.$$

From here,

$$\theta_r = \frac{\pi}{180} \approx 0.017.$$

So, a 1-degree angle approximately equals to 0.017 radians. ∎

Problem 10.13. Traveling Professor Smartman purchased a bike in which the middle wheel has a radius of 10 cm, the inner wheel has a radius of 3 cm, and the large wheels have a radius of 30 cm:

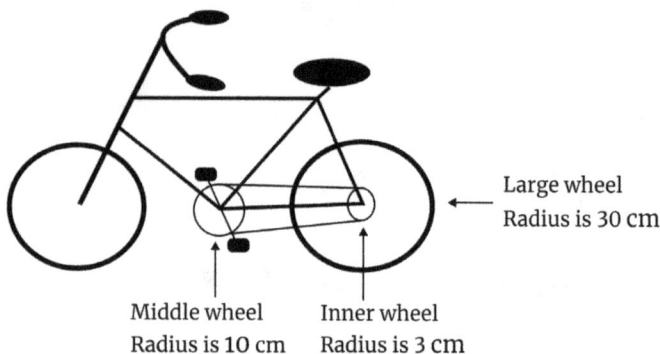

Large wheel
Radius is 30 cm

Middle wheel   Inner wheel
Radius is 10 cm   Radius is 3 cm

The professor determined that he could pedal 1 revolution/s. How fast can the professor ride a bicycle in miles/h?

Solution. We determine the speed of the bike in three steps. First, we convert the angular speed[2] of the middle wheel to linear speed. The relationship between linear speed v and angular speed w of the wheel with radius r is the same as the relationship of the arc with the central angle: $v = rw$. The middle wheel's linear speed is transmitted by the chain to the inner wheel and remains the same. In the second step, we convert the linear speed of the inner wheel into its angular speed. This speed is equal to the angular speed of the large wheel. Finally, we convert the angular speed of the large wheel to linear speed.

> **Step 1.** Since the middle wheel makes 1 revolution per second, and 1 revolution corresponds to 2π radians, the angular speed w_m of the middle wheel is $w_m = 2\pi$ rad/s. Its radius $r_m = 10$ cm, therefore the linear speed $v_m = r_m w_m = 10 \cdot 2\pi \approx 62.83$ cm/s.
>
> **Step 2.** The linear speed of the inner wheel is equal to the linear speed v_m of the middle wheel. Its radius $r_i = 3$ cm, therefore the angular speed
> $$w_i = \frac{v_m}{r_i} = \frac{62.83}{3} = 20.94 \frac{\text{rad}}{\text{s}}.$$

---

[2]The angular speed is the angle by which the wheel turns per unit of time.

Step 3. The angular speed of the large wheel is equal to the angular speed w_i of the inner wheel. Its radius $r = 30$ cm, therefore the linear speed $v = rw_i = 30 \cdot 20.94 = 628.2 \frac{\text{cm}}{\text{s}}$. Finally, we convert this speed into miles/h. Since there are 160,934 cm in 1 mile and 3,600 s in 1 h, we have $628.2 \frac{\text{cm}}{\text{s}} = 628.2 \cdot \frac{3,600}{160,934} \approx 14 \text{miles/h}$.

So, the professor was riding his bike at 14 miles/h. ∎

Solve the problem below on your own.

Problem 10.14. Prove that special angles in degrees and radians are related according to the following table. Show that sine, cosine, and tangent have the indicated values.

Degrees	0°	30°	45°	60°	90°	180°	270°	360°
Radians	0	$\frac{\pi}{6}$	$\frac{\pi}{4}$	$\frac{\pi}{3}$	$\frac{\pi}{2}$	π	$\frac{3\pi}{2}$	2π
sine	0	$\frac{1}{2}$	$\frac{\sqrt{2}}{2}$	$\frac{\sqrt{3}}{2}$	1	0	−1	0
cosine	1	$\frac{\sqrt{3}}{2}$	$\frac{\sqrt{2}}{2}$	$\frac{1}{2}$	0	−1	0	1
tangent	0	$\frac{\sqrt{3}}{3}$	1	$\sqrt{3}$	Undefined	0	Undefined	0

∎

Let's mark the quadrantal angles in degrees and radians on the circle. Compare the left and right figures.

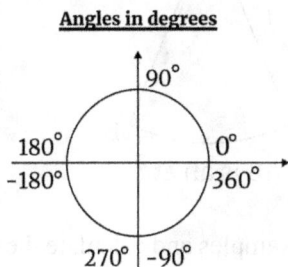

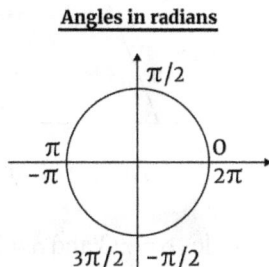

Angles in degrees · Angles in radians

In the next chapters, we will mostly use radians.

Exercises for Solving on Your Own

Exercise 10.1. How many

(a) Radians are there in 50°?
(b) Degrees are there in 1.7 radians?

Exercise 10.2. In a circle, the central angle θ is 30° and the corresponding sector area is 3.6 feet^2. Find the radius of the circle.

Exercise 10.3. The angles of a triangle are in the ratio of 3:4:5. Express these angles in radians.

Exercise 10.4. The diameter of a Ferris wheel is 16 m. Its spokes connecting each cab to the center of the wheel make an angle of $\pi/8$. How many cabs are on the wheel? What is the length of the arc between two consecutive cabs?

Exercise 10.5. Nick is running around a circular track of a 30-m radius. Michelle is staying at the center and observing him. She found that she turned by 6 radians in 1 min. Find how many meters does Nick run in 1 min?

Exercise 10.6. A fly sat on the top of the second hand of a large clock and rode 48 cm. How long was the fly riding if the length of the second hand is 1.5 m?

Exercise 10.7. The length of the arc of the central angle is proportional to its size. Find out if the same proportionality property remains between the chord AB and the central angle α.

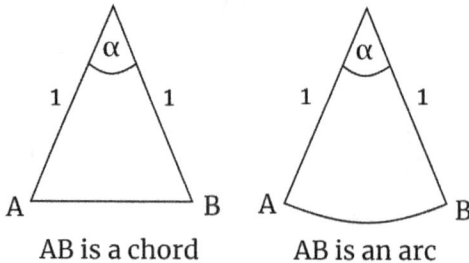

AB is a chord   AB is an arc

Take angles $\alpha = \pi/3$ and $\alpha = 2\pi/3$ as examples and calculate the lengths of chord AB and arc AB for these angles (see figure above).

Exercise 10.8. Express the length of the chord c through the length of its arc s and the diameter of the circle.

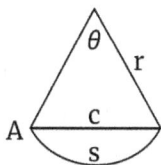

Exercise 10.9. Prove that $x = 0$ is the only solution of the equation $\sin x = x$ in the interval $[0, \pi/2]$.

Exercise 10.10. In what quadrant does a 100-radian angle lie?

Exercise 10.11. Traveling Professor Smartman is leisurely floating aboard his boat at the center of a round lake. Suddenly a huge beast approaches the lake and threatens the professor. Fortunately, the beast cannot swim. Also, when on land, the professor can run faster than the beast. To save himself, it is enough for the professor to reach the shore before the beast gets there and run away. However, the boat travels 4 times slower than the beast.

How can the professor get to the shore before the beast?

Let's find out if the professor can simply travel in the direction opposite of the beast's original location. Since the speed of the beast is 4 times greater, in order to escape, the distance from the boat to the shore (which is the radius R of the lake) should be 4 times less than the distance that the beast must run (this is half the circle $= \pi R$). So, R should be less than $\pi R/4$. However, this is not so, since $R > \pi R/4 = 0.8R$. With this tactic, the professor will not be able to get ashore before the beast. Is it possible for the professor to escape?

Entertainment Problems

Problem E10.1. Test your intuition with this thought experiment. Imagine our planet as an ideal sphere. Let's brace it tightly around the equator with a rope or a wire. Extend the length of the wire by only 1 m. As a result, there will be some gap between the wire and the surface of the earth. Do you think this gap will be microscopically small? Calculate the gap to verify your intuition.

Problem E10.2. Professor Smartman's son dreamed of buying a rather expensive mountain bike for $1,500. The professor said that his son should try to earn at least part of the money himself. He promised to square the sum of all the money that his son earns in a week and give this money towards the bike purchase. Unfortunately, the son was so busy at school that he could only earn $4. Nevertheless, he was able to do something, such that his father was forced to give him all the needed money to buy the bike, as promised. What did the son do?

Hint: Remember that in this chapter, we discussed converting one unit of measure to another.

Chapter 11

Graphs of Sine and Cosine

In this chapter, we will show how to present sine and cosine functions visually using a graph. Graphs of functions allow not only to visualize them but also to show some of their properties.

We use the notation $y = f(x)$ for an arbitrary function f and call the variable x the **argument** of f. A possible method for constructing the graph of f is to identify (mark) several points on the graph, and then try to imagine the entire graph's view on the coordinate plane.

As we already mentioned, sine and cosine functions have odd/even properties. Let's consider these properties in general.

A function $f(x)$ is called **even** if $f(-x) = f(x)$ for any x from the domain of f. If function $f(x)$ is a polynomial with all terms having even powers, then $f(x)$ is an even function, so the word *even* is used. Since both points $(x, f(x))$ and $(-x, f(x))$ are on the graph of the function $f(x)$, and these points are symmetric to each other over the y-axis, the entire graph of any even function is also symmetric over the y-axis. This means that the graph remains unchanged after reflection about the y-axis. So, the graph of an even function has what is called a **line symmetry property**.

A function $f(x)$ is called **odd** if $f(-x) = -f(x)$ for any x from the domain of f. If a function $f(x)$ is a polynomial with all terms having odd powers, then $f(x)$ is an odd function, so the word *odd* is used. Both points $(x, f(x))$ and $(-x, -f(x))$ are on the graph of the function $f(x)$, and these points are symmetric to each other over the origin $(0, 0)$. Therefore, the entire graph of any odd function is also symmetric over the origin. This means that the graph remains unchanged after 180° rotation about the origin. So, the graph of an odd function has what is called a **point symmetry property**.

Other properties are increasing/decreasing. We say that the function $f(x)$ **increases** if the bigger the argument x, the bigger the function (more exactly, the value of function), and the function **decreases**, if the bigger the x, the smaller the function.

Before we look at graphs of trig functions, we will first practice with some simpler functions. Namely, we consider the graphs of functions $y = x^2$ and $y = x^3$. Function $y = x^2$, and its graph, is called a **parabola**, and function $y = x^3$, and its graph, is called a **cubic parabola**.

Problem 11.1. Sketch the graph of the parabola $y = x^2$.

Solution. Note that this function returns the same values for x and $-x$ since $(-x)^2 = x^2$. So, the parabola is an even function, and its graph is symmetric over the y-axis. Therefore, if we draw the graph only for positive x, then we can simply reflect (flip) this graph over the y-axis to get the entire picture. Also, for positive x, the bigger the x, the bigger the x^2, so the parabola increases (for positive x). However, this function is **not linear.** It means that its graph is not a straight line. Instead, the graph is a **curve**. To draw this curve, we calculate several values of the parabola for some values of x. The following table represents some sample values.

x	0	1	2	3
$y = x^2$	0	1	4	9
(x, y)	$(0, 0)$	$(1, 1)$	$(2, 4)$	$(3, 9)$

If we plot points (x, y) on the coordinate plane and connect them with a smooth curve, we get the figure:

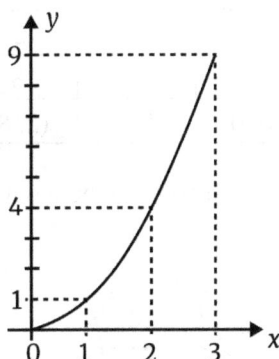

Graph of the parabola $y = x^2$ for positive x.

To get the entire parabola (to include negative x), we reflect this graph over the y-axis. Here is the final figure.

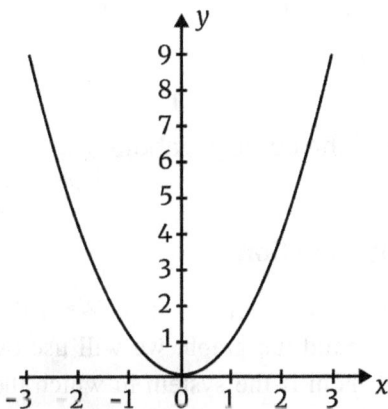

Graph of the parabola $y = x^2$. ■

Problem 11.2. Sketch the graph of the cubic parabola $y = x^3$.

Solution. Unlike the parabola $y = x^2$, the cubic parabola is an odd function, since $(-x)^3 = -x^3$. Its graph is symmetric with respect to the origin $(0, 0)$. Therefore, if we draw a graph only for positive x, we can reflect this graph over the origin to get the entire picture. Also, the bigger the x, the bigger the x^3, so our cubic parabola increases along the entire number line. To sketch its graph, we again calculate several values of y for some values of x. The following table represents some sample values.

x	0	1	2	3
$y = x^3$	0	1	8	27
(x, y)	$(0, 0)$	$(1, 1)$	$(2, 8)$	$(3, 27)$

If we plot points (x, y) and connect them with a smooth curve, we will get the figure:

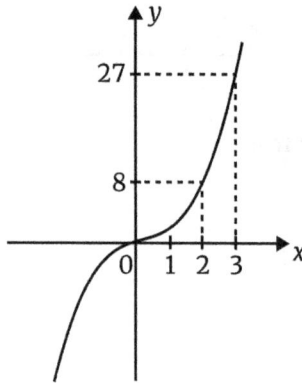

Graph of the cubic parabola $y = x^3$. ∎

Graphing of the Sine Function

Now let's start graphing the function $y = \sin \theta$. We will take the argument (angle) θ in radians. To build the graph, we will use two different coordinate systems. One of them is the system in which the sine function is defined. Recall that this system contains the unit circle. Recall also that sine of the central angle θ is defined as the vertical coordinate of the point on the unit circle (point of its intersection with the terminal side of angle θ). This coordinate system will be used as a source of information for drawing a graph. Depict this system:

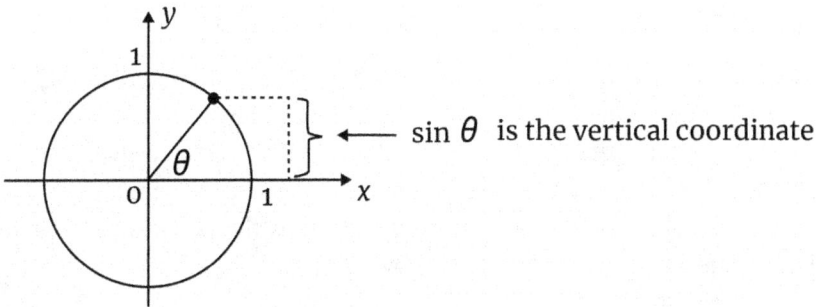

Figure 1.

The second coordinate system will be used directly for drawing the sine graph. In this system, we mark angles on the horizontal x-axis (we will use the letter x for the angle instead of θ) and mark $\sin x$ on the vertical y-axis.

Warning. Do not confuse these two coordinate systems. In the first system, where the sine is defined, the sine arguments are shown as angles in the unit circle. But in the second system, where we built the graph, the sine arguments are shown as points on the x-axis.

As an example, let's look at $\sin\dfrac{\pi}{4} = \dfrac{\sqrt{2}}{2}$ in each system:

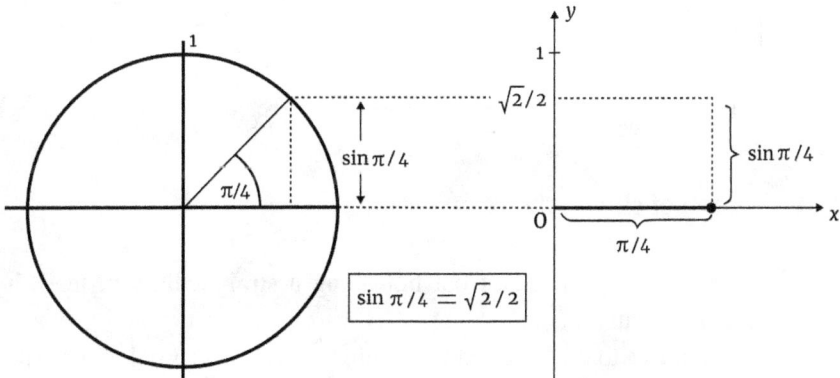

In this system,
$\pi/4$ is shown as an angle

In this system,
$\pi/4$ is shown as a point on the x–axis

To graph the sine function, we will move from quadrant to quadrant and observe how the graph will look.

Problem 11.3. Sketch the graph of sine in the 1st quadrant.

Solution. In the 1st quadrant, argument x belongs to the interval $[0, \pi/2]$. If we increase the angle θ in the system shown in Figure 1 and observe how the vertical coordinate of the point on the unit circle changes, we see that the sine increases from 0 to 1. To imagine the shape of the graph, we select those values of the argument x for which we know the exact values of sine. These values are presented in the following table:

x	0	$\pi/6$	$\pi/4$	$\pi/3$	$\pi/2$
$\sin x$	0	$1/2$	$\dfrac{\sqrt{2}}{2}$	$\dfrac{\sqrt{3}}{2}$	1

Now plot these points in the system of coordinates and connect them with a smooth curve:

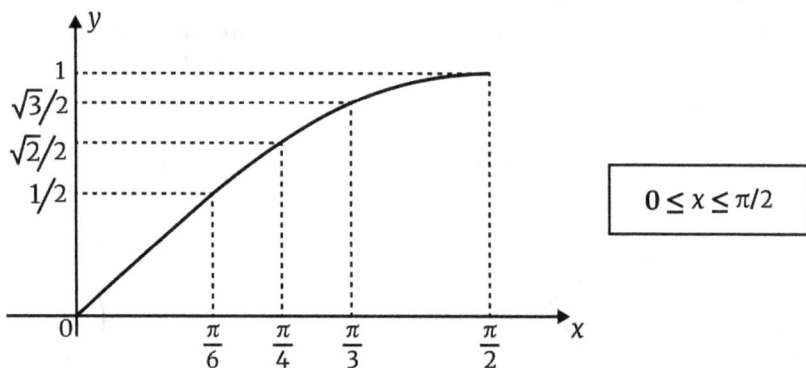

Graph of sine in the 1st quadrant ∎

$$0 \le x \le \pi/2$$

We see that sine increases but not along a straight line. Instead, it increases along a curve.

Now we move to the 2nd quadrant. Here the argument x belongs to the interval $[\pi/2, \pi]$. If we look at Figure 1, we see that while angle θ increases from $\pi/2$ to π, sine decreases from 1 to 0. The following problem shows how to get an explicit graph of the sine in the 2nd quadrant.

Problem 11.4. Prove that in the interval $[0, \pi]$ (i.e., in the 1st and 2nd quadrants), the graph of sine is symmetric about the vertical line $x = \pi/2$.

Solution. On the x-axis, points $\pi/2 - \theta$ and $\pi/2 + \theta$ are symmetric to each other with respect to point $\pi/2$. Therefore, it is enough to show that sine at these points takes the same values:

$$\sin\left(\frac{\pi}{2} - \theta\right) = \sin\left(\frac{\pi}{2} + \theta\right).$$

It follows from the reduction formulas (see Chapter 5):

$$\sin\left(\pi/2 - \theta\right) = \cos\theta \text{ and } \sin\left(\pi/2 + \theta\right) = \cos\theta. \quad \blacksquare$$

Using this property, the graph in the 2nd quadrant can be obtained by reflecting the graph from the 1st quadrant over the vertical line $x = \pi/2$:

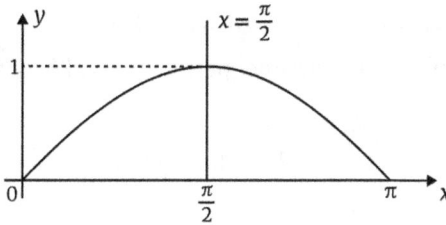

Graph of sine in the 1st and 2nd quadrants

Let's continue moving angle θ along the unit circle from π to 2π (i.e., in the 3rd and 4th quadrants). From Figure 1, you can see that in the 3rd quadrant, sine decreases from 0 to -1, and in the 4th quadrant, sine increases from -1 to 0. We can get the exact graph using the following point symmetry property.

Problem 11.5. Prove that in the interval $[0, 2\pi]$, the sine graph is symmetrical about the point $(\pi, 0)$ on the x-axis.

Solution. On the x-axis, points $\pi - \theta$ and $\pi + \theta$ are symmetric to each other with respect to point π. Therefore, it is enough to show that

$$\sin(\pi - \theta) = -\sin(\pi + \theta).$$

It follows from the reduction formulas:

$$\sin(\pi - \theta) = \sin\theta \text{ and } \sin(\pi + \theta) = -\sin\theta. \quad \blacksquare$$

Using this property, we can get the graph of sine in the 3rd and 4th quadrants (i.e., in the interval $[\pi, 2\pi]$) by reflecting the graph located in the 1st and 2nd quadrants (i.e., in the interval $[0, \pi]$) over the point $(\pi, 0)$. It is the same as moving the graph horizontally to the right from the interval $[0, \pi]$ to $[\pi, 2\pi]$, and then reflecting it over the x-axis. As a result, we get the graph of sine for one full cycle (we can also say, for a one-period interval $[0, 2\pi]$):

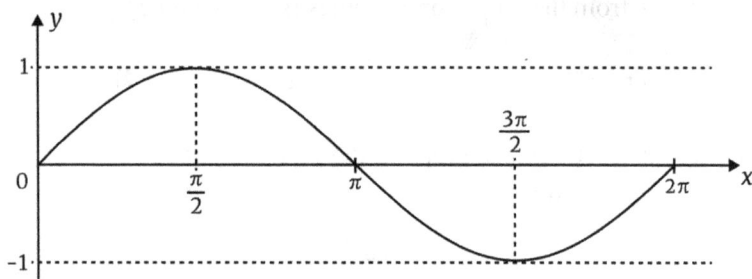

Graph of sine in a one-period interval $[0, 2\pi]$

Problem 11.6. Sketch the graph of the function $y = \sin x$ on the entire x-axis.

Solution. Since sine is a periodic function with a period of 2π, we can extend its graph in either direction (to the left and to the right) on the x-axis, starting from the interval $[0, 2\pi]$. By doing this, we get the graph of sine on the entire number line, i.e., for all values of x from $-\infty$ to $+\infty$:

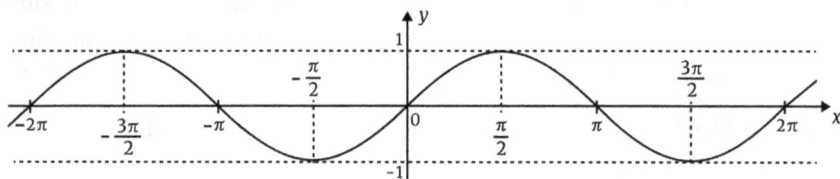

Entire graph of the function $y = \sin x$

It is interesting that sinusoidal lines can be observed in wildlife. For example, snakes have four ways of moving, one of which is a wave-like motion. This wavy motion is called a serpentine and allows the snake to move at the greatest velocity. A serpent advances like a wave, through the sinusoidal movement of its body.

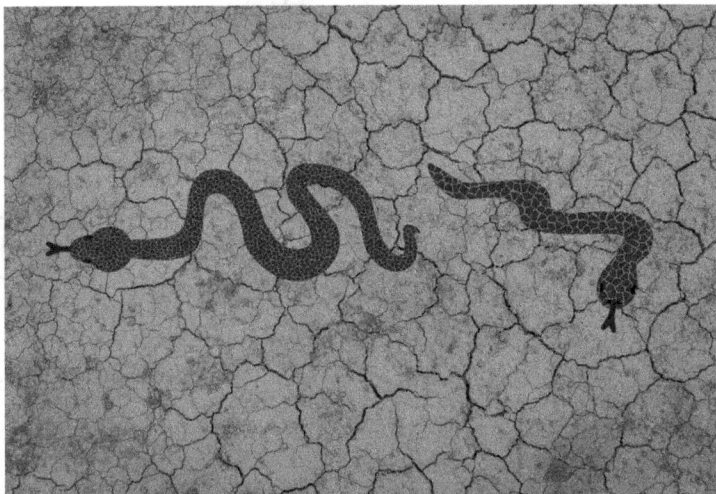

The graph of the sine function allows us to see a wide range of its properties. Let's list some of them.

(1) sin x is defined for all values of x. So, the domain of sine function is the interval $(-\infty, +\infty)$.

(2) Sine changes between -1 and 1, so its range is the interval $[-1, 1]$.

(3) Sine is symmetric about the origin. Therefore, sine is an odd function:

$$\sin(-x) = -\sin x.$$

(4) Sine is a periodic function with a period of 2π. We can say that sine "repeats itself" at consecutive intervals of the length 2π. It means that for any x,

$$\sin(x + 2\pi) = \sin x.$$

The number 2π represents one full period. If we replace it with k full periods, i.e., with the number $2k\pi$, then again $\sin(x + 2k\pi) = \sin x$. Here, k is any integer that may be positive or negative: $k = 0, \pm1, \pm2,$ ... . Below we will use this property in the following way. If some specific property holds in the interval $[a, b]$, then the same property holds for all intervals that we can get by shifting $[a, b]$ by any number of full periods, i.e., by adding $2k\pi$ to both a and b. In other words, the property holds at an infinite number of intervals $[a + 2\pi, b + 2\pi]$,

$[a + 4\pi, b + 4\pi]$, and so on. We can describe this infinite family of intervals in a parametric form as $[a + 2k\pi, b + 2k\pi]$.[1]

(5) In a one-period interval $[0, 2\pi]$, sine returns the largest (maximum) value, equaling to 1, at $x = \dfrac{\pi}{2}$. Therefore, on the entire number line, sine takes this value at an infinite number of points that we can get by adding any number of periods $2k\pi$ to $\dfrac{\pi}{2}$. As a result, all the points of the maxima can be described in the parametric form: $x = \dfrac{\pi}{2} + 2k\pi$.

Similarly, in a one-period interval $[0, 2\pi]$, sine returns the smallest (minimum) value, equaling to -1, at $x = \dfrac{3\pi}{2}$. Therefore, on the entire number line, sine returns the minimum value at an infinite number of points $x = \dfrac{3\pi}{2} + 2k\pi$. The same infinite family can also be described as $x = -\dfrac{\pi}{2} + 2k\pi$.

(6) Sine is positive (i.e., the graph is located above the *x*-axis) in the interval $[0, \pi]$, and negative (the graph is located below the *x*-axis) in the interval $[\pi, 2\pi]$. Therefore, on the entire number line, sine is positive in the intervals $[2k\pi, \pi + 2k\pi]$, and is negative in the intervals $[\pi + 2k\pi, 2\pi + 2k\pi]$.

(7) In a one-period interval $[0, 2\pi]$, sine produces the value 0 at three points. These are the points of intersection of the graph with the *x*-axis: 0, π, and 2π, which are located at a distance of π one after the other. Such points are called **zeros** of the function. Therefore, all zeros of the sine function can be described by the parametric formula $x = k\pi$.

(8) Sine increases at the interval $[-\pi/2, \pi/2]$. On the entire number line sine increases at intervals

$$[-\pi/2 + 2k\pi, \pi/2 + 2k\pi].$$

Sine decreases at the interval $[\pi/2, 3\pi/2]$. Therefore, on the entire number line sine decreases at the intervals $[\pi/2 + 2k\pi, 3\pi/2 + 2k\pi]$.

---

[1] By parametric form, we mean that the last interval represents an infinite number of intervals, assigning any integer value to the number k, which is called the parameter.

Graphing of the Cosine Function

To graph the cosine function, we can proceed in the same way as we built the sine graph. Namely, using the definition of cosine as a horizontal coordinate, we can construct its graph moving along the unit circle quadrant by quadrant. However, there is a simpler way. It is based on the relationship between sine and cosine. According to the reduction formulas, we have

$$\cos x = \sin(x + \pi/2).$$

Using this formula, we can figure out how sine and cosine graphs are related and then get the graph of cosine from the graph of sine. It is useful to consider these questions in a general form.

Problem 11.7. How do we get the graph of the function $g(x) = f(x + a)$ from the graph of function $f(x)$, where a is a fixed number?

Solution. It may be intuitively clear that the graph of the function $g(x)$ is obtained from the $f(x)$ graph by shifting along the horizontal x-axis by the value a. Yes, that's how it is. It might also seem that if $a > 0$, then the shift will be to the right, and if $a < 0$, then it is to the left. **However, this is a mistake.** In fact, the opposite is true: if $a > 0$, then the shift will be to the **left**, and if $a < 0$, then to the **right**. To understand why it is, let's denote $x' = x + a$. Then $x = x' - a$ and

$$g(x) = f(x + a) = f(x' - a + a) = f(x').$$

If, for example, $a > 0$, then $x < x' = x + a$. The above equality

$$g(x) = f(x')$$
$$(x < x')$$

shows that the function g takes at point x the same value as f at point x', and the point x is to the left of the point x' ($x < x'$). So, in order to get graph g, we should shift graph f to the left. ∎

Let's illustrate this result with the examples of the parabolas $y = x^2$, $y = (x + 5)^2$ and $y = (x - 5)^2$. Compare their graphs:

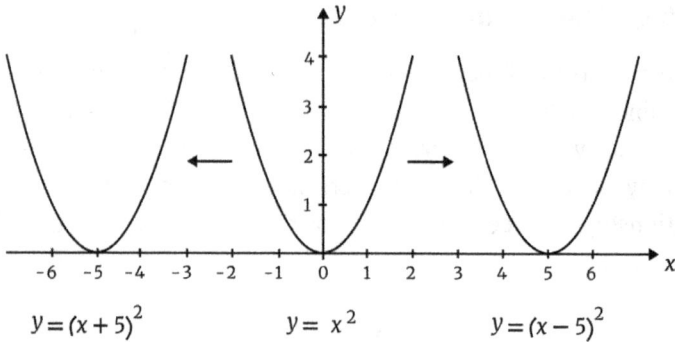

$$y = (x+5)^2 \qquad y = x^2 \qquad y = (x-5)^2$$

Now we return to the graph of cosine.

Problem 11.8. Sketch the graph of the function $y = \cos x$.

Solution. Since $\cos x = \sin(x + \pi/2)$, we can get the graph of the function $y = \cos x$ by shifting the graph of $y = \sin x$ by $\pi/2$ to the left:

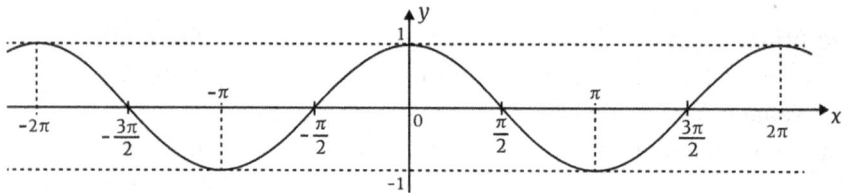

Graph of the function $y = \cos x$

If we ignore the location of the y-axes on the coordinate plane, then the graphs for sine and cosine represent the same wavy curve.

Looking at the graph of the cosine function, we can list the following properties (number k below is any integer):

(1) Cosine is defined for all values of x. The domain of cosine is the interval $(-\infty, +\infty)$.
(2) Cosine changes between -1 and 1, so its range is the interval $[-1, 1]$.
(3) Cosine is a periodic function with a period of 2π.
(4) In a one-period interval $[-\pi/2, 3\pi/2]$, cosine returns the largest (maximum) value, equaling to 1, at $x = 0$. By adding any number of periods, which is $2k\pi$, we get the parametric description of the infinite number of all maximum points: $x = 2k\pi$.

In a one-period interval $[0, 2\pi]$, cosine returns the smallest (minimum) value, equals to -1, at $x = \pi$. Just as for maximum points, we can describe all minimum points on the entire number line in the parametric form: $x = \pi + 2k\pi$.

(5) Cosine is symmetric about the y-axis. Therefore, it is an even function:

$$\cos (-x) = \cos x.$$

(6) Cosine is positive at the interval $[-\pi/2, \pi/2]$. On the entire number line, cosine is positive at each interval

$$[-\pi/2 + 2k\pi, \pi/2 + 2k\pi].$$

Cosine is negative at the interval $[\pi/2, 3\pi/2]$. On the entire number line, it is negative at intervals

$$[\pi/2 + 2k\pi, 3\pi/2 + 2k\pi].$$

(7) In a one-period interval $[0, 2\pi]$, cosine has two zeros: $x = \pi/2$ and $x = 3\pi/2$. These points are located at a distance of π from each other. Therefore, we can describe all zeros of the cosine on the entire number line by one parametric formula $x = \pi/2 + k\pi$.

(8) Cosine increases at the interval $[-\pi, 0]$ and decreases at $[0, \pi]$. On the entire number line cosine increases at intervals $[-\pi + 2k\pi, 2k\pi]$ and decreases at $[2k\pi, \pi + 2k\pi]$.

Graphical Illustration of Solutions for Some Equations

If we need to solve the equation $f(x) = g(x)$, in some cases it is useful to consider a graphical interpretation of solutions. It allows us to confirm the correctness of solutions, or even get an idea of possible solutions. Graphically, to get solutions to the above equation, we graph both functions $f(x)$ and $g(x)$ in the same system of coordinates and consider points of their intersection. Then the x-coordinates of these points give us the solutions. Let's consider some examples.

Problem 11.9. Solve the equation $\sin x = x - \pi$.

Solution. This equation belongs to the so-called **transcendental** equations. These are equations that contain functions of different nature. In our example, the left side is a trig function, and the right side is a polynomial. In general, it is not possible to get solutions of such equations in exact form. It is not even possible to predict the number of solutions.

To get an idea about the solutions of a given equation, let's graph both functions $\sin x$ and $x - \pi$, and find points of intersections:

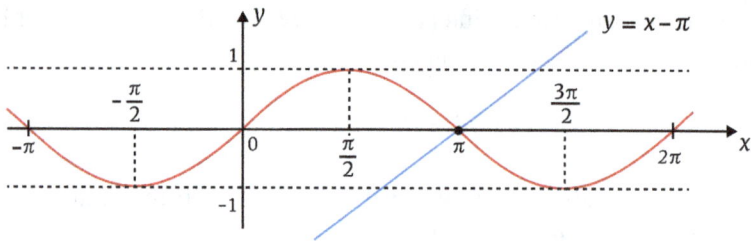

The figure clearly shows that there is only one point of intersection at $x = \pi$, so this is the only solution to the given equation. ■

Problem 11.10. Determine the number of solutions for the equation $\cos x = (x - \pi/2)^2$.

Solution. We graph both functions: $\cos x$ and $(x - \pi/2)^2$:

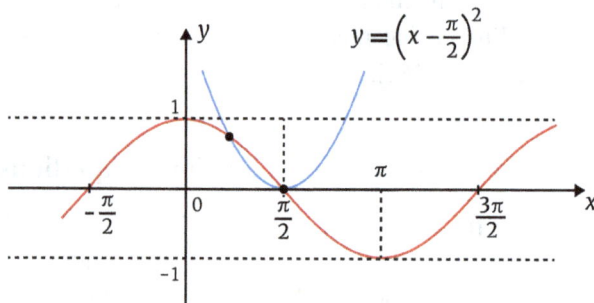

As we see, these graphs intersect at two points. The given equation has two solutions. ■

Problem 11.11. Solve the equation $\sin x = (x - \pi/2)^2 + 1$.

Solution. Graph both functions $\sin x$ and $(x - \pi/2)^2 + 1$:

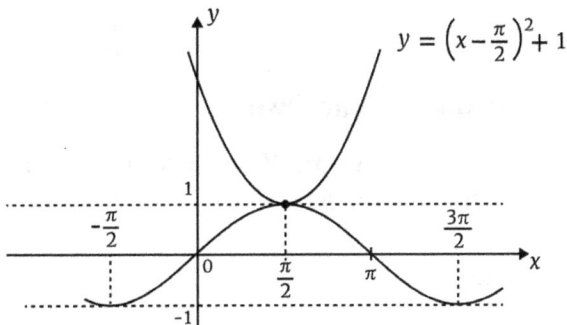

$$y = \left(x - \frac{\pi}{2}\right)^2 + 1$$

We can see from the graphs that they have only one common point at $x = \dfrac{\pi}{2}$. We can also get this result algebraically. Indeed, the maximum value of sine is 1, and this value occurs at $x = \dfrac{\pi}{2} + 2k\pi$ (see item 5 above in the list of properties of sine). The parabola $(x - \pi/2)^2 + 1$ produces values from 1 and up since it equals the nonnegative number $(x - \pi/2)^2$ plus 1. The minimum value of the parabola is 1, and it occurs at $x = \dfrac{\pi}{2}$. The last number is one of the numbers at which sine returns the maximum value. Thus, the only solution is $x = \dfrac{\pi}{2}$. ∎

Problem 11.12. Find all values of parameter t, for which the equation $\sin x - 1 = t$ has at least one solution.

Solution. The graph of function $\sin x - 1$ is obtained from the graph of $\sin x$, by shifting one unit down:

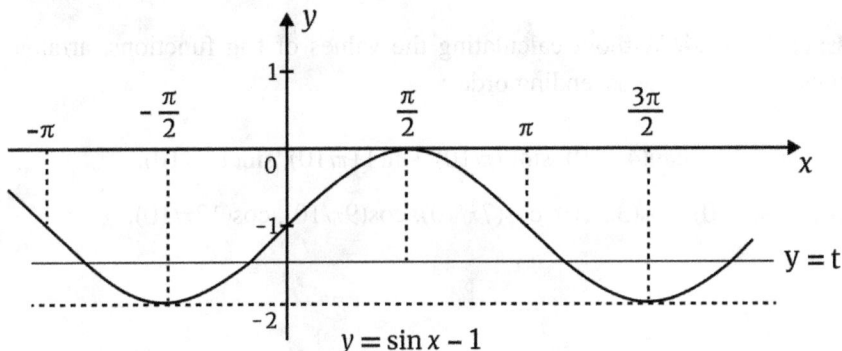

$$y = \sin x - 1$$

For the equation $\sin x - 1 = t$ to have a solution, the horizontal line $y = t$ must intersect the graph of the function $\sin x - 1$. As can be seen from the figure, this happens only when the parameter t is in the interval $[-2, 0]$. ■

Exercises for Solving on Your Own

Exercise 11.1. Let $f(x) = 5\sin x + x^2$. Write the formula for a new function $g(x)$, if the graph of the function $f(x)$ is

(a) shifted 7 units to the right;
(b) shifted 4 units down;
(c) reflected over the y-axis;
(d) reflected over the origin.

Exercise 11.2. Let a function $f(x)$ have the following graph:

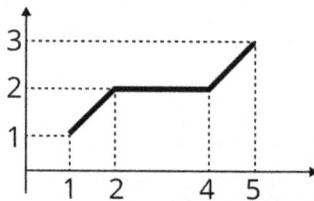

Sketch the graph of the function $g(x) = 2f(x + 5)$.

Exercise 11.3. Consider the equation $\cos x = x^2 + t$, where t is a parameter. Determine the possible number of solutions to this equation depending on the values of the parameter t.

Exercise 11.4. Without calculating the values of trig functions, arrange these values in the ascending order:

(a) $\sin(\pi/10)$, $\sin(4\pi/10)$, $\sin(7\pi/10)$, $\sin(11\pi/10)$, $\sin(13\pi/10)$.

(b) $\cos(\pi/10)$, $\cos(3\pi/10)$, $\cos(7\pi/10)$, $\cos(9\pi/10)$, $\cos(12\pi/10)$.

Exercise 11.5. In solving a trig problem, Sofia figured out that angles α and β are in the interval $(0, \pi/4)$. She calculated $\sin \alpha = 0.8$ and $\cos \beta = 0.6$. Both answers are incorrect. Why?

Exercise 11.6. Find all values of x in the interval $[0, 2\pi]$ for which $\cos x > \sin x$.

Entertainment Problem

Problem E11.1. Traveling Professor Smartman met his old friend Mike and asked about his children. Mike said that he has three kids. "How old are they now?" enquired Smartman.

"You are a professor. Why don't you try to figure it out? The product of their years is 36."

"I'm sorry, but this information is not sufficient," replied the professor.

"Well, the sum of their years is equal to the number of windows in that house over there," said Mike, pointing to a building across the street. The professor counted the windows and said, "But this information is also not sufficient."

"Okay, my oldest daughter is really good at drawing caricatures," Mike replied.

"Oh, now everything is clear!" exclaimed the professor.

Question: So, how old are Mike's children?

Chapter 12

Sinusoidal Functions

Many processes and natural phenomena turn out to be periodic fluctuations. Such processes and phenomena include vibration of strings, the motion of a pendulum, pulsation of stars, musical tones, radio waves, and much more. Since the trig functions sine and cosine are periodic waves, they are useful objects to model such processes and phenomena.

The following problem leads us to a transformation of sine function from $y = \sin x$ to the more general $y = a \sin (bx + c) + d$, where a, b, c, and d are some constants.

Problem 12.1. The radius of a Ferris wheel is a m, the height from the center of the wheel to the ground is d m, and the wheel rotates at a speed of f revolutions per hour. Find the function $y(t)$, showing the height of the cabin to the ground at any time t, assuming that at the initial moment the cabin is at a point corresponding to an angle with the horizon equal to φ radians:

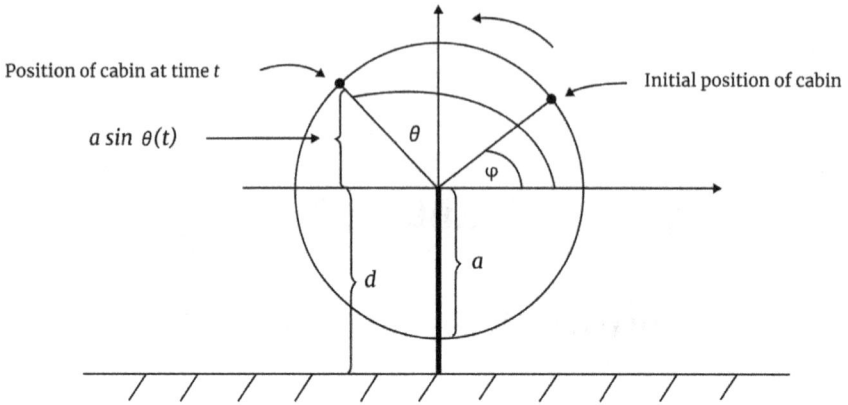

Solution. If we look at the position of the cabin at time t, we will see that function $y(t)$ is the sum of two parts: $y(t) = a \sin \theta(t) + d$. To get the explicit formula for $y(t)$, what remains is to express angle θ as a function of time t. It is given that a cabin makes f revolutions per hour. One full revolution corresponds to an angle of 2π radians, so f revolutions correspond to an angle of $2\pi f$. Since the wheel rotates uniformly, we can set up a proportion for angle $\overline{\theta}$ that the cabin sweeps over time t:

$$\frac{1h}{2\pi f} = \frac{t}{\overline{\theta}}.$$

Solving for $\overline{\theta}$, we get $\overline{\theta} = 2\pi f t$. Since the cabin starts moving at an angle φ, the angle θ at which the cabin will be at time t is the sum: $\theta = \overline{\theta} + \varphi = 2\pi f t + \varphi$. So, the height $y(t)$ of the cabin at time t is defined by the formula $y(t) = a \sin (2\pi f t + \varphi) + d$. ∎

Note. The expression $2\pi f$ is the angle that a cabin makes in 1 h. We can treat this expression as the rate (or speed) of the cabin in terms of the angle. Such speed is called the **angular speed** or **angular frequency**. The value of f, which is the number of revolutions per hour, is called **frequency**. We will say more about the frequency below.

Based on the above problem, we give the following definition:

> Sinusoid or sinusoidal function is a function of the form
> $$y = a \sin (bt + c) + d.$$

Here, t is a variable (argument) of the function. In many applications, this variable means time (in seconds, minutes, days, etc.). Letters a, b, c, and d are constant numbers (parameters).

Note. A periodic fluctuation can also be defined by the cosine function. It can be easily transformed into a sinusoid using the reduction formulas: $\cos t = \sin(\pi/2 - t)$ or $\cos t = \sin(t + \pi/2)$.

Let's analyze the meaning of the parameters a, b, c and d for the sinusoid.

It's easy to understand the meaning of the parameter d. Indeed, it shows how to transform the graph of the function $y = a \sin(bt + c)$ to $y = a \sin(bt + c) + d$: just shift it in the vertical direction by the value of d. If d is positive, shift up, and if d is negative, shift down.

Amplitude, Period, and Phase Shift

To simplify further analyses of parameters, we assume that $d = 0$. In other words, we consider the sinusoid

$$y = a \sin(bt + c).$$

Let's start with the parameter a. Since function $y = \sin(bt + c)$ returns its values between -1 and 1, the function $y = a \sin(bt + c)$ returns values between $-|a|$ and $|a|$. So, $|a|$ is the highest value (or peak) of this function. The number $A = |a|$ is called the **amplitude** of the sinusoid $y = a \sin(bt + a)$:

$$\boxed{A = |a| \text{ is the amplitude of the function } y = a \sin(bt + c).}$$

Amplitude A tells us how high and how low the graph of the sinusoid $a \sin(bt + c)$ extends in the vertical direction. Compare the graphs of $\sin t$, $2\sin t$ and $-\dfrac{1}{2}\sin t$ with amplitudes 1, 2 and $\dfrac{1}{2}$:

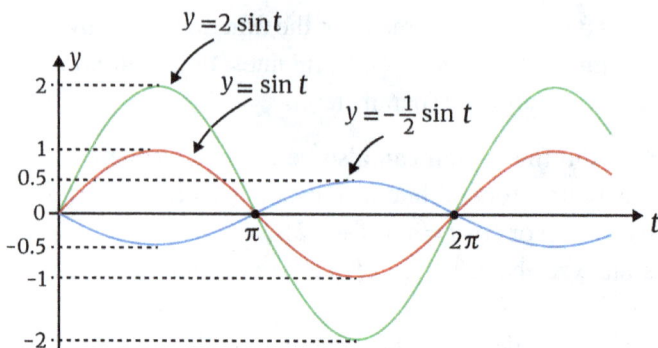

Next, we analyze parameter b. Let's look at the general case.

Problem 12.2. Let $f(t)$ be a periodic function with period P. Prove that the period of the function $g(t) = f(bt + c)$ is equal to P/b.

Solution. By definition of period P, $f(t + P) = f(t)$ for any t. We have

$$g\left(t + \frac{P}{b}\right) = f\left[b\left(t + \frac{P}{b}\right) + c\right] = f(bt + P + c) = f(bt + c) = g(t).$$

The equality $g\left(t + \frac{P}{b}\right) = g(t)$ shows that $\dfrac{P}{b}$ is the period of $g(t)$. ■

Let's return to the sinusoid $y = a\,\sin(bt + c)$. Since the function $y = a\sin t$ has a period of 2π, then, using the result from the previous problem, we come up to the following statement:

> Period P of the sinusoid $y = a\,\sin(bt + c)$ is
> $$P = 2\pi/b.$$

Compare the graphs of $\sin t$, $\sin 2t$ and $\sin\dfrac{1}{2}t$ with periods 2π, π and 4π:

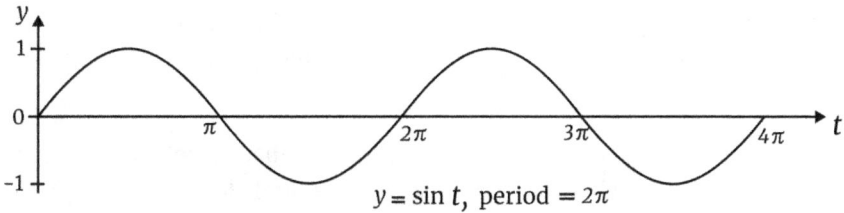

$y = \sin t$, period $= 2\pi$

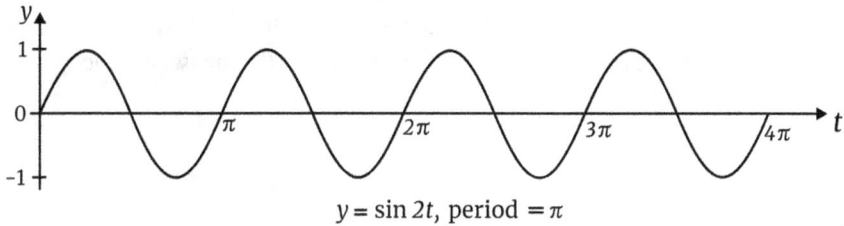

$y = \sin 2t$, period $= \pi$

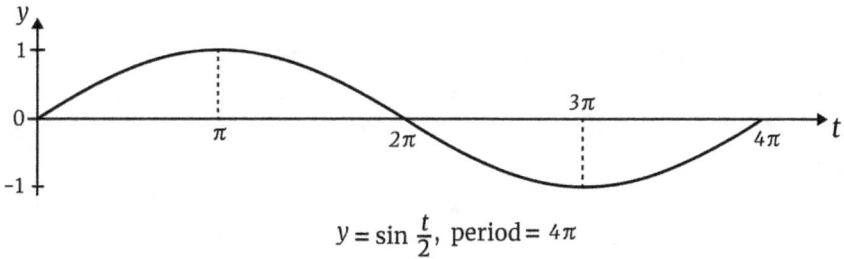

$y = \sin \frac{t}{2}$, period $= 4\pi$

In Problem 12.1, we represented the sinusoid in the form $a \sin(2\pi f t + \varphi) + d$. Comparing it with $a \sin(bt + c)$, we conclude that $b = 2\pi f$. As we indicated in the note after Problem 12.1, this is the angular speed. Thus:

> Parameter b of the sinusoid
> is the angular speed.

If a sinusoid describes the rotation of an object, parameter b measures the rotation rate, i.e., the angle by which the object turns in a unit of time.

Now let's find out the effect of the remaining parameter c. Again, we consider the general case.

Problem 12.3. Assume that we can plot the graph of the function $g(t) = f(bt)$. How would we transform it to the graph of the function $h(t) = f(bt + c)$?

Solution. We already discussed in Chapter 11 that to get the graph of $f(t + c)$ from the graph of $f(t)$, we make a horizontal shift of this graph by the value of c. If c is positive, shift left, and if c is negative, shift right. At first glance, it may seem that the same answer holds for the problem at hand: we need to shift by the value of c. However, this answer is incorrect. To see why, let us modify the function $h(t)$:

$$h(t) = f(bt + c) = f\left[b\left(t + \frac{c}{b}\right)\right] = g\left(t + \frac{c}{b}\right).$$

From here we conclude that to get the graph of $h(t)$ from the graph of $g(t)$, we shift the latter graph in the horizontal direction not by the value c, but by the value $\frac{c}{b}$ instead. If $\frac{c}{b}$ is positive, shift left, and if $\frac{c}{b}$ is negative, shift right. ■

Returning to the sinusoid $y = a \sin(bt + c)$, we conclude that its graph can be obtained from the graph of $y = a \sin(bt)$ by horizontally shifting by the value $\frac{c}{b}$.

Since for positive $\frac{c}{b}$ the shift goes to the left, and for negative $\frac{c}{b}$ it goes to the right, this result may seem counterintuitive. Perhaps we would feel more comfortable if the result was to be just the opposite. Thus, to make it more intuitive, a special value is introduced, which is called the **phase shift** and is often denoted by the letter Φ (Greek letter Phi). Here is its definition:

$$\boxed{\text{Phase Shift } \Phi = -c/b.}$$

Using the phase shift Φ, we can say that to get the graph of the sinusoid $y = a \sin(bt + c)$ from the graph of $y = a \sin(bt)$, we shift this graph in the horizontal direction by Φ: if Φ is positive, shift to the right, and if Φ is negative, shift to the left. This result agrees with our intuition. Using the phase shift Φ, we can represent sinusoid in the form: $y = a \sin[b(t - \Phi)]$.

Let's summarize our discussion about parameters a, b, and c for the sinusoid.

$$y = a \sin(bt + c)$$

Amplitude $A = |a|$. It shows peaks and troughs (maximum and minimum values) of the sinusoid: maximum value is A, and minimum value is $-A$.

Period $P = 2\pi/b$. It shows the smallest interval after which the sinusoid "repeats itself."

Phase shift $\Phi = -c/b$. It shows the value by which the graph of the function $y = a \sin(bt)$ is shifted in the horizontal direction to get the graph of the above sinusoid: if Φ is positive, shift to the right, and if Φ is negative, shift to the left.

Graphing the Sinusoid $y = a \sin(bt + c)$ at a One-Period Interval

Sometimes we want to graph the sinusoid $y = a \sin(bt + c)$ at a one-period interval and mark its essential points. The essential points are points of intersection with the t-axis (zeros of the sinusoid: points at which the sinusoid is equal to zero), and points of maxima and minima (points at which the function has biggest and smallest values). Below are the steps to sketch the graph of the sinusoid at a one-period interval. We consider the case when parameter a is positive. We can graph the sinusoid in three steps according to the diagram:

$$\sin(t) \implies a \sin(bt) \implies a \sin(bt + c).$$

Here are the steps:

(1) Graph the basic function $y = \sin(t)$ on a one-period interval $[0, 2\pi]$, and label the quadrantal angles 0, $\dfrac{\pi}{2}$, π, $\dfrac{3\pi}{2}$ and 2π on the horizontal t-axis.

(2) Divide all labels on the t-axis by b. Labels become:

$$0, \frac{\pi}{2b}, \frac{\pi}{b}, \frac{3\pi}{2b}, \frac{2\pi}{b}.$$

Also, mark the labels $A = |a|$ and $-A$ on the vertical y-axis. As a result, we will get the graph of the sinusoid $y = a \sin(bt)$.

(3) Add the phase shift $\Phi = -\dfrac{c}{b}$ to all labels on the t-axis and shift the graph along the t-axis by Φ: to the right, if $\Phi > 0$ and to the left, if $\Phi < 0$. Thus, we get the final graph at a one-period interval.

You can organize the calculation of labels in the above steps in a table:

Quadrantal angles	0	$\dfrac{\pi}{2}$	π	$\dfrac{3\pi}{2}$	2π
Divide by b	0	$\dfrac{\pi}{2b}$	$\dfrac{\pi}{b}$	$\dfrac{3\pi}{2b}$	$\dfrac{2\pi}{b}$
Add Φ	Φ	$\dfrac{\pi}{2b} + \Phi$	$\dfrac{\pi}{b} + \Phi$	$\dfrac{3\pi}{2b} + \Phi$	$\dfrac{2\pi}{b} + \Phi$

If needed, adjust the scale on the coordinate axes.

Note. If parameter a is negative, then in step 1, start with the graph of $y = -\sin(t)$ (reflect the graph of $y = \sin(t)$ over the t-axis).

Problem 12.4. For the sinusoid $y = 3 \sin(2t - \pi/3)$,
(a) Calculate the amplitude, period, and phase shift.
(b) Graph the sinusoid in a one-period interval and on the t-axis label all maxima, minima, and zero points.

Solution.
(a) For the given sinusoid, $a = 3$, $b = 2$, $c = -\pi/3$. Thus, the amplitude is $A = |a| = |3| = 3$, period $P = \dfrac{2\pi}{b} = \dfrac{2\pi}{2} = \pi$ and phase shift $\Phi = -\dfrac{c}{b} = -\dfrac{-(\pi/3)}{2} = \dfrac{\pi}{6}$.

(b) To graph the sinusoid, we follow the above steps. Graph of the basic function $y = \sin(t)$ on the interval $[0, 2\pi]$ is already drawn in the previous chapter. Following step 2, we divide labels $0, \dfrac{\pi}{2}, \pi, \dfrac{3\pi}{2}, 2\pi$ on the *t*-axis by $b = 2$. Labels become $0, \dfrac{\pi}{4}, \dfrac{\pi}{2}, \dfrac{3\pi}{4}, \pi$. Also, we label 3 and –3 on the *y*-axis. We get the graph of the sinusoid $y = 3 \sin(2t)$:

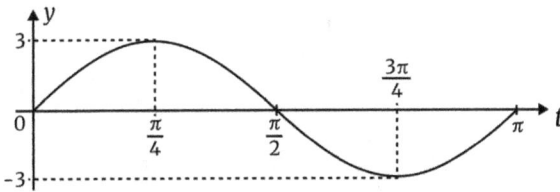

Graph of $y = 3 \sin(2t)$

Next, we add the phase shift $\varPhi = \dfrac{\pi}{6}$ to all labels on the *t*-axis. Labels become

$$\frac{\pi}{6}, \frac{5\pi}{12}, \frac{2\pi}{3}, \frac{11\pi}{12}, \frac{7\pi}{6}.$$

Finally, we shift the above graph by $\varPhi = \dfrac{\pi}{6}$ to the right. We get this figure

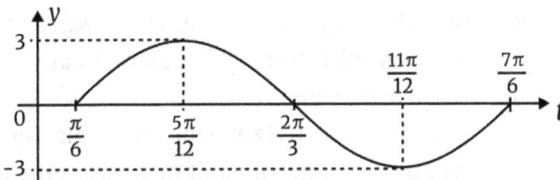

Graph of $y = 3 \sin(2t - \pi/3)$ ∎

Frequencies of Sinusoidal Signals

As we already mentioned, sinusoids are useful models to describe many periodic oscillations (the repetitive variations). A branch of physics and engineering called Signal Processing studies such phenomena. It is well known that radio waves can reach your radio set with different frequencies.

Generally speaking, frequency tells us how often something happens per unit of time, for example, in one second. For the sinusoid $y = a \sin(bt + c)$, if we treat the variable *t* as time in seconds, the frequency means the number of full cycles in one second. Period *P* of the sinusoid is related to frequency f. Indeed, since P is the time for 1 rotation, then $1/P$ rotations occur per 1 unit of time, which is frequency f. So, period and frequency are **reciprocal** to each other:

$$f = \frac{1}{P} \text{ and } P = \frac{1}{f}.$$

Since the period $P = \frac{2\pi}{b}$, it follows that $b = \frac{2\pi}{P} = 2\pi f$. From here, we get an expression of the sinusoid in terms of frequency (the same expression that we have seen in Problem 12.1):

$$y = a \sin(2\pi ft + c).$$

In electronics, frequencies are usually measured in hertz (Hz). They get this name in honor of a German physicist Heinrich Hertz (1857–1894). He was the first who experimentally proved the existence of electromagnetic waves. However, Hertz did not realize the practical importance of his experiment. When he was asked about the applications of his discovery, Hertz replied, "Nothing, I guess." He died at the age of 36 from granulomatosis (a very rare disease). Before Hertz, Scottish mathematical physicist James Clerk Maxwell (1831–1879) theoretically predicted the existence of electromagnetic waves.

By definition, a frequency of 1 Hz means one cycle per second. For example, 30 Hz means 30 cycles/s. This unit of measurement was established by the International Electrotechnical Commission in 1930 and was adopted by the General Conference on Weights and Measures in 1960. Consider, for example, sound waves. The frequency of the sound affects how high the audio tone is. The human ear can hear sounds at frequencies from about 20 to 20,000 Hz (20 kHz). The best range for hearing is from 2 to 5 kHz. Some animals, for example, dogs, can hear in a much wider frequency range.

An important property of sinusoids is that the sum of any finite number of sinusoids with the same frequency is also a sinusoid with the same frequency. We establish this fact in the following two problems.

Problem 12.5. Show that the expression $a_1 \sin x + a_2 \cos x$ can be written as a single sinusoid.

Solution. We can assume that $a_2 > 0$ (otherwise, we replace the original expression with $-(-a_1 \sin x + (-a_2)\cos x)$). First, we modify the given expression like this

$$a_1 \sin x + a_2 \cos x = \sqrt{a_1^2 + a_2^2}\left(\frac{a_1}{\sqrt{a_1^2 + a_2^2}} \sin x + \frac{a_2}{\sqrt{a_1^2 + a_2^2}} \cos x \right).$$

Now we make the following observation: since $\left| a_1 / \sqrt{a_1^2 + a_2^2} \right| < 1$, there exists an angle φ such that $\cos \varphi = a_1 / \sqrt{a_1^2 + a_2^2}$. Let's calculate $\sin^2 \varphi$:

$$\sin^2 \varphi = 1 - \cos^2 \varphi = 1 - \frac{a_1^2}{a_1^2 + a_2^2} = \frac{a_2^2}{a_1^2 + a_2^2}.$$

From here, $\sin \varphi = a_2 / \sqrt{a_1^2 + a_2^2}$. This is exactly the coefficient for $\cos x$ in the parentheses above. Using the Sine Sum formula (see Chapter 8), the original expression can be written as a single sinusoid:

$$a_1 \sin x + a_2 \cos x = \sqrt{a_1^2 + a_2^2}\, (\sin x \cos \varphi + \cos x \sin \varphi)$$
$$= \sqrt{a_1^2 + a_2^2}\, \sin(x + \varphi). \ \blacksquare$$

Problem 12.6. Show that the sum of two sinusoids with the same frequency can be written as a single sinusoid with the same frequency. Calculate the amplitude of the sum.

Solution. Let two sinusoids be

$$y = a_1 \sin(bt + c_1) \text{ and } y = a_2 \sin(bt + c_2).$$

Note that the coefficient b is the same for both sinusoids. This is because the sinusoids have the same frequency $f = b/2\pi$. Using the Sine Sum formula, we have

$$a_1 \sin(bt + c_1) + a_2 \sin(bt + c_2)$$
$$= a_1 \left[\sin(bt)\cos(c_1) + \cos(bt)\sin(c_1) \right]$$
$$+ a_2 \left[\sin(bt)\cos(c_2) + \cos(bt)\sin(c_2) \right]$$
$$= \left[a_1 \cos(c_1) + a_2 \cos(c_2) \right] \sin(bt)$$
$$+ \left[a_1 \sin(c_1) + a_2 \sin(c_2) \right] \cos(bt).$$

Let's denote the coefficients for $\sin(bt)$ and $\cos(bt)$ as

$$A_1 = a_1 \cos(c_1) + a_2 \cos(c_2) \text{ and } A_2 = a_1 \sin(c_1) + a_2 \sin(c_2).$$

Then

$$a_1 \sin(bt + c_1) + a_2 \sin(bt + c_2) = A_1 \sin(bt) + A_2 \cos(bt).$$

According to the result from the previous problem, we conclude that last expression can be transformed into a single sinusoid with the same frequency and the amplitude $A = \sqrt{A_1^2 + A_2^2}$. ■

Sinusoids are useful tools to model smooth periodic oscillations. However, in digital electronics and digital signal processing some non-smooth periodical signals are used like the following square wave:

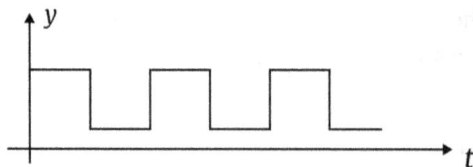

It turns out that the combination of an infinite number of sinusoids with multiple frequencies can describe and approximate almost any periodic wave, including non-smooth one. This discovery was made by a French mathematician Joseph Fourier (1768–1830) when he worked on a specific problem having to do with the distribution of heat in a metal plate. His work led to an important area of mathematics, which is called Fourier analysis, in particular Fourier series. This theory found numerous applications in many areas of physics and engineering. Alas, this interesting topic is beyond the scope of this book.

Exercises for Solving on Your Own

Exercise 12.1. Find the amplitude, period and phase shift for the following sinusoidal functions.

$$\text{(a) } y = 2 \sin(3t + \pi/6); \text{ (b) } y = -3 \sin(2t - \pi/4).$$

Sketch the graphs of these functions in a one-period interval.

Exercise 12.2. Write the equations of sinusoids in the form $y = a \sin(bx + c)$ that match the given graphs.

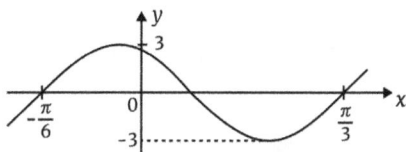

(a)

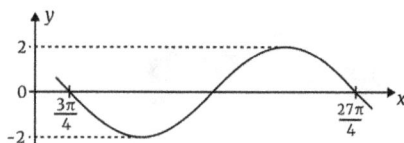

(b)

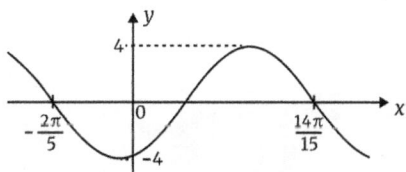

(c)

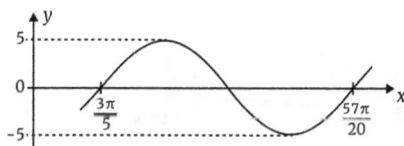

(d)

Exercise 12.3. For the following function, describe

(a) domain, (b) range, (c) amplitude, (d) period, (e) phase shift, (f) vertical shift.

$$-\frac{1}{4}(y+12) = \sin\left(5t - \frac{\pi}{4}\right) + 3.$$

Exercise 12.4. Represent the given expressions as a single sinusoid.

$$\text{(a) } \sin x + \cos x; \text{ (b) } \sqrt{3} \sin x - \cos x.$$

Exercise 12.5. Prove that for any x, a and b,

$$\left| a \sin x + b \cos x \right| \le \sqrt{a^2 + b^2}.$$

Exercise 12.6. Solve the equation $\sin x + \cos x = \sqrt{2}$ in the interval $[0, 2\pi]$.

Entertainment Problem

Problem E12.1. You have three keys to three suitcases with different locks. Each key fits only one suitcase. Are three attempts sufficient to determine which key opens which suitcase?

Chapter 13

Applications of Sinusoids

In the previous chapter, we introduced the sinusoidal function $y(t) = a \sin(2\pi f t + c) + d$ and discussed some of its basic properties. Here we consider the application of sinusoids in solving problems related to periodic fluctuations from different areas of science and nature.

In some practical cases, it is possible to determine (empirically) the maximum and minimum values of the periodic process. These data can be used to determine parameters of the sinusoidal function, which model the process.

Problem 13.1. Let y_{max} and y_{min} be the maximum and minimum values of the sinusoidal function $y = a \sin(2\pi f t + c) + d$. Assuming that $a > 0$, prove that parameters a and d are determined by the formulas:

$$a = \frac{y_{max} - y_{min}}{2}, \quad d = \frac{y_{max} + y_{min}}{2}. \tag{1}$$

Solution. The function $y = a \sin(2\pi f t + c) + d$ reaches its maximum and minimum values when the sine function reaches its maximum and minimum values that are 1 and -1. So, we have

$$\begin{cases} y_{max} = a + d \\ y_{min} = -a + d. \end{cases}$$

Solving this system for a and d, we get the above result. ∎

The next problem considers the process of changing blood pressure during a circulation cycle (pumping and relaxing the heart). When our heart pumps the blood throughout the body, the blood pressure has the highest value (it is called systolic). When our heart relaxes between beats, then the blood pressure has the lowest value (it is called diastolic). Experiments show that blood pressure with a good approximation can be expressed through sinusoidal functions. The problem below derives a sinusoidal function that expresses blood pressure as a function of time for the average person.

Problem 13.2. The average blood pressure for adults is about 120 (highest) over 80 (lowest). It is usually written as 120/80, and it is measured in millimeters of mercury (mmHg). The average pulse (heart rate) is 80 beats/min. Determine the sinusoidal function that expresses the dependence of blood pressure on time in minutes. Assume that the function begins at the time when the blood pressure reaches the average value.

Solution. The problem is to find the parameters a, f, c, and d for the sinusoid $y(t) = a \sin(2\pi f t + c) + d$, where t is time in minutes. Actually, two models could be used: when the blood pressure increases, starting from the average value, or when it decreases. Here we consider the case when the pressure increases. For case when the pressure decreases, see Exercise 13.2. Using the notations from the previous problem, we have $y_{max} = 120$ and $y_{min} = 80$. From here, according to formulas (1),

$$a = (y_{max} - y_{min})/2 = (120 - 80)/2 = 20,$$
$$d = (y_{max} + y_{min})/2 = (120 + 80)/2 = 100.$$

Therefore, the sinusoid takes the form: $y(t) = 20 \sin(2\pi f t + c) + 100$. Parameter f is the frequency (the number of cycles per unit of time). In our case, f is the number of beats per minute (pulse), so $f = 80$. Therefore, the sinusoid becomes $y(t) = 20 \sin(160\pi t + c) + 100$. Finally, to find c, we set $t = 0$ (initial value of time) and use the condition that at the initial moment the blood pressure has the average value, which is $d = 100$. Therefore,

$$y(0) = 100 = 20 \sin(2\pi f \cdot 0 + c) + 100 = 20 \sin c + 100.$$

From here, $\sin c = 0$ and $c = 0$. The final answer is

$$y(t) = 20 \sin(160\pi t) + 100. \blacksquare$$

Note. Another solution to the equation $\sin c = 0$ is $c = \pi$. This case corresponds to the model when blood pressure decreases after the average value.

The following problem demonstrates the creation of a model for describing the tidal movement. It is known that tides are caused mostly by the gravity of the moon (the sun's gravitation also has some influences). Observations show that sinusoidal functions provide a good approximation of the motion of the tides.

Problem 13.3. We've been tasked in determining the height of a tide along a dam at any given time, especially at 3:00 pm. It is known that on a particular day, the high tide reaches 14 feet at 11:00 am, and the low tide drops to 6 feet. This periodic process occurs every 12 h. The problem is to create a function to determine the water level.

Solution. We will use a sinusoid as a model for the water level considering it as a function of time t (in hours), so $y(t) = a \sin(2\pi ft + c) + d$. We will treat 11:00 am as a starting time and assign $t = 0$ to it. It is given that $y(0) = 14$. We have $y_{max} = 14$, $y_{min} = 6$. Using formulas (1),

$$a = (y_{max} - y_{min})/2 = (14-6)/2 = 4, \ d = (y_{max} + y_{min})/2 = (14+6)/2 = 10.$$

Since the periodic process occurs every 12 h (it is one cycle), in 1 h, it occurs $\dfrac{1}{12}$ part of a cycle. This is the frequency f, so $f = \dfrac{1}{12}$. Therefore, the sinusoid takes the form

$$y(t) = 4 \sin(\pi t/6 + c) + 10.$$

To find parameter c, we use the condition that $y(0) = 14 = 4 \sin c + 10$. Solving for $\sin c$, we get $\sin c = 1$. From here, $c = \pi/2$ and the sinusoid is

$$y(t) = 4 \sin(\pi t/6 + \pi/2) + 10.$$

It can be simplified by the reduction formula, so $y(t) = 4 \cos(\pi t/6) + 10$.

In order to determine the water level at 3:00 pm, we set $t = 4$ (4 h after the initial time 11:00 am). We have

$$y(4) = 4 \cos(\pi \cdot 4/6) + 10 = 4 \cos(2\pi/3) + 10 = -4/2 + 10 = 8.$$

So, the water level at 3:00 pm will be 8 feet. ■

The problem below determines days of the year with acceptable daylight hours.

Problem 13.4. To carry out important construction work in New York, we plan to hire two teams of workers who will work in two shifts with a total duration of 14 h/day (so that the work duration for one team would not be too long). They can only work in daylight, and the work has to be completed in 3 months. To start the work, it is necessary to assess what days of the year daylight hours are at least 14 h. Take into account that in New York, the shortest day is 9 h and 15 min, and the longest day is 15 h and 6 min. Also use the fact that the spring equinox[1] is on March 21. The countdown of days must begin on January 1.

Solution. Since the change of seasons is a periodic process, we will model it by a sinusoidal function $y(t) = a \sin(2\pi f t + c) + d$. Here, t is the number of days starting from January 1. So, $t = 0$ means January 1 and $t = 365$ means December 31. For March 21 (the day of the spring equinox), $t = 80$ ($31 + 28 + 21$). The value $y(t)$ means the number of daylight hours on day t. Thus, $y(80) = 12$ h (on March 21, day and night are equal in length). As in the previous problems, we will use formulas (1):

$$y_{max} = 15\text{h } 6 \text{ min} = 15.1 \text{ h}, \ y_{min} = 9 \text{ h } 15 \text{ min} = 9.25 \text{ h},$$
$$a = (y_{max} - y_{min})/2 = (15.1 - 9.25)/2 = 3,$$
$$d = (y_{max} + y_{min})/2 = (15.1 + 9.25)/2 = 12.$$

We rounded a and d to the whole number of hours. Value $d = 12$ corresponds to the spring equinox on March 21. The sinusoid $y(t)$ makes one cycle per year (365 days), so the frequency f (number of cycles in one day) is $f = \dfrac{1}{365}$. The sinusoid $y(t)$ takes the form

$$y(t) = 3 \sin(2\pi t/365 + c) + 12 = 3 \sin(0.0172 \cdot t + c) + 12.$$

To find parameter c, we use the condition that $y(80) = 12$. So,

---

[1] On the day of the equinox, daytime and nighttime are approximately equal.

$$y(80) = 3\sin(0.0172 \cdot 80 + c) + 12$$
$$= 3\sin(1.376 + c) + 12 = 12 \Rightarrow \sin(1.376 + c) = 0.$$

From here, we get two equations for c: $1.376 + c = 0$ and $1.376 + c = \pi$. Below we will show that the second equation should be rejected. From the first equation, we get $c = -1.376$, and our model function is found:

$$y(t) = 3\sin(0.0172 \cdot t - 1.376) + 12.$$

To detect days when daylight hours are at least 14 h, we need to solve the inequality: $y(t) \geq 14$. If we denote $\theta = 0.0172 \cdot t - 1.376$, then the inequality $y(t) \geq 14$ implies:

$3\sin(\theta) + 12 \geq 14 \Rightarrow \sin\theta \geq 0.667$. To solve this inequality, we first solve the equation $\sin\theta = 0.667$. It has two solutions in the interval $[0, 2\pi]$: $\theta_1 = \sin^{-1} 0.667$ and $\theta_2 = \pi - \theta_1$. Using a calculator, $\theta_1 = 0.73$ and $\theta_2 = 2.41$ (angles are in radians). Let's mark these angles on the graph of sine:

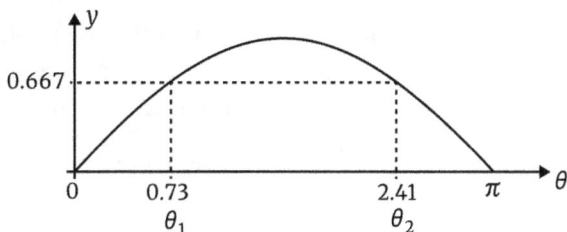

We can see that solutions of the inequality $\sin\theta \geq 0.667$ are angles between θ_1 and θ_2: $\theta_1 \leq \theta \leq \theta_2 \Rightarrow 0.73 \leq \theta \leq 2.41 \Rightarrow 0.73 \leq 0.0172 \cdot t - 1.376 \leq 2.41$. Solving this inequality for t, we get $122 \leq t \leq 220$. The number 122 corresponds to May 2, and 220 corresponds to August 8. So, the days with at least 14 h of daylight will be from May 2 to August 8. There are 99 such days, which is more than the required 3 months.

Now consider the second equation for c: $1.376 + c = \pi$. We get $c = 1.766$ and the function $y(t)$ becomes $y(t) = 3\sin(0.0172 \cdot t + 1.766) + 12$. Using the notation $\theta = 0.0172 \cdot t + 1.766$, we have the same inequality for θ as before: $0.73 \leq \theta \leq 2.41$ or $0.73 \leq 0.0172 \cdot t + 1.766 \leq 2.41$. From here $-60 \leq t \leq 37$. In the current year (starting with $t = 0$), there are not enough days with the required daylight hours, so we reject the second equation. ■

The next problem is related to determining the brightness of the star Delta Cephei. In 1784, a young English deaf-mute astronomer John Goodrick (1764–1786, born in the Netherlands), discovered that this star periodically changes its brightness and suggested an explanation of this effect. For this discovery and others, he was elected as a Fellow of the Royal Society. He never learned of this honor, however, as he died four days later from pneumonia at age 21.

An important contribution to the study of variable stars was made by Henrietta Leavitt (1868–1921). She was born in Cambridge, Massachusetts. After a severe illness, she became deaf. Working at Harvard College Observatory (with a salary of 30 cents/h) she studied photographic plates. Based on that, she discovered more than 2400 variable stars.

Problem 13.5. The star Delta Cephei is pulsating with a period of 5 days and 9 h: its radius alternately expands and contracts by millions of miles. As a consequence, its brightness varies from magnitude 3.48^m to 4.37^m (in astronomy, magnitude (m) is a measure of the brightness). Find a sinusoidal function that models the magnitude of Delta Cephei as a function of time in days. Select the initial time when the star is at maximum brightness.

Solution. According to the problem's condition, the sinusoid

$$y(t) = a \, \sin(2\pi f t + c) + d$$

has the period of 5 days and 9 hours, which is 5.375 days. Therefore,

$$f = 1/5.375 = 0.186, \, 2f = 0.372.$$

Since $y_{max} = 4.37$, $y_{min} = 3.48$, then

$$a = (4.37 - 3.48)/2 = 0.445,$$
$$d = (4.37 + 3.48)/2 = 3.925$$

At this point, the sinusoid is

$$y(t) = 0.445 \, \sin(0.372\pi t + c) + 3.925.$$

At initial time $t = 0$, $y(0) = y_{max} = 4.37$. It occurs for the maximum value of the sine function. Therefore, sin (c) = 1, and $c = \pi/2$. Finally, the sinusoid is defined as

$$y(t) = 0.445 \sin(0.372\pi t + \pi/2) + 3.925.$$

Using the reduction formula $\sin(x + \pi/2) = \cos x$, we can present this result as

$$y(t) = 0.445 \cos(0.372\pi t) + 3.925. \blacksquare$$

The next problem demonstrates the modeling of an alternating current (AC). Alternating current occurs when a rotor turns around its axis in a magnetic field. During the rotation, the wire enters a different magnetic polarity that causes the current to reverse the direction periodically. It also causes the voltage to alternate. The unit of voltage measurement is the volt (V), named after the Italian physicist Alessandro Volta (1745–1827). The voltage can be modeled with the sinusoidal function $y = a \sin(bt + c) + d$. Usually, the notation ω is used instead of parameter b. Parameter ω is called the angular frequency. Also, for an alternating current, parameter c is often denoted as ψ, and it is called the phase angle. So, instead of $bt + c$, the expression $\omega t + \psi$ is used.

Problem 13.6. It is established that ordinary household alternating current can be described by a sinusoidal function that oscillates from +155 V to –155 V with a frequency $f = 60$ Hz (60 cycles/s). Find the sinusoidal function $V(t)$, which describes the dependence of the voltage change on time t. Use the maximum voltage when $t = 0$.

Solution. We have $V(t) = a \sin(\omega t + \psi) + d$. It is given that $V_{max} = 155$, and $V_{min} = -155$. From here,

$$a = \left(V_{max} - V_{min}\right)/2 = \left(155 - (-155)\right)/2 = 155,$$
$$d = \left(V_{max} + V_{min}\right)/2 = \left(155 + (-155)\right)/2 = 0,$$
$$\omega = 2\pi f = 2\pi \cdot 60 = 120\pi.$$

Voltage $V(t)$ becomes $V(t) = 155 \sin(120\pi t + \psi)$. To determine ψ, we set $t = 0$. It is given that $V(0) = V_{max} = 155$. We have

$V(0) = 155 = 155 \sin(\psi)$. From here, $\sin(\psi) = 1$ and $\psi = \pi/2$. The final answer is $V(t) = 155 \sin(120\pi t + \pi/2) = 155 \cos(120\pi t)$. ∎

Exercises for Solving on Your Own

Exercise 13.1. A mass hanging on the end of a vertical spring is pulled down so that its distance from the floor is 4 in. The mass is then released and springs up and down. Assuming that there are no resistive and external forces, the mass will periodically reach the constant maximum and minimum heights above the floor, which are 8 and 4 in, respectively. Also, assume that the mass is moving from minimum to maximum heights (and vice versa) in 0.6 s. Determine the sinusoidal function that describes the motion of the mass. At what distance from the floor will the mass be after 1 s?

Exercise 13.2. Bill has a blood pressure of 124/72 and a pulse rate of 75 beats/min. Using a sinusoidal function as a model, find the formula for this function. Assume that the blood pressure decreases, starting from the average value.

Exercise 13.3. The Ferris wheel's diameter is 32 m and the distance from the bottom of the wheel to the ground is 3 m. The wheel makes one full rotation in 1.5 min. A rider gets onto the wheel at the lowest point. How high off the ground will the rider be in 1 min?

Exercise 13.4. A boat must leave the pier in the daytime (before 6 pm). To sail safely, it needs at least 12 feet of depth in the sea. In this area, the high tide reaches 20 feet at 11 am, and at low tide it drops to 4 feet at 5 pm. This periodic process occurs every 12 h. Between what times is it safe for the boat to leave the pier?

Exercise 13.5. In a certain area, the monthly average temperature changes periodically throughout the year. The maximum average temperature occurs in August and equals to 35°C. The minimum average temperature is −5°C. In what months is the average temperature below 25°C? Assume that the sinusoidal model can be used to describe the monthly average temperature.

Entertainment Problem

Problem E13.1. Bandits captured a group of tourists and involved them in a dangerous game — to play for their chance of survival. They said that the next morning they would take all the tourists into the courtyard. The bandits will place a black or a white hat on each tourist, such that none of them would know the color of their own hat. After that, the bandits would ask each tourist, one by one, the color his/her hat. If the correct answer is given, the tourist would be released. If not, the tourist will be executed. The bandits warned that the tourists would not be able to speak, signal or give any kind of cues to others. However, everyone will be able to see everyone else and hear their answers. Throughout the night, the tourists discussed a possible strategy. And they came up with a wonderful idea which would allow almost all of them to survive. What did the tourists come up with?

Chapter 14

Graphs of Tangent and Cotangent

As we saw in previous chapters, sines and cosine as well as sinusoidal functions are presented with wavy graphs. In contrast, the graphs of tangent and cotangent do not have a wave-like shape.

Before starting with these graphs, consider a simpler function $f(x) = \dfrac{1}{x}$. It is called the **hyperbola**. This function is odd:

$$f(-x) = \frac{1}{-x} = -f(x).$$

Therefore, its graph is symmetric about the origin $(0, 0)$. To construct its graph, we can plot the graph for positive x, and then reflect it over the origin. To plot the graph for positive x, we create a table of values, for example, like this

x	1/3	1/2	1	2	3
y = 1/x	3	2	1	1/2	1/3
(x, y)	(1/3, 3)	(1/2, 2)	(1, 1)	(2, 1/2)	(3, 1/3)

By smoothly connecting these points in the coordinate system, we get the graph.

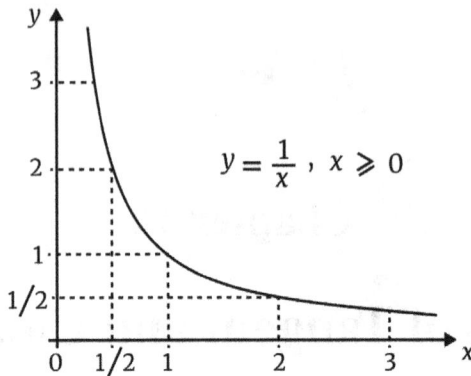

Reflecting this graph over the origin, we get the complete graph along the entire x-axis:

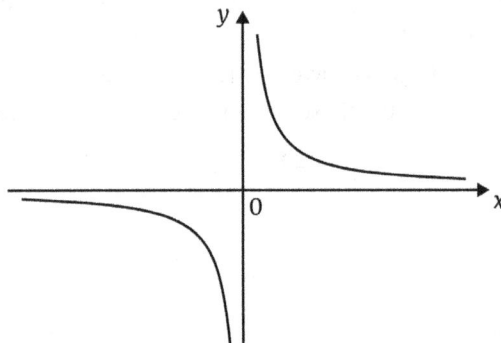

Graph of the hyperbola $y = \dfrac{1}{x}$

Note some specific features of this function:

(1) The function is discontinuous: its graph consists of two separate branches: for $x > 0$ and for $x < 0$.
(2) When x approaches zero from the right side, the graph increases to infinity, and if x approaches zero from the left side, the graph decreases to negative infinity. We say that the function $y = \dfrac{1}{x}$ has a **vertical asymptote** $x = 0$ (this is the y-axis).
(3) When x approaches positive or negative infinity, the graph approaches the x-axes, but never touches it. We say that the function has a **horizontal asymptote** $y = 0$ (this is the x-axis).

Below we will see that tangent and cotangent graphs have features similar to (1) and (2).

Graphing the Tangent Function

Recall that on the unit circle in the coordinates system, we can interpret the tangent like this: On the right side of the unit circle, draw a vertical line and extend the terminal side of the angle to meet this line. Then the tangent is the vertical coordinate of the point of intersection.

Problem 14.1. Draw figures in the unit circle for tangent for angle θ located in each quadrant.

Solution.

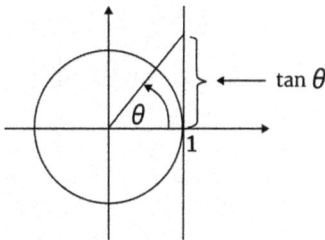

Angle is in 1st quadrant

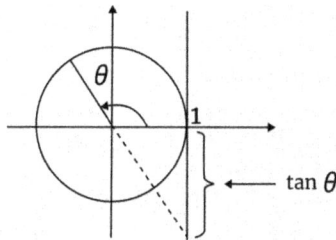

Angle is in 2nd quadrant

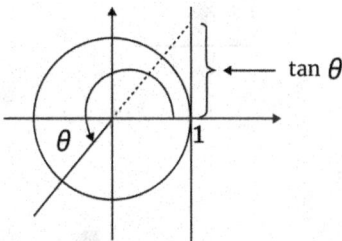

Angle is in 3rd quadrant

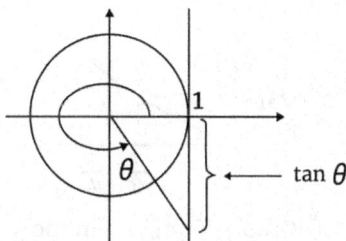

Angle is in 4th quadrant

Now we draw a graph of the tangent in another coordinate system, where the tangent argument will be presented as points on the x-axes. In this system, we denote the angle as x instead of θ. As in the cases of sine and cosine, we will use radians.

Problem 14.2. Sketch the graph of the function $y = \tan x$ for x in the interval $\left[0, \dfrac{\pi}{2}\right]$.

Solution. This interval corresponds to the 1ˢᵗ quadrant. Moving a point along the unit circle in this quadrant, we can see that tangent increases from zero to infinity. To detect more details, we select several values for the special angles for which we know the tangent's exact values. The following table presents such values:

x	0	$\dfrac{\pi}{6}$	$\dfrac{\pi}{4}$	$\dfrac{\pi}{3}$
$\tan x$	0	$\dfrac{\sqrt{3}}{3}$	1	$\sqrt{3}$

Now plot these points in the system of coordinates and connect them with a smooth curve:

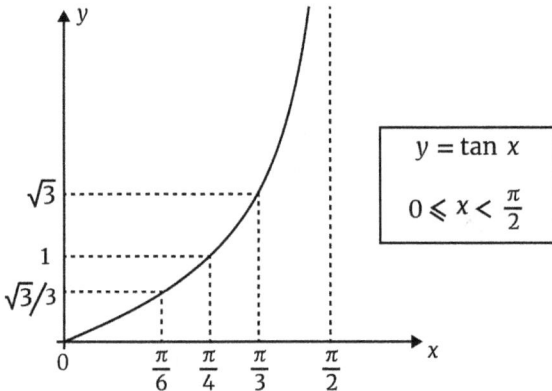

Graph of tangent in the 1ˢᵗ quadrant

As we see, when x approaches $\dfrac{\pi}{2}$, the graph of tangent increases to infinity and becomes close to the vertical line $x = \dfrac{\pi}{2}$, but does not touch it. This line, similar to the case of the function $y = \dfrac{1}{x}$, is called a vertical asymptote.

As the next interval for graphing, we will not use the 2nd quadrant (as you may expect), but the 4th. The reason is the symmetric property of tangent that you will see shortly. The 4th quadrant corresponds to the interval $\left[-\dfrac{\pi}{2}, 0\right]$.

Problem 14.3. Sketch the graph of tangent in the interval $\left[-\dfrac{\pi}{2}, \dfrac{\pi}{2}\right]$.

Solution. This interval corresponds to the 1st and 4th quadrants. In the 4th quadrant, we can also draw the graph using the values of special angles as we did in the 1st quadrant. However, there is a better way. We can use the odd property of the tangent function. As we indicated in Chapter 5, $\tan(-x) = -\tan x$. As any odd function, the graph of tangent is symmetric about the origin. Therefore, using the graph in the interval $\left[0, \dfrac{\pi}{2}\right]$ (1st quadrant), we can immediately plot tangent in the interval $\left[-\dfrac{\pi}{2}, 0\right]$ (4th quadrant) by reflecting the graph over the origin. Here is the figure of the tangent in the 1st and 4th quadrants:

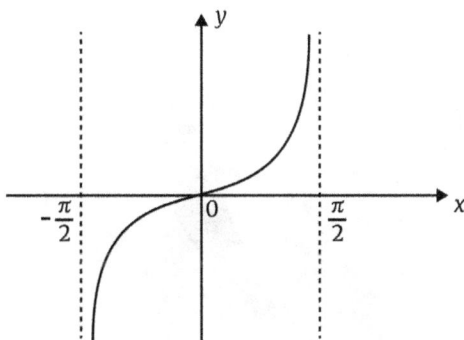

Graph of tangent in the interval $\left(-\dfrac{\pi}{2}, \dfrac{\pi}{2}\right)$ ■

Problem 14.4. Sketch the graph of tangent for the entire x-axis.

Solution. Since tangent is a periodic function with a period of π, we can extend its graph from the interval $\left[-\dfrac{\pi}{2}, \dfrac{\pi}{2}\right]$ in either direction (to the left and to the right) on the x-axis. As a result, we get the graph of tangent across the entire number line, i.e., for all values of x from $-\infty$ to $+\infty$:

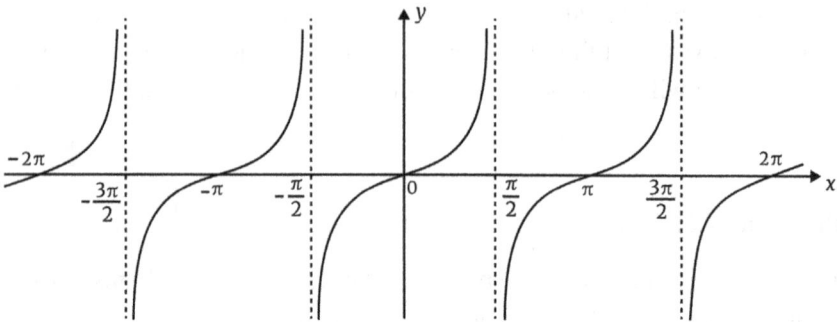

Entire graph of the function $y = \tan x$

We see that the graph consists of an infinite number of separate branches (so, this function is discontinuous), and it has an infinite number of vertical asymptotes. ∎

An interesting observation was made in biology. If you keep track of a fixed point on the tail of a fish, while it swims, this point moves along a sinusoid. And the movement of the body of the fish itself is similar to a graph of the tangent.

The graph of the tangent function allows us to see a wide range of properties. Here is a list of some of them.

(1) Tangent is a periodic function with a period of π:

$$\tan(x + \pi) = \tan x.$$

Let k be any integer ($k = 0, \pm1, \pm2, \ldots$). If we replace the period π with k periods $k\pi$, we will also have $\tan(x + k\pi) = \tan x$. Therefore, if some property holds for the number x on the interval $[a, b]$, then the same property holds for all numbers $x + k\pi$ on intervals $[a + k\pi, b + k\pi]$.

(2) Tangent is defined for all values of x, except $x = \dfrac{\pi}{2} + k\pi$. The domain of the tangent is the union of intervals $\left(-\dfrac{\pi}{2} + k\pi, \dfrac{\pi}{2} + k\pi\right)$ for all integer k.

(3) Tangent returns any values, so its range is the interval $(-\infty, \infty)$. Also, tangent has no maximum and no minimum.

(4) Tangent is symmetric about the origin. Also, it is an odd function:

$$\tan(-x) = -\tan x.$$

(5) Tangent is positive in the interval $\left(0, \dfrac{\pi}{2}\right)$. By adding $k\pi$ to both endpoints of this interval, we get that the tangent is positive at an infinite number of intervals

$$\left(k\pi, \dfrac{\pi}{2} + k\pi\right).$$

Similarly, tangent is negative at all intervals $\left(-\dfrac{\pi}{2} + k\pi, k\pi\right)$.

(6) Tangent equals 0 at $x = 0$, therefore, it also equals 0 at all points $x = k\pi$.

(7) Tangent increases at the interval $\left(-\dfrac{\pi}{2}, \dfrac{\pi}{2}\right)$. Therefore, it increases at the infinite number of intervals $\left(-\dfrac{\pi}{2} + k\pi, \dfrac{\pi}{2} + k\pi\right)$.

(8) Tangent has infinitely many vertical asymptotes $x = \pi/2 + k\pi$.

Solve the problems below on your own.

Problem 14.5. Describe all intervals in which tangent decreases. ∎

Problem 14.6. Describe all values of x at which the tangent graph is discontinuous. ■

In some cases, graphical representation helps to find the number of solutions for specific equations.

Problem 14.7. Find the number of roots of the equation $\tan x = \dfrac{1}{x}$ in the interval $[0, 100\pi]$.

Solution. Roots of this equation are the x-coordinates of points of intersection of the two graphs: $\tan x$ and $\dfrac{1}{x}$. Thus, to solve the problem, we need to count the number of common points of these two graphs. The following figure shows both graphs:

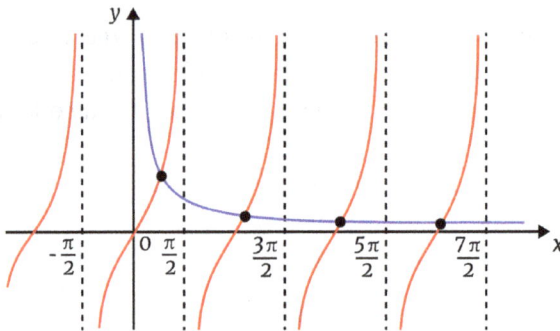

Here, we draw the graph of $y = \tan x$ in red, and the graph of $y = \dfrac{1}{x}$ in blue.

As we see, each interval between vertical asymptotes contains one common point. And tangent increases in each interval. Let's count the number of such intervals on the x-axis, taking into account the condition $x \leq 100\pi$. We use letter k for the last interval in the sequence of intervals:

$$\left[-\frac{\pi}{2}, \frac{\pi}{2}\right], \left[\frac{\pi}{2}, \frac{3\pi}{2}\right], \left[\frac{3\pi}{2}, \frac{5\pi}{2}\right], \dots, \left[\frac{(2k-1)\pi}{2}, \frac{(2k+1)\pi}{2}\right],$$

Taking the right endpoint from each interval, we get

$$\frac{\pi}{2}, \frac{3\pi}{2}, \frac{5\pi}{2}, \dots, \frac{(2k+1)\pi}{2}.$$

To calculate the number of elements in this sequence, we add π to each numerator, and then divide each fraction by π. As a result, we get the sequence $1, 2, 3, \ldots, (k+1)$ with $k+1$ terms. Hence, there are $k+1$ of the above intervals.

In the given interval $[0, 100\pi]$, number 100π should be inside of the last interval $\left[\dfrac{(2k-1)\pi}{2}, \dfrac{(2k+1)\pi}{2}\right]$ (see above). Therefore, k is the smallest integer such that

$$100\pi \le \frac{(2k+1)\pi}{2}.$$

Solving this inequality for k, we get $k \ge 99.5$. The smallest integer satisfying this inequality is $k = 100$, so the number of the above intervals is 101.

In each of these intervals, except the last, there is one point of intersection of the two graphs. Since our equation has a limitation that solution x should belong to the interval $[0, 100\pi]$, then the last interval for $k = 100$ is $[99.5\pi, 100\pi)$.

In this interval, the tangent has value zero at the endpoint $x = 100\pi$. Since the tangent increases, it is negative for all x in the $[99.5\pi, 100\pi)$ interval. On the other hand, the value of the function $y = \dfrac{1}{x}$ is greater than zero for positive x.

Therefore, there are no points of intersection in the last interval, and the total number of solutions of a given equation is $101 - 1 = 100$. ∎

Graphing the Cotangent Function

The easiest way to do this is to express cotangent through tangent, using the reduction formulas from Chapter 5:

$$\tan\left(x - \frac{\pi}{2}\right) = \frac{\sin\left(x - \dfrac{\pi}{2}\right)}{\cos\left(x - \dfrac{\pi}{2}\right)} = -\frac{\sin\left(\dfrac{\pi}{2} - x\right)}{\cos\left(\dfrac{\pi}{2} - x\right)} = -\frac{\cos x}{\sin x} = -\cot x$$

From here,

$$\cot x = -\tan\left(x - \frac{\pi}{2}\right).$$

This formula suggests the following two steps for plotting the cotangent graph:

(1) Reflect (flip) the tangent graph over the x-axis (it will take into account the negative sign before tangent). For the interval $\left(-\frac{\pi}{2}, \frac{\pi}{2}\right)$ this transformation looks like this

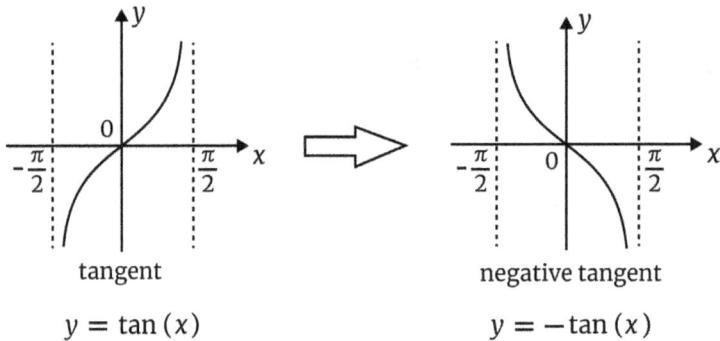

tangent

$y = \tan(x)$ negative tangent

$y = -\tan(x)$

(2) Shift the transformed graph to the right by $\frac{\pi}{2}$. In the figure below, the graph from the interval $\left(\frac{-\pi}{2}, \frac{\pi}{2}\right)$ moves to the interval $(0, \pi)$:

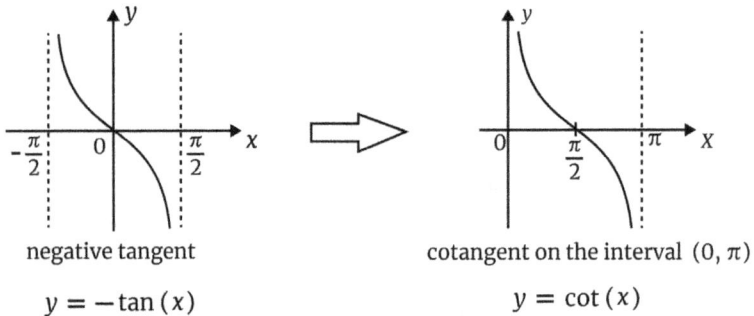

negative tangent   cotangent on the interval $(0, \pi)$

$y = -\tan(x)$ $y = \cot(x)$

By making two mentioned transformations for the entire tangent graph, we get a complete graph of cotangent:

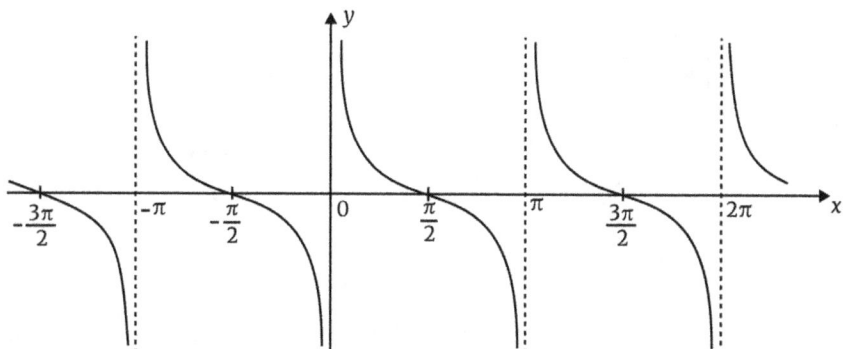

Graph of the function $y = \cot(x)$

Solve the following problem on your own.

Problem 14.8. Looking at the graph of the cotangent function, determine the following properties:

1. What is the period of cotangent?
2. What is its domain?
3. What is its range?
4. What kind of symmetry does cotangent have?
5. In which intervals does cotangent take positive values, and in which negative?
6. At what values of x is cotangent equal to 0?
7. In which intervals does cotangent increase and decrease?
8. What are the asymptotes of cotangent? ■

Problem 14.9. Sketch the graph of the function

$$y = \frac{\cot x - \tan x}{2}.$$

Solution. At first glance, it looks quite complicated. However, it is possible to simplify it. We have

$$\frac{\cot x - \tan x}{2} = \frac{1}{2}(\cot x - \tan x) = \frac{1}{2}\left(\frac{\cos x}{\sin x} - \frac{\sin x}{\cos x}\right)$$

$$= \frac{1}{2}\frac{(\cos^2 x - \sin^2 x)}{\cos x \cdot \sin x} = \frac{\cos 2x}{\sin 2x} = \cot 2x.$$

So, we get the identity

$$\frac{\cot x - \tan x}{2} = \cot 2x.$$

Now it is much easier to graph this function. According to Problem 12.2, Chapter 12, the period of the function $\cot 2x$ is $\dfrac{\pi}{2}$ (half of the period of $\cot x$). To plot this function, we can just compress the graph of $\cot x$ along the x-axis in half. Here is the figure:

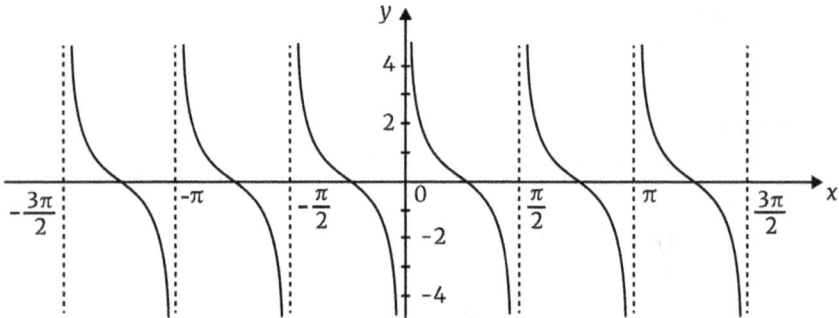

Graph of the function $y = \cot 2x$ ∎

Exercises for Solving on Your Own

Exercise 14.1. Sketch the graph of the function $y = 2 \tan x$.

Exercise 14.2. Sketch the graph of the function $y = \tan(x + \pi/3)$.

Exercise 14.3. What is the period of the function $f(x) = \tan(3x + 4)$?

Exercise 14.4. Sketch the graph of the function
$y = (\tan x - \cot x)/(\tan x + \cot x)$.
Hint: Try to modify this function. It can be significantly simplified.

Exercise 14.5. Without calculating the values of trig functions, arrange the values in ascending order:

(a) $\tan (\pi/10)$, $\tan (4\pi/10)$, $\tan(7\pi/10)$, $\tan (9\pi/10)$, $\tan (13\pi/10)$.
(b) $\cot (\pi/10)$, $\cot(3\pi/10)$, $\cot (7\pi/10)$, $\cot (9\pi/10)$, $\cot (12\pi/10)$.

Exercise 14.6. In solving a trig problem, Sofia found that the angle θ is in the interval $(\pi/4, \pi/2)$. She calculated $\tan \theta = 0.7$. This answer is incorrect. Why?

Entertainment Problems

Problem E14.1. There are three baskets with balls. The first basket is labeled "BLACK" (B), the second "WHITE" (W), and the third "BLACK AND WHITE" (BW). One of these baskets contains only white balls, the other only black balls, and the remaining basket contains both black and white balls. All basket labels are knowingly false. You are allowed to get one ball from only one basket to determine which basket is which.

Problem E14.2. Out of eleven precious coins, there is one counterfeit, which differs in weight from the genuine ones. Using a balance scale, we need to determine whether it is lighter or heavier than a genuine coin using only two weighings. Is it possible to solve the same problem (do only two weighings) for any number of coins?

Chapter 15

Basic Sine Equation and Inverse Sine Function

In the previous chapters, several times we have used $\sin^{-1}$, $\cos^{-1}$, and $\tan^{-1}$ buttons on a calculator to get angles from the values of trig functions. In other words, these buttons allow us to solve trig equations $\sin x = A$, $\cos x = A$, and $\tan x = A$ for a given number A. We call these equations basic (or simplest), since the solution of many more complex trig equations can be reduced to them. In this chapter, we examine in detail the equation $\sin x = A$ and a related, so-called, inverse sine function.

As we know, using the above buttons, calculators return only one root, while the equation $\sin x = A$ may have an infinite number of roots. The problem below demonstrates it. We will use the notation t for the variable instead of x, so as not to mix up with points on the x-axis.

Problem 15.1. Solve the equation $\sin t = \dfrac{1}{2}$.

Solution. Depict value $\dfrac{1}{2}$ for sine in the unit circle as a vertical coordinate:

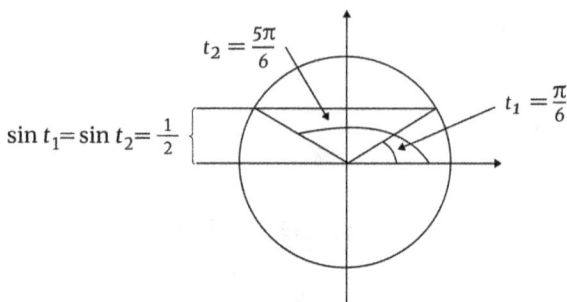

This figure shows two roots: $t_1 = \dfrac{\pi}{6}$ and $t_2 = \pi - \dfrac{\pi}{6} = \dfrac{5\pi}{6}$.

If we add any integer number n of full rotations to these angles, then new angles will be coterminal to these and will also be solutions to this equation. Since n full rotations are represented by angles $2\pi n$, all solutions of the equation $\sin t = \dfrac{1}{2}$ are defined by the formulas:

$$
\left[
\begin{array}{l}
t_1 = \dfrac{\pi}{6} + 2\pi n \\[2mm]
t_2 = \dfrac{5\pi}{6} + 2\pi n,
\end{array}
\right.
$$

where n is any integer number.

As we see, the equation $\sin t = \dfrac{1}{2}$ has an infinite number of solutions. ∎

Solve the problem below on your own.

Problem 15.2. Find all solutions of the equation $\sin t = -\dfrac{\sqrt{3}}{2}$. ∎

Problem 15.3. Solve the equation $\sin t = A$ for the following values of A: 2, 1, 0, –1.

Solution.

(1) Equation $\sin t = 2$ has no solutions since the maximum value of sine is 1.

(2) Equation $\sin t = 1$ has one solution $t = \dfrac{\pi}{2}$ (in one cycle) on the unit circle:

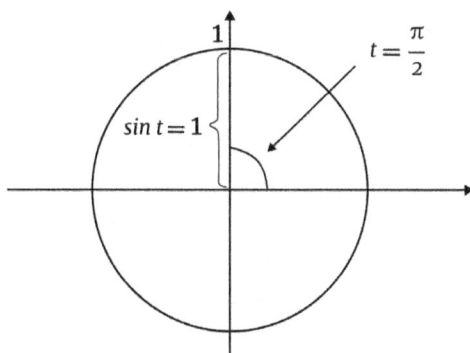

Adding n full rotations, we get a complete set of solutions: $t = \dfrac{\pi}{2} + 2\pi n$, where n is any integer number.

(3) Equation $\sin t = 0$ has two solutions in the unit circle: $t_1 = 0$ and $t_2 = \pi$. Adding n full rotations, we get $t_1 = 2n\pi$ and $t_2 = \pi + 2\pi n = (2n+1)\pi$. The expressions for t_1 and t_2 can be combined into one formula: $t = \pi n$.

(4) Equation $\sin t = -1$ has one solution in the unit circle: $t = -\dfrac{\pi}{2}$. Adding n full rotations, we get a full set of solutions: $t = -\dfrac{\pi}{2} + 2\pi n$. ∎

In the problems above, we solved the equation $\sin t = A$ for well-known values of sine. But how do we solve this equation in a general case for an arbitrary A?

The answer may look strange: to solve the equation $\sin t = A$, we just denote its solutions by some symbols, and we're done. At first glance, such a way may seem to be a trick: we avoid solving the equation by merely denoting its solutions. And after that, we claim that we have solved it. However, such a conclusion is unjustified.

To see the benefit of using mere notations, let's look at something well-known, a quadratic equation:

$$x^2 = A \ (A > 0).$$

Its solutions are $x_1 = \sqrt{A}$ and $x_2 = -\sqrt{A}$. We can stop at this point, saying that the equation is solved. Note that we just use the **notation** $\sqrt{\ }$ (square root or radical) to claim that the equation is solved. How to calculate $\sqrt{A}$ is a separate problem.

One of the efficient approaches to calculating $\sqrt{A}$ for the **constant** number A is to investigate the **function** $y = \sqrt{x}$ instead. Doing this, we "unfreeze" the number A, making it a variable. When considering the number $\sqrt{A}$, we only think about how to calculate it. When considering the function $y = \sqrt{x}$, we extend our thinking and may explore the various properties of this function. As a result, we may develop methods for approximating square roots (this is beyond the scope of this book). Numerous math studies show that the "functional" point of view is very useful.

However, the functional approach has a subtle point. To see it, let's compare the operation of raising a number to a second power with the inverse operation of extracting the square roots (it is used when solving the equation $x^2 = A$). The operation of raising to a power is the function $y = x^2$. However, the inverse operation of extracting the square roots **is not a function**. The reason is that it produces **two** values for $A > 0$: $\sqrt{A}$ and $-\sqrt{A}$. But, as we know, a function should produce only **one** value as an output. In order to make this operation be a function, we must restrict its output. The natural way to do this is to allow it to produce only positive numbers.[1] In this case, the operation will represent the function $y = \sqrt{t}$. The mentioned subtle point is that the "functional" approach to solving the equation $x^2 = A$ allows us to directly get only one root x_1 from the value of the inverse function $y = \sqrt{t}$ for $t = A$: $x_1 = \sqrt{A}$. Let us call it the **main** root. The second root x_2 can be obtained from the main: $x_2 = -x_1$.

In order for the process of raising to a second power to be **mutually** inverse to extracting the root, we should also limit it's input to positive numbers (including zero). It means that we need to cut off negative numbers from the domain of the function $y = x^2$. In this case, we get a one-to-one correspondence between arguments x of $y = x^2$ (which are now positive numbers) and values of the function $y = \sqrt{x}$: function $y = x^2$ maps number $s \geq 0$ to $A = s^2$, and the function $y = \sqrt{x}$ returns A back to $s = \sqrt{A}$.

Functions $y = \sqrt{x}$ and the truncated $y = x^2$ (i.e. $x \geq 0$) are called **inverse** to each other.

---

[1] Theoretically, there is nothing wrong with using negative numbers as the output of the square root extraction operation. However, positive numbers are more straightforward and easier to use.

For $y = f(x)$, the inverse function is usually denoted as $y = f^{-1}(x)$. Note that this is **not** the same as $\dfrac{1}{f(x)}$. The figure below illustrates the action of function $f(x)$ and its inverse $f^{-1}(x)$. In this figure, function f moves value a to b, and inverse function f^{-1} returns b to a:

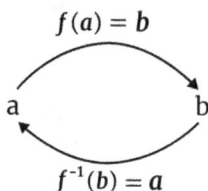

$$f(a) = b$$

a   b

$$f^{-1}(b) = a$$

Now we discuss in general terms the way to construct the inverse function for a given function $y = f(x)$. Let the original function $f(x)$ map some x_1 to y_1. The inverse function should map y_1 back to x_1. Since we want to make both functions mutually inverse, we use the condition that the inverse function should map its values to either the entire domain of the original function, or some part of the domain. The following figure shows an unsuitable part of the graph of the original function for the construction of an inverse function.

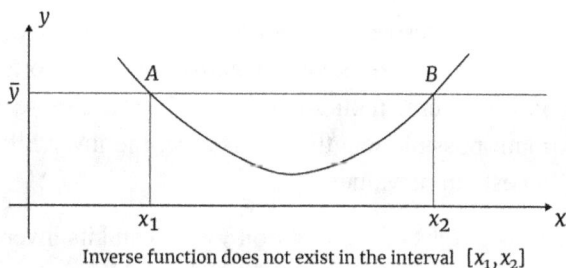

Inverse function does not exist in the interval $[x_1, x_2]$

As we see, this function maps two different x_1 and x_2 to the same $\bar{y}$. However, the inverse function should map $\bar{y}$ back to only **one** x. Thus, either x_1 or x_2 will remain uncovered by the value $\bar{y}$. Geometrically, it means that a horizontal line going from $\bar{y}$ intersects the graph of $f(x)$ at more than one point (they are A and B). To avoid such a case, the graph within the selected interval from the domain, should go either up or down, but not in both directions. Otherwise, a horizontal line will intersect the graph at more than one point.[2] If the function either goes up or down, we say that the function is **monotonic** in this interval. Here are figures of monotonic intervals:

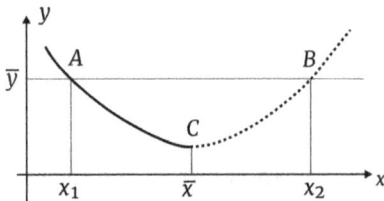

Truncated graph. Function decreases.
Inverse function exists in the interval $[x_1, \bar{x}]$

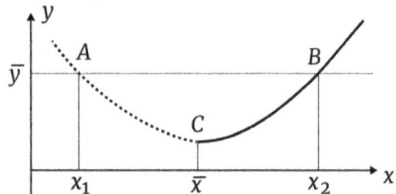

Truncated graph. Function increases.
Inverse function exists in the interval $[\bar{x}, x_2]$

On the left figure, the function is monotonic in the interval $[x_1, \bar{x}]$, and on the right — it is monotonic on the interval $[\bar{x}, x_2]$. Each of them allows us to get an inverse function.

Thus, we come up with the following method of constructing the inverse function. We should choose a monotonic interval for the graph of the original function by restricting its domain. This is exactly what we did with the function $y = x^2$ by removing negative x values from its domain. Since there can be several monotonic intervals, we choose the interval with the maximum possible length. In this case, the inverse function will become the "richest" in its values.

Let's see how graphs of the function $y = f(x)$ and its inverse $y = f^{-1}(x)$ are related. Let point (x_0, y_0) lie on the graph of f. It means that function f maps x_0 to y_0. The inverse function f^{-1} returns y_0 to x_0. Therefore, point (y_0, x_0) lies on the graph of the inverse function. Points (x_0, y_0) and (y_0, x_0) are symmetric to each other with respect to the line $y = x$, which is the bisector of the 1st and 3rd quadrants. Therefore, graphs of f and f^{-1} are

---

[2] Here we assume that the considered part of the graph is a continuous (unbroken) line. We do not give a formal definition of a continuous line here, leaving its understanding on an intuitive level.

symmetric to each other over the line $y = x$. The figure below demonstrates this for graphs of the truncated $y = x^2$ and its inverse $y = \sqrt{x}$:

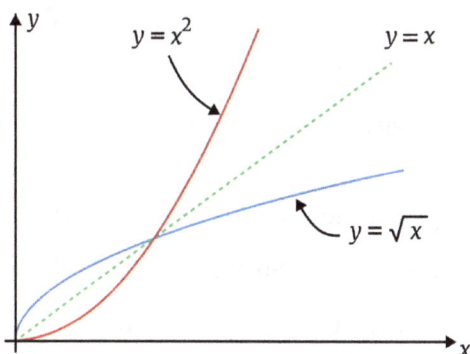

Construction of the Inverse Sine Function

Let's apply the above ideas to the function $y = \sin x$. First, we consider its graph on the entire number line:

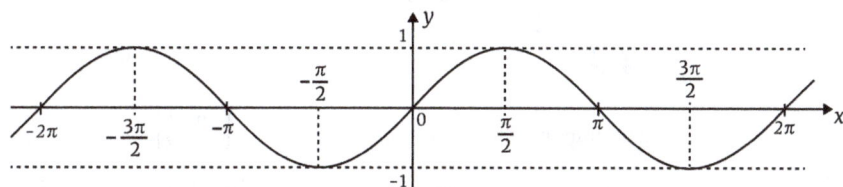

Graph of the function $y = \sin x$

This graph is not monotonic, so the sine does not have an inverse on the entire x-axes. To get the inverse, we need to cut (truncate) this function such that its part becomes monotonic. Take a look at the graph above and try to find an interval on the x-axis with the maximum possible length, where the graph is monotonic (increases or decreases). You can see many such intervals. For example, in the interval $\left[-\dfrac{3\pi}{2}, -\dfrac{\pi}{2} \right]$ sine decreases, in the next interval $\left[-\dfrac{\pi}{2}, \dfrac{\pi}{2} \right]$ it increases, in $\left[\dfrac{\pi}{2}, \dfrac{3\pi}{2} \right]$ decreases, and so on. Theoretically, any of these intervals could be taken for values (range) of the inverse sine. We choose $\left[-\dfrac{\pi}{2}, \dfrac{\pi}{2} \right]$ interval:

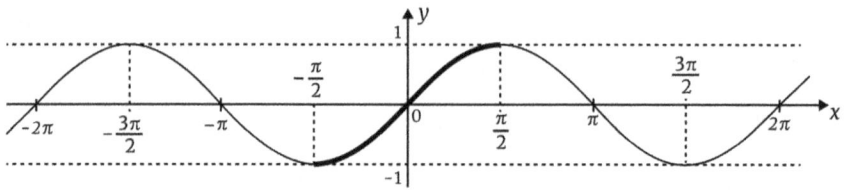

Interval $\left[-\frac{\pi}{2},\frac{\pi}{2}\right]$ on the graph of sine is used for construction of the inverse function

Here is the reason for choosing this interval. It is natural to require that the interval contain acute angles. So, we first choose the interval $\left[0,\frac{\pi}{2}\right]$ on the graph as a part. Then we expand it as much as possible while maintaining monotony. As can be seen from the graph, the only interval that satisfies this condition is $\left[-\frac{\pi}{2},\frac{\pi}{2}\right]$.

Thus, the truncated sine looks like this:

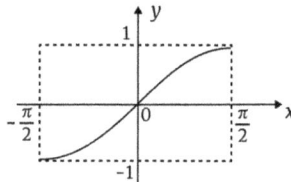

Graph of the truncated function $y = \sin x$ on the interval $\left[-\frac{\pi}{2},\frac{\pi}{2}\right]$.

From the truncated sine, we can get the inverse function. To give it a name, it was decided to add an "arc" prefix to the notation of the original sine function. So, the notation for the inverse sine is this: $y = $ **arcsin** x. Values y are angles located in the 1st and 4th quadrants. While there are many solutions to the equation $\sin x = A$, calculators produce only values for arcsine. Calculators (as well as some textbooks) use a different notation: $y = \sin^{-1} x$. Actually, such a notation may be confusing with the notation of the reciprocal. Some computer programming languages use the abbreviation asin. In 2009, the International Organization for Standardization (ISO) accepted the prefix "arc" for the inverse trig functions.

By selecting the interval $\left[-\frac{\pi}{2},\frac{\pi}{2}\right]$ for the truncated sine function, and introducing the notation for its inverse, we may say that the inverse function $y = $ arcsin x is wholly defined. Its graph can be obtained from the above graph of the truncated sine by reflecting it over the line $y = x$ (bisector of the 1st and 3rd quadrants):

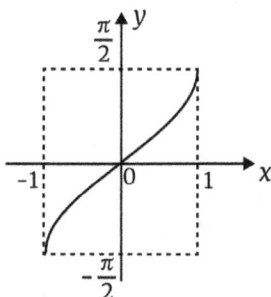

Graph of the inverse function $y = \arcsin x$

Note that the graph of the inverse sine function is symmetric (similar to the sine function) over the origin. We call such functions odd. Algebraically, it means that

$$\arcsin(-x) = -\arcsin x.$$

Let's rephrase the definition of arcsin A for any number A from the interval $[-1, 1]$.

Definition 1

arcsin A is an angle with two properties:
(1) Sine of this angle is equal to A:
$$\sin(\arcsin A) = A$$
(2) This angle lies in the 1st or 4th quadrants and is acute by absolute value

Definition 2

arcsin A is a root of the equation $\sin x = A$ located in the interval $\left[-\dfrac{\pi}{2}, \dfrac{\pi}{2}\right]$

There is a method of how to recall the range of the arcsine function. First, always include the 1st quadrant in the range. In it, sine produces values from 0 to 1. Then, ask yourself: "What is the contiguous (adjacent) interval for the 1st quadrant where sine produces the remaining values from −1 to 0?" Since the sine values are presented by vertical coordinates, we need to select the 4th quadrant. The 1st and 4th quadrants make up the entire range of arcsine, which is the interval $\left[-\dfrac{\pi}{2}, \dfrac{\pi}{2}\right]$. The figure below illustrates this.

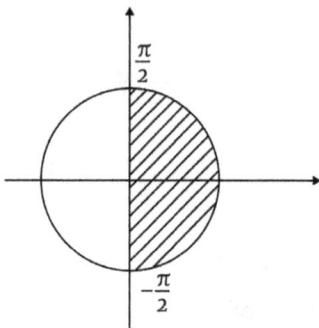

Range of the function $y = \arcsin x$ is $\left[-\dfrac{\pi}{2}, \dfrac{\pi}{2}\right]$

Recall again that truncated sine maps angles from the interval $\left[-\dfrac{\pi}{2}, \dfrac{\pi}{2}\right]$ to the numbers in the interval $[-1, 1]$. The arcsine function acts exactly in the opposite direction: it maps numbers from the interval $[-1, 1]$ to the angles in the interval $\left[-\dfrac{\pi}{2}, \dfrac{\pi}{2}\right]$.

Problem 15.4. Determine the values for

$$\arcsin 1, \quad \arcsin \frac{1}{2}, \quad \arcsin\left(-\frac{\sqrt{3}}{2}\right), \quad \arcsin(-1), \quad \arcsin 0.$$

Solution. To do this, we can ask ourselves the following question: what are the angles for which the sine has values $1, \dfrac{1}{2}, -\dfrac{\sqrt{3}}{2}, -1, 0$. Also, take into account that the range of arcsine is $\left[-\dfrac{\pi}{2}, \dfrac{\pi}{2}\right]$. Then, looking at the unit circle, we get:

$$\arcsin 1 = \frac{\pi}{2}, \arcsin \frac{1}{2} = \frac{\pi}{6}, \arcsin\left(-\frac{\sqrt{3}}{2}\right) = -\frac{\pi}{3},$$

$$\arcsin(-1) = -\frac{\pi}{2}, \arcsin 0 = 0.$$

■

Using the arcsine function, we can describe all solutions to the equation $\sin t = A$. According to the definition of arcsine, it is one of its roots. We call this root as the **main or principal root**. We will show that any other root can be expressed through the main root.

Problem 15.5. Solve equation $\sin t = A$ where A is a nonnegative number $A \leq 1$.

Solution. The figure below shows two roots in the unit circle: t_1 and $t_2 = \pi - t_1$ (A < 1):

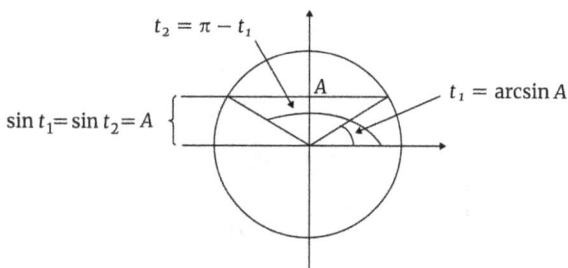

According to the definition of arcsine, $t_1 = \arcsin A$. It is the main solution. The second solution t_2 is expressed through the main: $t_2 = \pi - t_1 = \pi - \arcsin A$. For $A = 1$ we have one solution $t = \arcsin 1 = \pi/2$. If we add any integer number n of full rotations $2\pi n$ to these angles, then we get a full set of solutions for the equation $\sin t = A$, where $0 \leq A \leq 1$:

$$\begin{cases} t_1 = \arcsin A + 2\pi n \\ t_2 = \pi - \arcsin A + 2\pi n = -\arcsin A + \pi(2n+1). \end{cases} \blacksquare \qquad (1)$$

Solve the problem below on your own.

Problem 15.6. Prove that all solutions of the equation $\sin t = A$ for negative $A \geq -1$ are determined by the same formulas (1). $\blacksquare$

As we see, all solutions of the equation $\sin t = A$ are expressed through the main root $t = \arcsin A$ by two formulas (1). It turns out that both formulas could be combined into one.

To do this, note that the first formula (1) is the sum of arcsin A and an even multiple of π, while the second formula is the sum of negative arcsin A and an odd multiple of π. In other words, the difference between the two formulas is that when we add odd multiples of π, we need to include a minus sign before arcsin A. It can be done by introducing a special "device" $(-1)^n$. This device switches plus and minus signs depending on whether n is even or odd: if n is even, $(-1)^n = 1$, and if n is odd, $(-1)^n = -1$. Using this device, we can combine the above two sets into one, and write the following final result:

> All solutions of the equation
> $$\sin x = A, \quad -1 \le A \le 1$$
> are defined by the formula:
> $$x = (-1)^n \text{ arcsin } A + \pi n,$$
> where n is any integer.

In three special cases, when $A = 1, -1$, and 0, as we showed in Problem 15.3, the above formula can be written in a simpler form:

$$\sin x = 1 \implies x = \pi/2 + 2n\pi,$$
$$\sin x = -1 \implies x = -\pi/2 + 2n\pi,$$
$$\sin x = 0 \implies x = n\pi.$$

Solution to the Equation $\sin x = A$, $-1 \le A \le 1$ in a One-Period Interval $[0, 2\pi)$

In some cases, we need to solve the equation $\sin x = A$ only in a one-period interval $[0, 2\pi)$, and not for the entire x-axes. The angle 2π is not included since it corresponds to the same point on the unit circle as angle 0. We will assume that $-1 < A < 1$ (solutions for $A = \pm 1$ are clearly seen in the unit circle and are $\pi/2$ for $A = 1$, and $3\pi/2$ for $A = -1$). If $|A| < 1$, the equation $\sin x = A$ has two solutions in the interval $[0, 2\pi)$. We describe them by using the reference angle and the definition of sine on the unit circle: sine is the second (vertical) coordinate of points on the unit circle. Here are the steps to do that.

(1) Set up the equation for the reference angle x_r : $\sin x_r = |A|$. In other words, ignore the sign of the number A (always take a positive sign). Solve this equation for the reference angle: $x_r = \arcsin(|A|) = \sin^{-1}(|A|)$. To calculate x_r, you may use the $\sin^{-1}$ button on a calculator.

(2) Determine the quadrants in which angles are located based on the sign of the number A:

If $A > 0$, then angles are located in the 1st and 2nd quadrants.

If $A < 0$, then angles are located in the 3rd and 4th quadrants.

(3) Using quadrants and a reference angle, find two solutions x_1 and x_2 of the equation $\sin x = A$.

If $A > 0$, then $x_1 = x_r$ and $x_2 = \pi - x_r$, solutions are in the 1st and 2nd quadrants:

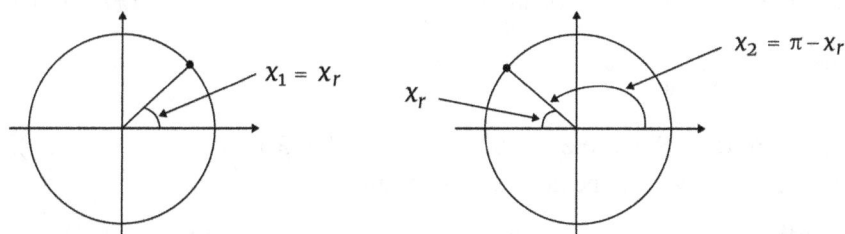

If $A < 0$, then $x_1 = \pi + x_r$ and $x_2 = 2\pi - x_r$, solutions are in the 3rd and 4th quadrants:

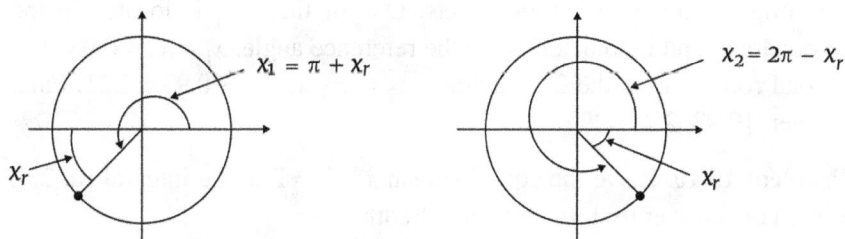

Problem 15.7. Solve the equation $2 \sin x + 4 = 5$ in the interval $[0, 2\pi)$.

Solution. We can reduce this equation to the basic one by solving for $\sin x$: $2 \sin x = 1$, so, $\sin x = \dfrac{1}{2}$. Here, $A = \dfrac{1}{2} > 0$. Therefore, the equation for reference angle x_r is the same as for x : $\sin x_r = 1/2$. From here $x_r = \sin^{-1}(1/2)$. We can find x_r with a calculator or using the special value

$1/2 : x_r = \sin^{-1}(1/2) = 30° = \pi/6$. The original equation has two roots. One of them, x_1, is located in the 1st quadrant and it coincides with the reference angle: $x_1 = x_r = \pi/6$. The second root x_2 is in the 2nd quadrant:

$$x_2 = \pi - x_r = \pi - \pi/6 = 5\pi/6. \text{ Final answer: } \left\{\frac{\pi}{6}, \frac{5\pi}{6}\right\}. \ \blacksquare$$

Problem 15.8. Solve the equation $-2 \sin x = \sqrt{2}$ in the interval $[0, 2\pi)$.

Solution. Solving for $\sin x$, we get the basic equation $\sin x = -\sqrt{2}/2$. Here $A = -\sqrt{2}/2 < 0$. The equation for reference angle x_r is $\sin x_r = |A| = \sqrt{2}/2$. From here, using a calculator or the special value $\sqrt{2}/2$, we can find that $x_r = \sin^{-1}(\sqrt{2}/2) = \pi/4$. The original equation has two roots. One of them, x_1, is located in the 3rd quadrant: $x_1 = \pi + x_r = \pi + \pi/4 = 5\pi/4$. The second root x_2 is in the 4th quadrant: $x_2 = 2\pi - x_r = 2\pi - \pi/4 = 7\pi/4$. Final answer: $\left\{\frac{5\pi}{4}, \frac{7\pi}{4}\right\}. \ \blacksquare$

Problem 15.9. Solve the equation $5\sin x - 1 = 3$ in the interval $[0, 2\pi)$. Round the answer to the nearest hundredth.

Solution. Solving for $\sin x$, we get the basic equation $\sin x = 4/5$. Here, $A = 4/5 > 0$. The equation for reference angle x_r is the same as for x: $\sin(x_r) = 4/5$. The value $4/5$ is not a special value. To find x_r, use a calculator (make sure the calculator is in radian mode): $x_r = \sin^{-1}(4/5) = 0.93$. The original equation has two roots. One of them, x_1 is located in the 1st quadrant, and it coincides with the reference angle: $x_1 = x_r = 0.93$. The second root x_2 is in the 2nd quadrant: $x_2 = \pi - x_r = \pi - 0.93 = 2.21$. Final answer: $\{0.93, 2.21\}$. $\blacksquare$

Problem 15.10. Solve the equation $6\sin x + 7 = 2$ in the interval $[0, 2\pi)$. Round the answer to the nearest hundredth.

Solution. Solving for $\sin x$, we get the basic equation $\sin x = -5/6$. Here $A = -5/6 < 0$. The equation for reference angle x_r is $\sin x_r = |-5/6| = 5/6$. The value $5/6$ is not a special value, and, to find x_r, we use a calculator in the radian mode: $x_r = \sin^{-1}(5/6) = 0.99$. The original equation has two roots. One of them, x_1, is located in the 3rd quadrant: $x_1 = \pi + x_r = \pi + 0.99 = 4.13$. The second root x_2 is in the 4th quadrant: $x_2 = 2\pi - x_r = 2\pi - 0.99 = 5.29$. Final answer: $\{4.13, 5.29\}$. $\blacksquare$

Exercises for Solving on Your Own

In all exercises, use the radian measure.

In Exercises 15.1 and 15.2, solve the given equations in the interval $[0, 2\pi)$.

Exercise 15.1. Write the answers in terms of π (do not round).

(a) $2\sqrt{2} \sin x - 1 = 1$; (b) $2\sqrt{3} \sin x + 4 = 1$; (c) $2\sqrt{3} \sin x - 3 = 1$.

Exercise 15.2. Round the answers to the nearest hundredth.

(a) $4 \sin x - 1 = 2$; (b) $6 \sin x + 5 = 3$.

Exercise 15.3. Solve the equations for the entire number line. If possible, write the answers in exact form in terms of π. If not, round the answers to the nearest hundredth.

(a) $2 \sin x - 3 = -2$;   (b) $2\sqrt{3} \sin x + 5 = 2$; (c) $3\sqrt{2} \sin x + 8 = 2$;
(d) $5 \sin x - 2 = 1$;   (e) $7 \sin x + 6 = 2$;   (f) $2 \sin 3x + 1 = 2$;
(g) $2 \sin(x - \pi/3) = -\sqrt{3}$; (h) $\sin^3 x = \sin x$;   (i) $2 \sin^2 x - 5 \sin x - 3 = 0$;
(j) $3 \sin^2 x - \cos^2 x = 0$;   (k) $\sin^3 2x + \sin x = 2$; (l) $\sin^2 2x \cdot \sin x = 1$.

Exercise 15.4. Sofia wanted to check whether arcsin(sin 2) equals to 2. According to her knowledge, the functions sine and arcsine are inverse to each other, so she concluded that these two values are equal. However, her conclusion is wrong. Why?

Exercise 15.5. Prove that

(a) If θ lies in the interval $[\pi/2, 3\pi/2]$, then arcsin(sin θ) = $\pi - \theta$.
(b) If θ lies in the interval $[3\pi/2, 2\pi]$, then arcsin(sin θ) = $\theta - 2\pi$.

Exercise 15.6. Sofia tried to calculate arcsin 1.2, but the calculator showed an error. She thought that something was wrong with the calculator. Can we conclude that the calculator is broken?

Entertainment Problem

Problem E15.1. Mike has two sons, Eli and Ben, who live in different cities. To visit them, Mike goes to a train station and travels by train. Trains that lead to Eli arrive at the station at intervals of 1 h with the duration of the stop being 1 min. The situation is exactly the same for the trains that lead to Ben: they also arrive at intervals of 1 h with a stop of 1 min. Since the trains travel in different directions, Mike can only visit one son per train. Mike comes to the station at any arbitrary time of the day and gets on the first train that arrives. It turns out that Mike visits Eli much more often than Ben. How can this be? After all, Mike gets to the station at a completely random time, and each train arrives at the same interval. So, it looks like the trains are "symmetrical" about Mike.

Chapter 16

Inverse Cosine and Tangent. Basic Trigonometric Equations for Cosine and Tangent

In this chapter, we discuss the inverse cosine and inverse tangent functions, and their application to the solution of the basic equations $\cos x = A$ and $\tan x = A$. As we mentioned, in order to construct the inverse function, we need to select a monotonic interval with the maximum length in the domain of the original function. Its graph helps to visualize the selection.

Inverse Cosine Function

Let's observe the graph of the cosine function on the entire number line.

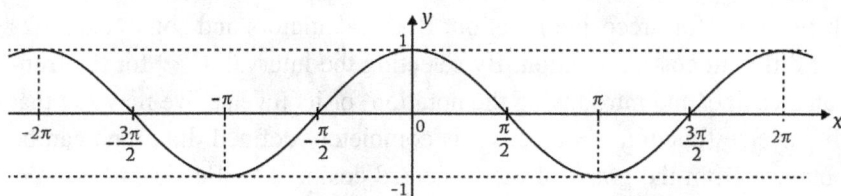

Graph of the function $y = \cos x$

This graph is not monotonic, so the cosine on the entire x-axes does not have an inverse. Similar to the sine function case, we need to restrict the domain of the cosine function to get a monotonic interval. Looking at the above graph, we can see many monotonic intervals of the length π (this is the maximum possible length). For example, in the interval $[-2\pi, -\pi]$ the function decreases, in the next interval $[-\pi, 0]$ the function increases, in the interval $[0, \pi]$ function decreases, and so on. It was decided to select the interval $[0, \pi]$.

The reason is quite similar to inverse sine. First of all, this interval should contain acute angles. Therefore, we select the interval $\left[0, \dfrac{\pi}{2}\right]$ as a part. Next, we expand it as much as possible while maintaining monotony. As can be seen from the graph, the result is the interval $[0, \pi]$.

In this interval, the graph of cosine goes down (decreases) and looks like this:

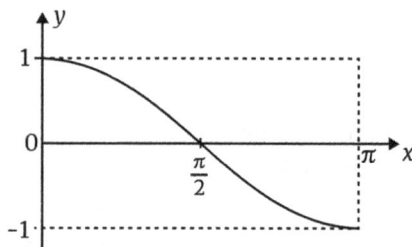

Graph of the truncated function $y = \cos x$ in the interval $[0, \pi]$

This truncated cosine function has an inverse. Like the inverse sine, it is denoted by adding an "arc" prefix to the notation of the original cosine function. So, the notation for the inverse cosine is this: $y = \mathbf{arccos}\ x$. Values y are angles located in the 1st and 2nd quadrants. Calculators produce values for arccosine function. Both calculators and some textbooks use different $\cos^{-1} x$ notation. By selecting the interval $[0, \pi]$ for the truncated cosine, and introducing the notation for its inverse, we may say that the inverse function $y = \arccos x$ is completely defined. Its graph can be obtained from the graph of the truncated cosine by reflecting it over the line $y = x$ (bisector of the 1st quadrant):

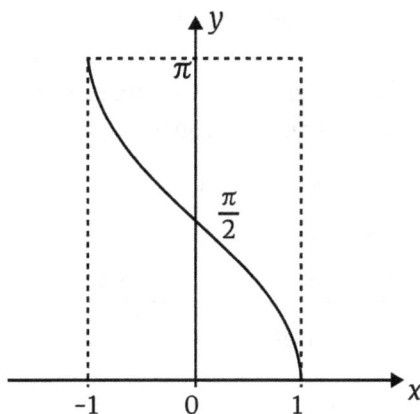

Graph of the inverse function $y = \arccos x$

Note again that the domain of the truncated cosine is the interval $[0, \pi]$, and the range is the interval $[-1, 1]$. For the inverse cosine, everything is just the opposite: its domain is the interval $[-1, 1]$, and the range is the interval $[0, \pi]$. Also, note that the graph of inverse cosine has a point of symmetry: it is symmetric with respect to the point $(0, \pi/2)$ on the y-axis. Let's describe this property algebraically. If we move this graph down by $\pi/2$, then the point $(0, \pi/2)$ goes to the origin. By this transformation, we get the function $f(x) = \arccos x - \pi/2$. Its graph becomes symmetric with respect to the origin, so the function $f(x)$ is odd:

$$f(-x) = -f(x) \quad \Rightarrow \quad \arccos(-x) - \pi/2 = -(\arccos x - \pi/2).$$

Solving the last equation for $\arccos(-x)$, we get the following property of the inverse cosine:

$$\arccos(-x) = \pi - \arccos x.$$

Let's rephrase the definition of $\arccos A$ for number A from the interval $[-1, 1]$ in two versions:

Definition 1

arccos A is an angle with two properties:
1) cosine of this angle is equal to A:

$$\cos(\arccos A) = A$$

2) This angle lies in the 1st or 2nd quadrants. It is nonnegative and does not exceed π.

Definition 2

arccos A is a root of the equation $\cos x = A$ located in the interval $[0, \pi]$.

There is a way of how to recall the range of the arccosine function. It is similar to arcsine. First, always include the 1st quadrant in the range. In it, cosine produces values from 0 to 1. Then, ask yourself: "What is the contiguous (adjacent) interval for the 1st quadrant where cosine takes the remaining values from -1 to 0?" Since cosine values are presented by horizontal coordinates, then we need to select the 2nd quadrant. The 1st and 2nd quadrants make up the entire range of arccosine, which is the interval $[0, \pi]$. The figure below illustrates this.

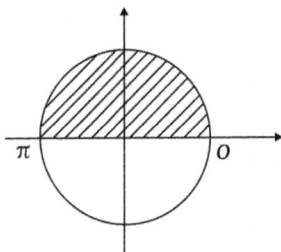

Range of the function $y = \arccos x$ is $[0, \pi]$

Problem 16.1. Calculate $\arccos 1$, $\arccos \dfrac{1}{2}$, $\arccos\left(-\dfrac{\sqrt{3}}{2}\right)$, $\arccos(-1)$, $\arccos 0$.

Solution. We need to find angles from interval $[0, \pi]$ whose cosines are 1, $\dfrac{1}{2}$, $-\dfrac{\sqrt{3}}{2}$, $-1, 0$. Actually, we are familiar with these special angles.

To recall, we can look at the unit circle (draw it on your own). We will see that these angles are

$$\arccos 1 = 0, \ \arccos \frac{1}{2} = \frac{\pi}{3}, \ \arccos\left(-\frac{\sqrt{3}}{2}\right) = \pi - \frac{\pi}{6} = \frac{5\pi}{6},$$

$$\arccos(-1) = \pi, \ \arccos 0 = \frac{\pi}{2}.$$

∎

According to the definition of arccosine, it is one of the roots of the equation $\cos x = A$. We call this root the **main** or **principal** root. We will show that all other roots of the equation can be expressed through the main root.

Problem 16.2. Solve the equation $\cos x = A$ for $-1 \le A \le 1$.

Solution. The figures below show two roots x_1 and x_2 for $A \ne \pm 1$ in the unit circle:

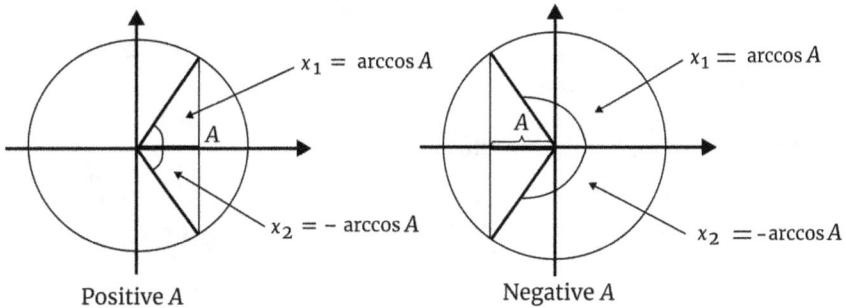

Positive A Negative A

As we see, the 2nd solution x_2 is expressed through the main x_1: $x_2 = -x_1$. All other solutions to the equation $\cos x = A$ can be obtained by adding any number n of full rotations (i.e., $2n\pi$) to x_1 and x_2 where n is any integer. One set of solutions is $x = x_1 + 2n\pi = \arccos A + 2n\pi$ and another is $x = x_2 + 2n\pi = -\arccos A + 2n\pi$. We can combine both sets into one formula by using $\pm$ notation: $x = \pm\arccos A + 2n\pi$.

We obtain the following result.

> All solutions of the equation
>
> $$\cos x = A, \quad -1 \le A \le 1$$
>
> are described by the formula:
>
> $$x = \pm\arccos A + 2\pi n,$$
>
> where n is any integer.

∎

Let's take a closer look at the solutions of the equation $\cos x = A$ for the extreme values $A = 1$, $A = -1$, and for $A = 0$ that correspond to the quadrantal points on the unit circle.

Problem 16.3. Solve the equations: (a) $\cos x = 1$; (b) $\cos x = -1$; (c) $\cos x = 0$.

Solution. In the equation $\cos x = A$, number A is the horizontal coordinate. We have:

(a) If $A = 1$, then the corresponding point on the unit circle is the rightmost point (the angle is 0), and all roots can be written as $x = 2n\pi$.

(b) If $A = -1$, then the corresponding point is the leftmost point (the angle is π), and all roots can be described as $x = \pi + 2n\pi = (2n + 1)\pi$.

(c) If $A = 0$, then there are two corresponding points: the upper point (the angle is $\pi/2$), and the lower point (angle is $-\pi/2$). Since points $\pi/2$ and $-\pi/2$ are at a distance π from each other, then all the roots can be written as $x = \pi/2 + n\pi$.

Let's summarize the above three cases:

$$
\begin{array}{ll}
\cos x = 1 & \Rightarrow \quad x = 2n\pi \\
\cos x = -1 & \Rightarrow \quad x = (2n + 1)\pi \\
\cos x = 0 & \Rightarrow \quad x = \pi/2 + n\pi
\end{array}
\quad \blacksquare
$$

Some of the problems require us to solve the basic equation $\cos x = A$ only for a specified interval for x.

Problem 16.4. Find solutions to the equation $\cos x = 1/2$

(a) In a one-period interval $[-\pi, \pi]$.
(b) In a one-period interval $[0, 2\pi]$.
(c) On the entire number line.

Solution. Here $A = 1/2$ is a special value, and the principal root arccos A is

$$\arccos A = \arccos(1/2) = \pi/3.$$

(a) In the interval $[-\pi, \pi]$, the two solutions are

$$x_{1,2} = \pm \arccos A = \pm \arccos(1/2) = \pm\pi/3.$$

(b) In the interval $[0, 2\pi]$, the two solutions are

$$x_1 = \arccos A = \arccos(1/2) = \pi/3$$

and

$$x_2 = 2\pi - \arccos A = 2\pi - \arccos(1/2) = 2\pi - \pi/3 = 5\pi/3.$$

(c) On the entire number line, there are infinitely many solutions that are presented by the formula $x = \pm\pi/3 + 2n\pi$, where n is any integer. ■

Problem 16.5. Solve the equation $\cos x = -\sqrt{2}/2$.

(a) In a one-period interval $[-\pi, \pi]$.
(b) In a one-period interval $[0, 2\pi]$.
(c) On the entire number line.

Solution. Here $A = -\sqrt{2}/2$ and it can be reduced to the special value $\sqrt{2}/2$ by using the formula $\arccos(-x) = \pi - \arccos x$. The principal root $\arccos A$ is:

$$\arccos A = \arccos(-\sqrt{2}/2) = \pi - \arccos(\sqrt{2}/2) = \pi - \pi/4 = 3\pi/4.$$

(a) In the interval $[-\pi, \pi]$, two solutions are

$$x_{1,2} = \pm \arccos A = \pm\arccos(-\sqrt{2}/2) = \pm 3\pi/4.$$

(b) In the interval $[0, 2\pi]$, two solutions are

$$x_1 = \arccos A = \arccos(-\sqrt{2}/2) = 3\pi/4$$

and

$$x_2 = 2\pi - \arccos A = 2\pi - \arccos(-\sqrt{2}/2) = 2\pi - 3\pi/4 = 5\pi/4.$$

(c) On the entire number line, an infinite family of solutions can be represented by the formula $x = \pm 3\pi/4 + 2n\pi$, where n is any integer. ∎

Problem 16.6. Solve the equation $\cos x = -0.23$.

(a) In one-period interval $[-\pi, \pi]$.
(b) In one-period interval $[0, 2\pi]$.
(c) On the entire number line.

Solution. Here $A = -0.23$ is not a special value. Using a calculator,

$$\arccos A = \arccos(-0.23) = 1.8.$$

(a) In the interval $[-\pi, \pi]$, two solutions are

$$x_{1,2} = \pm\arccos A = \pm\arccos(-0.23) = \pm 1.8.$$

(b) In the interval $[0, 2\pi]$, two solutions are

$$x_1 = \arccos A = \arccos(-0.23) = 1.8$$

and

$$x_2 = 2\pi - \arccos A = 2\pi - \arccos(-0.23) = 2\pi - 1.8 = 4.5.$$

(c) On the entire number line, an infinite family of solutions can be represented by the formula $x = \pm 1.8 + 2n\pi$, where n is any integer. ∎

Inverse Tangent Function

Here is the graph of tangent on the entire number line

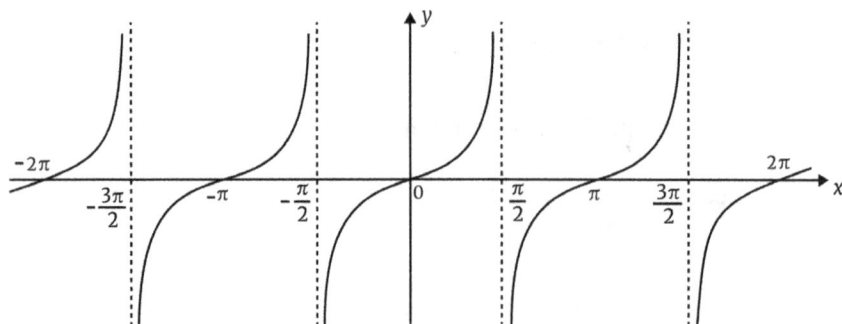

Graph of the function $y = \tan x$

Similar to the function $y = \sin x$, we select $\left(-\dfrac{\pi}{2}, \dfrac{\pi}{2} \right)$ as a monotonic interval for the truncated tangent. In this interval, tangent increases and has the graph:

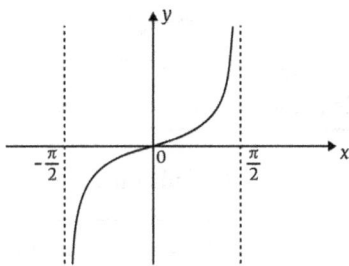

Graph of the truncated function $y = \tan x$ in the interval $\left(-\dfrac{\pi}{2}, \dfrac{\pi}{2} \right)$.

We denote the inverse function for this truncated tangent by arctan x and construct its graph by reflecting the above graph over the line $y = x$:

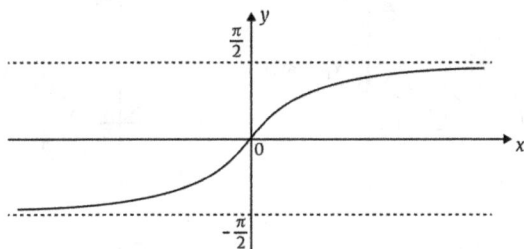

Graph of the inverse function $y = \arctan x$

Like for inverse sine, this graph is also symmetric over the origin $(0, 0)$, so the inverse tangent is an odd function:

$$\arctan(-x) = -\arctan x.$$

Also, it has two horizontal asymptotes: $y = \pi/2$ and $y = -\pi/2$. Two possible definitions of the inverse tangent are these:

Definition 1

arctan A is an angle with two properties:

1) Tangent of this angle is equal to A:

$$\tan(\arctan A) = A.$$

2) This angle lies in the 1st or 4th quadrants and is acute by absolute value.

Definition 2

arctan A is a root of the equation $\tan x = A$ located in the interval $(-\pi/2, \pi/2)$.

Problem 16.7. Solve the equation $\tan x = A$ on the entire number line.

Solution. Depict $\tan x$ in the unit circle:

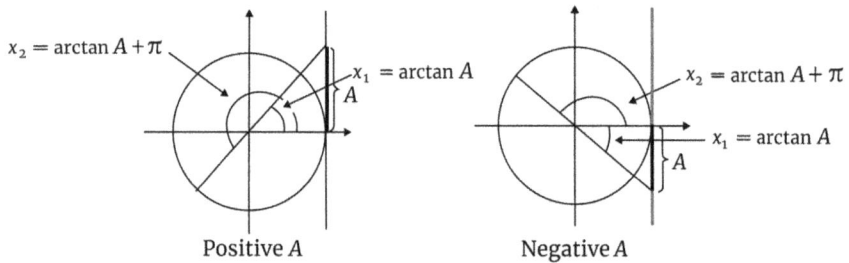

Positive A Negative A

These figures show two roots: $x_1 = \arctan A$ and $x_2 = \arctan A + \pi$. Since the tangent function period is π, then all other roots can be obtained by adding $n\pi$ to x_1. Therefore, we get the following result:

All solutions of the equation

tan $x = A$

are described by the formula:

$$x = \textbf{arctan } A + \pi n,$$

where n is any integer ∎

Problem 16.8. Solve the equation $\tan x = A$ in the interval $[0, 2\pi)$.

Solution. Let $A > 0$. Look at the left figure above for positive A. You can see two solutions and both are positive: $x_1 = \arctan A$ (in the 1ˢᵗ quadrant) and $x_2 = \arctan A + \pi$ (in the 3ʳᵈ quadrant).

If $A = 0$, then $x_1 = 0$ and $x_1 = \pi$.

Let $A < 0$. In the right figure above for negative A, you can see two roots: $x_1 = \arctan A$ (negative number in the 4ᵗʰ quadrant) and $x_2 = \arctan A + \pi$ (positive number in the 2ⁿᵈ quadrant). Since root x_1 is negative, we replace it with the coterminal angle by adding 2π to it. So, there are two roots: $x_1 = \arctan A + 2\pi$ and $x_2 = \arctan A + \pi$.

Final answer:

If $A \geq 0$, then $x_1 = \arctan A$, $x_2 = \arctan A + \pi$;

If $A < 0$, then $x_1 = \arctan A + 2\pi$ and $x_2 = \arctan A + \pi$. ■

Problem 16.9. Solve the equation $\tan x = \sqrt{3}$.

(a) In a one-period interval $(-\pi/2, \pi/2)$.
(b) On the entire number line.
(c) In the interval $[0, 2\pi)$.

Solution. Here $A = \sqrt{3}$. This is a special value, and $\arctan A = \arctan \sqrt{3} = \pi/3$.

(a) In the interval $(-\pi/2, \pi/2)$, the only solution is
$x = \arctan A = \arctan \sqrt{3} = \pi/3$.
(b) On the entire number line, there are infinitely many solutions that are described by the formula $x = \pi/3 + n\pi$, where n is any integer.
(c) In the interval $[0, 2\pi)$, there are two solutions that correspond to the values $n = 0$ and $n = 1$. The solutions are:
$x_1 = \pi/3$ and $x_2 = \pi/3 + \pi = 4\pi/3$. ■

Problem 16.10. Solve the equation $\tan x = -1$.

(a) In one-period interval $(-\pi/2, \pi/2)$.
(b) On the entire number line.
(c) In the interval $[0, 2\pi)$.

Solution. Here $A = -1$. This is also a special value and

$$\arctan A = \arctan(-1) = -\arctan 1 = -\pi/4.$$

(a) In the interval $(-\pi/2, \pi/2)$, the only solution is

$$x = \arctan A = \arctan(-1) = -\pi/4.$$

(b) On the entire number line, there are infinitely many solutions that are described by the formula $x = -\pi/4 + n\pi$, where n is any integer.
(c) In the interval $[0, 2\pi)$, there are two solutions that correspond to the values $n = 1$ and $n = 2$. The two solutions are:

$$x_1 = -\pi/4 + \pi = 3\pi/4 \text{ and } x_2 = -\pi/4 + 2\pi = 7\pi/4. \ \blacksquare$$

Problem 16.11. Solve the equation $\tan x = 4/3$.

(a) In one-period interval $(-\pi/2, \pi/2)$.
(b) On the entire number line.
(c) In the interval $[0, 2\pi)$.

Solution. Here $A = 4/3$. This is not a special value. Using a calculator, we have

$$\arctan A = \arctan(4/3) = 0.927.$$

(a) In the interval $(-\pi/2, \pi/2)$, the only solution is $x = \arctan A = 0.927$.
(b) On the entire number line, there are infinitely many solutions that are described by the formula $x = 0.927 + n\pi$, where n is any integer.
(c) In the interval $[0, 2\pi)$, there are two solutions that correspond to the values $n = 0$ and $n = 1$. The solutions are: $x_1 = 0.927$ and $x_2 = 0.927 + \pi = 4.069. \ \blacksquare$

Note. Angle $0.927 \approx 53°$ is an angle in the Egyptian triangle with sides 3, 4, and 5. This angle is opposite to side 4.

Exercises for Solving on Your Own

For all exercises, use radian measure.

In Exercises 16.1 and 16.2, solve the given equations in the interval $[0, 2\pi)$.

Exercise 16.1. Write the answers in terms of π (do not round).

(a) $2\sqrt{2} \cos x + 3 = 5$; (b) $2\sqrt{3} \cos x + 5 = 2$; (c) $3\sqrt{2} \cos x - 7 = 2$;

(d) $\sqrt{3} \tan x + 2 = 3$; (e) $\sqrt{3} \tan x + 6 = 3$.

Exercise 16.2. Round the answers to the nearest hundredth.

(a) $5 \cos x - 2 = 1$; (b) $7 \cos x + 6 = 2$; (c) $3 \tan x - 1 = 5$; (d) $4 \tan x + 7 = 5$.

Exercise 16.3. Solve the equations for the entire number line. Write the answers in the exact form in terms of π.

(a) $2 \cos^2 x + \cos x - 1 = 0$; (b) $\sin^2 x - 2 \cos x + 2 = 0$; (c) $\cos x = \sin 2x$;

(d) $\tan^2 x = 1$; (e) $\sin x - \cos 2x = 2$.

Exercise 16.4. Sofia solved the equation $\sin x + \cos x = 1$ in the following way: she squared both sides:

$(\sin x + \cos x)^2 = 1^2 \implies \sin^2 x + 2 \sin x \cos x + \cos^2 x = 1 \implies$
$1 + 2 \sin x \cos x = 1 \implies \sin x \cos x = 0 \implies$
$\sin x = 0$ or $\cos x = 0$. From the first equation,
$x = n\pi$ and from the second, $x = \pi/2 + n\pi$.

Together, both sets of solutions describe the quadrantal angles, and Sofia combined them into one set, getting the final answer
$x = n\pi/2$. However, this answer is wrong. Why?

Exercise 16.5. Sofia wanted to derive a formula for $\sin(\arccos x)$. She denoted $\theta = \arccos x$. Now her problem is to calculate $\sin \theta$. She knew that by the definition of the inverse cosine, $\cos \theta = x$. Also, she knew that $\sin^2 \theta + \cos^2 \theta = 1$, so $\sin^2 \theta = 1 - \cos^2 \theta = 1 - x^2$. Trying to solve this equation for $\sin \theta$, she got $\sin \theta = \pm\sqrt{1 - x^2}$. At this point, Sofia was stuck: which sign, $+$ or $-$, should she choose in front of the square root? Could you help her?

Exercise 16.6. Prove that for any x from the interval $[-1, 1]$, $\cos(\arcsin x) = \sqrt{1 - x^2}$.

Exercise 16.7. Prove that for any x,

$$\sin(\arctan x) = \frac{x}{\sqrt{1+x^2}} \quad \text{and} \quad \cos(\arctan x) = \frac{1}{\sqrt{1+x^2}}.$$

Exercise 16.8. Prove that if θ lies in the interval $[\pi, 2\pi]$, then $\arccos(\cos \theta) = 2\pi - \theta$.

Exercise 16.9. Prove that if θ lies in the interval $\left(\frac{\pi}{2}, \frac{3\pi}{2}\right)$ then $\arctan (\tan \theta) = \theta - \pi$.

Entertainment Problem

Problem E16.1. A bridge, 7 miles long, cannot support loads weighing more than 7,000 pounds. A truck approaches the bridge, weighing exactly 7,000 pounds. When it reaches the middle of the bridge, a 0.5 pounds bird suddenly lands on it. Why doesn't the bridge collapse? Note that the driver and the passengers did not take any actions while crossing the bridge.

Chapter 17

Trigonometric Identities, Inequalities, and Equations

In this chapter, we present various results that follow from the study of trigonometry in the previous chapters. Some of the problems are challenging. Nevertheless, we decided to include them to demonstrate a wide range of methods for solving trig problems.

Trig Identities

In Problems 17.1–17.8, angles A, B, and C belong to an arbitrary $\triangle ABC$, and a, b, c are opposite sides to these angles.

Problem 17.1. Prove that $\tan A + \tan B + \tan C = \tan A \tan B \tan C$.

Solution. Since A, B, and C are angles in a triangle, then

$$A + B + C = \pi \Rightarrow A + B = \pi - C \Rightarrow \tan(A + B) = \tan(\pi - C) = -\tan C.$$

From the other side, $\tan(A + B) = \dfrac{\tan A + \tan B}{1 - \tan A \tan B}$ (see Exercise 8.4 from Chapter 8). From here,

$\tan A + \tan B = \tan(A + B)(1 - \tan A \tan B) = -\tan C(1 - \tan A \tan B) =$
$-\tan C + \tan A \tan B \tan C$. Therefore, $\tan A + \tan B + \tan C$
$= \tan A \tan B \tan C$. ∎

Problem 17.2. Prove that $\sin 2A + \sin 2B + \sin 2C = 4\sin A \sin B \sin C$.

Solution. $\sin 2A + \sin 2B + \sin 2C = \sin 2A + 2\sin(B + C)\cos(B - C)$
$= 2\sin A \cos A + 2\sin(\pi - A)\cos(B - C) = 2\sin A \cos A + 2\sin A \cos(B - C)$
$= 2\sin A [\cos A + \cos(B - C)] = 4\sin A \cdot \cos(A + B - C)/2 \cdot \cos(A - B + C)/2$.

Since A, B and C are angles in a triangle, then

$$(A + B - C)/2 = \pi/2 - C \text{ and } (A - B + C)/2 = \pi/2 - B.$$

From here

$$\sin 2A + \sin 2B + \sin 2C = 4\sin A \cos\left(\frac{\pi}{2} - C\right)\cos\left(\frac{\pi}{2} - B\right)$$

$$= 4\sin A \sin B \sin C. \blacksquare$$

Problem 17.3. Prove identity: $\dfrac{\sin A - \sin B}{\sin A + \sin B} = \dfrac{a - b}{a + b}$.

Solution. We can use the following property of proportions:

$$\text{If } \frac{x}{a} = \frac{y}{b}, \text{ then } \frac{x - y}{x + y} = \frac{a - b}{a + b}.$$

This property is easy to check by cross multiplication in each proportion.

Let $x = \sin A$ and $y = \sin B$, By the Law of Sines, $\dfrac{\sin A}{a} = \dfrac{\sin B}{b}$.

Therefore, $\dfrac{\sin A - \sin B}{\sin A + \sin B} = \dfrac{a - b}{a + b}$. $\blacksquare$

Problem 17.4. Prove the Law of Tangents: $\dfrac{\tan\frac{A-B}{2}}{a - b} = \dfrac{\tan\frac{A+B}{2}}{a + b}$.

Solution. Let's present the above formula in the form $\dfrac{\tan\frac{A-B}{2}}{\tan\frac{A+B}{2}} = \dfrac{a - b}{a + b}$, and

modify the left part: $\dfrac{\tan\frac{A-B}{2}}{\tan\frac{A+B}{2}} = \dfrac{\sin\frac{A-B}{2}}{\cos\frac{A-B}{2}} \div \dfrac{\sin\frac{A+B}{2}}{\cos\frac{A+B}{2}} = \dfrac{\sin\frac{A-B}{2}}{\cos\frac{A-B}{2}} \cdot \dfrac{\cos\frac{A+B}{2}}{\sin\frac{A+B}{2}}$.

Now apply the Product formulas (see Problem 9.2 from Chapter 9) to the numerator and denominator:

$$\sin\frac{A-B}{2}\cdot\cos\frac{A+B}{2}=\frac{1}{2}(\sin A-\sin B) \quad \text{and}$$

$$\cos\frac{A-B}{2}\cdot\sin\frac{A+B}{2}=\frac{1}{2}(\sin A+\sin B).$$

Therefore,

$$\frac{\tan\frac{A-B}{2}}{\tan\frac{A+B}{2}}=\frac{\sin\frac{A-B}{2}}{\cos\frac{A-B}{2}}\cdot\frac{\cos\frac{A+B}{2}}{\sin\frac{A+B}{2}}=\frac{\sin A-\sin B}{\sin A+\sin B}.$$

Using the result from the previous problem, we get $\dfrac{\tan\frac{A-B}{2}}{\tan\frac{A+B}{2}}=\dfrac{a-b}{a+b}.$ ■

Problem 17.5. Prove that

$$a=b\cos C+c\cos B, \quad b=a\cos C+c\cos A, \quad c=a\cos B+b\cos A.$$

Solution. We'll prove only the last formula. Proof for others is similar. In $\triangle ABC$, draw the height to the base c. We can see that

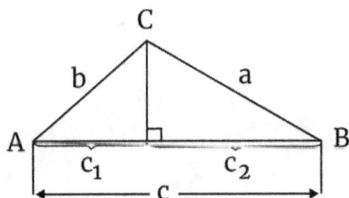

$$\cos A=c_1/b \;\Rightarrow\; c_1=b\cos A.$$
$$\cos B=c_2/a \;\Rightarrow\; c_2=a\cos B.$$
$$c=c_1+c_2=b\cos A+a\cos B. \;■$$

Problem 17.6. Prove that $\sin A+\sin B+\sin C=\dfrac{p}{d}$, where p is the perimeter of the triangle, and d is the diameter of the circumscribed circle.

Solution. According to Problem 6.3 from Chapter 6, $\dfrac{a}{\sin A} = \dfrac{b}{\sin B} = \dfrac{c}{\sin C} = d.$ From here, $\sin A = a/d$, $\sin B = b/d$, $\sin C = c/d$. To finish the proof, add all three expressions. ∎

Solve the following problem on your own.

Problem 17.7. Prove that $\sin A \sin B \sin C = \dfrac{abc}{d^3}$, where d is the diameter of the circumscribed circle. ∎

Problem 17.8. Prove that $\sin 2A + \sin 2B + \sin 2C = \dfrac{2s}{r^2}$, where s is the area of the triangle, and r is the radius of the circumscribed circle.

Solution. We will use the following result from geometry: the central angle in a circle is twice the inscribed angle (see the left figure below).

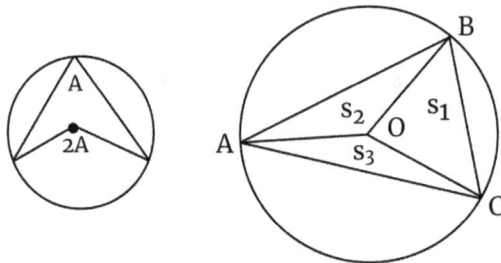

We have $AO = BO = CO = r$. Denote:

s_1 is the area of $\triangle BOC$, s_2 is the area of $\triangle AOB$, s_3 is the area of $\triangle AOC$. We have

$\angle BOC = 2A$, $s_1 = (1/2)r^2 \sin 2A$.
$\angle AOB = 2C$, $s_2 = (1/2)r^2 \sin 2C$.
$\angle AOC = 2B$, $s_3 = (1/2)r^2 \sin 2B$.

Summarize all three areas. The result is the area s of the entire $\triangle ABC$, so $s = (1/2)\, r^2\, (\sin 2A + \sin 2B + \sin 2C) \Rightarrow \sin 2A + \sin 2B + \sin 2C = 2s/r^2$. ∎

Problem 17.9. Prove that $\arcsin x + \arccos x = \pi/2$.

Solution. Denote $\alpha = \arcsin x$, $\beta = \arccos x$. We need to prove that $\alpha + \beta = \pi/2$.

By the definition of arcsine, $\sin \alpha = x$ and $-\pi/2 \le \alpha \le \pi/2$. Similarly, $\cos \beta = x$ and $0 \le \beta \le \pi$. We conclude that $\sin \alpha = \cos \beta$ (since both are equal to x). Using the reduction formula $\sin \alpha = \cos(\pi/2 - \alpha)$, we get $\cos(\pi/2 - \alpha) = \cos \beta$.

Let's find the interval in which the angle $(\pi/2 - \alpha)$ is located. To do this, multiply inequalities $-\pi/2 \le \alpha \le \pi/2$ by -1. We get: $-\pi/2 \le -\alpha \le \pi/2$. Then add $\pi/2$ to all parts of this inequality. We get $0 \le (\pi/2 - \alpha) \le \pi$.

As we see, both angles $(\pi/2 - \alpha)$ and β belong to the interval $[0, \pi]$, and the cosines of these angles are equal. Since the cosine function is monotonic in this interval (it decreases), we conclude that $(\pi/2 - \alpha) = \beta$, or $\alpha + \beta = \pi/2$. Therefore, $\arcsin x + \arccos x = \pi/2$. ∎

In Problems 17.10–17.12, we will use the following reasoning. Let $\tan \alpha = \tan \beta$. Then $\alpha - \beta = n\pi$ for some integer n. Indeed,

$$0 = \tan \alpha - \tan \beta = \frac{\sin \alpha}{\cos \alpha} - \frac{\sin \beta}{\cos \beta} = \frac{\sin \alpha \cos \beta - \sin \beta \cos \alpha}{\cos \alpha \cos \beta} = \frac{\sin(\alpha - \beta)}{\cos \alpha \cos \beta}.$$

From here, $\sin(\alpha - \beta) = 0 \Rightarrow \alpha - \beta = n\pi$. We can conclude that if $\tan \alpha = \tan \beta$ and $|\alpha - \beta| < \pi$, then $\alpha - \beta = 0$, i.e. $\alpha = \beta$. In other words, if angles α and β lie inside an interval of length less than π, then to prove that $\alpha = \beta$, it is enough to show that $\tan \alpha = \tan \beta$.

Problem 17.10. Prove that $\arctan 1 + \arctan 2 + \arctan 3 = \pi$.

Solution. Since $\arctan 1 = \pi/4$, it suffices to show that $\arctan 2 + \arctan 3 = 3\pi/4$. Denote $\alpha = \arctan 2$ and $\beta = \arctan 3$. Since the range of arctangent is the $(-\pi/2, \pi/2)$ interval, and both α and β are greater than 0, then $\alpha + \beta$ belongs to the $(0, \pi)$ interval. Note that angle $3\pi/4$ is also in the same interval. Therefore, to prove that $\alpha + \beta = 3\pi/4$, we just need to show that tangents of both sides are equal. On the right side, $\tan(3\pi/4) = -1$. To calculate the tangent on the left side, we can use the Tangent of the Sum formula (see Exercise 8.4 from Chapter 8) and take into account that $\tan \alpha = 2$ and $\tan \beta = 3$. We have
$$\tan(\alpha + \beta) = (\tan \alpha + \tan \beta)/(1 - \tan \alpha \cdot \tan \beta) = (2 + 3)/(1 - 2 \cdot 3) = -1.$$
Thus, $\alpha + \beta = 3\pi/4$. ∎

Solve problems below on your own.

Problem 17.11. Prove that $\arctan(1/2) + \arctan(1/3) = \pi/4$. ■

Problem 17.12. (Generalization of the previous problem). Prove that

$$\arctan x + \arctan y = \arctan \frac{x+y}{1-xy}, \; xy \neq 1.$$

Assume that the value of the left side is between $-\pi/2$ and $\pi/2$, so the right side makes sense. ■

Trig Inequalities

Solve the following problem on your own.

Problem 17.13. Prove that $|\sin x \cos x| \leq 1/2$. ■

Problem 17.14. Prove that $\sin^4 x + \cos^4 x \geq 1/2$.

Solution. According to the previous problem, $|\sin x \cos x| \leq 1/2 \Rightarrow \sin^2 x \cos^2 x \leq 1/4$.
From here,

$$1 = 1^2 = (\sin^2 x + \cos^2 x)^2 = \sin^4 x + 2\sin^2 x \cos^2 x + \cos^4 x$$
$$\leq \sin^4 x + \cos^4 x + \frac{1}{2} \Rightarrow \sin^4 x + \cos^4 x \geq 1 - \frac{1}{2} = \frac{1}{2}. \; ■$$

Problem 17.15. Prove that $\sin x + \cos x \leq \sqrt{2}$.

Solution. Using the Sine Sum formula from Chapter 8, expand $\sin(x + \pi/4)$:

$$\sin\left(x + \frac{\pi}{4}\right) = \sin x \cos \frac{\pi}{4} + \cos x \sin \frac{\pi}{4}$$
$$= \sin x \frac{\sqrt{2}}{2} + \cos x \frac{\sqrt{2}}{2} = \frac{\sqrt{2}}{2}(\sin x + \cos x).$$

From here, $\sin x + \cos x = \sqrt{2} \sin(x + \pi/4) \leq \sqrt{2}$. ■

Problem 17.16. Let angle C from $\triangle ABC$ be obtuse. Prove that $\tan A \tan B < 1$.

Solution. We use the identity: $\dfrac{\tan A + \tan B}{1 - \tan A \tan B} = \tan(A + B) = \tan(\pi - C) = -\tan C$. Because C is an obtuse angle, then $-\tan C > 0$. Therefore, the fraction on the left side of the above equality is positive. Since A and B are acute angles, the numerator of the fraction is positive: $\tan A + \tan B > 0$. Therefore, the denominator is also positive: $1 - \tan A \tan B > 0$. From here, $\tan A \tan B < 1$. ∎

In proving some trig inequalities, we can use the algebraic inequalities from the following problem.

Problem 17.17. Let m and n be two positive numbers. Prove the following inequalities

(a) Inequality connecting arithmetic and geometric means: $\dfrac{m + n}{2} \geq \sqrt{mn}$.

(b) $n + \dfrac{1}{n} \geq 2$.

Solution.

(a) Consider the obvious inequality $(\sqrt{m} - \sqrt{n})^2 \geq 0 \Rightarrow$ $m - 2\sqrt{mn} + n \geq 0 \Rightarrow m + n \geq 2\sqrt{mn} \Rightarrow (m + n)/2 \geq \sqrt{mn}$.

(b) In the inequality $m + n \geq 2\sqrt{mn}$, set $m = 1/n$. Then we get $1/n + n \geq 2$. ∎

Problem 17.18. Prove that $\tan^2 x + \cot^2 x \geq 2$.

Solution. In the inequality $n + \dfrac{1}{n} \geq 2$, set $n = \tan^2 x$. Then $\cot^2 x = 1/n$, and $\tan^2 x + \cot^2 x \geq 2$. ∎

Solve the following problem on your own.

Problem 17.19. Prove that for any natural number n, $\tan^n x + \cot^n x \geq 2$, if x is in the 1st quadrant. ∎

Problem 17.20. Prove that for any $\triangle ABC$, $\cos A \cos B \cos C \leq \dfrac{1}{8}$.

Solution. Without loss of generality, we may assume that the triangle is acute (otherwise one of the cosines is negative or zero and the inequality is obvious). So, we assume that all cosines are positive. Using the notations and the results from Problem 17.5, we have:

$$a = b \cos C + c \cos B, \quad b = a \cos C + c \cos A, \quad c = a \cos B + b \cos A.$$

Let's apply the inequality $m + n \geq 2\sqrt{mn}$ to each of these equations. We will get

$$a \geq 2\sqrt{bc \cos B \cos C}, \quad b \geq 2\sqrt{ac \cos C \cos A}, \quad c \geq 2\sqrt{ab \cos B \cos A}.$$

Now, multiply these inequalities:

$$abc \geq 8\sqrt{a^2 b^2 c^2 \cos^2 A \cos^2 B \cos^2 C} = 8abc \cos A \cos B \cos C.$$

Thus, $\cos A \cos B \cos C \leq \dfrac{1}{8}$. ∎

Trig Equations

In all problems below, n is any arbitrary integer.

Problem 17.21. Solve the equation $\sin x = \cos x$.

Solution. Take a look at the graphs of sine and cosine on the one-period interval $[0, 2\pi]$:

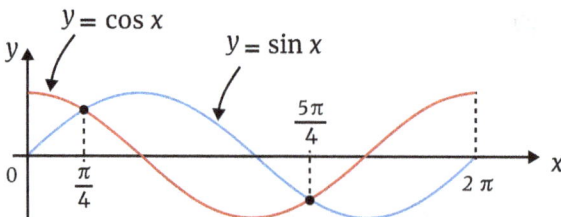

You can see two points of intersection at which $x = \pi/4$ and $x = \pi/4 + \pi = 5\pi/4$. Therefore, all solutions of the given equation can be described by the formula $x = \pi/4 + n\pi$. Another method of reaching the solution is to just look at sine and cosine in the unit circle:

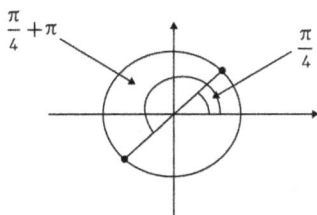

Problem 17.22. Solve the equation $\sin 5x = \sin 7x$.

Solution. We transform the equation to $\sin 7x - \sin 5x = 0$ and apply the Difference of the Sines formula from Chapter 9: $\sin \alpha - \sin \beta = 2 \sin \dfrac{\alpha - \beta}{2} \cos \dfrac{\alpha + \beta}{2}$. We have

$\sin 7x - \sin 5x = 2 \sin [(7x - 5x)/2] \cos[(7x + 5x)/2] = 0$, or $2 \sin x \cos 6x = 0$.

This equation is split into two basic equations: $\sin x = 0$ and $\cos 6x = 0$. The first equation has the solutions $x = n\pi$. From the second equation, we get $6x = \pi/2 + n\pi$ and $x = \pi/12 + n\pi/6 = \pi(2n + 1)/12$. Final answer:

$$x = n\pi \text{ and } x = \pi(2n + 1)/12. \blacksquare$$

Problem 17.23. Solve the equation $\sin x + \sin 2x + \sin 3x = 0$.

Solution. We will use the Sum of Sines formula

$\sin \alpha + \sin \beta = 2 \sin \dfrac{\alpha + \beta}{2} \cos \dfrac{\alpha - \beta}{2}$ from Chapter 9. We apply it to

the 1st and the 3rd terms of the equation by letting $\alpha = x$ and $\beta = 3x$. We have

$$\sin x + \sin 3x = 2 \sin[(x + 3x)/2] \cos[(3x - x)/2] = 2 \sin 2x \cos x.$$

Then the original equation becomes

$$2 \sin 2x \cos x + \sin 2x = 0,$$

or factoring $\sin 2x$, we get $\sin 2x(2 \cos x + 1) = 0$. This equation is split into two basics: $\sin 2x = 0$ and $\cos x = -1/2$. From them, we get the final answer: there are two infinite sets of solutions: $x = n\pi/2$ and $x = \pm 2\pi/3 + 2n\pi$. $\blacksquare$

Problem 17.24. Solve the equation $1 + \sin x + \cos x + \sin 2x + \cos 2x = 0$.

Solution. Let's group the terms: $(\sin 2x + 1) + (\sin x + \cos x) + \cos 2x = 0$.

Since $\sin 2x + 1 = (\sin x + \cos x)^2$ and $\cos 2x = \cos^2 x - \sin^2 x$, our equation takes the form:

$(\sin x + \cos x)^2 + (\sin x + \cos x) + (\cos^2 x - \sin^2 x) = 0,$

or $(\sin x + \cos x)^2 + (\sin x + \cos x) + (\cos x + \sin x)(\cos x - \sin x) = 0 \implies$

$(\sin x + \cos x)(\sin x + \cos x + 1 + \cos x - \sin x) = 0 \implies$

$(\sin x + \cos x)(1 + 2\cos x) = 0.$ From here we get two sets of solutions:

(1) $\sin x + \cos x = 0$. Note that $\cos x \neq 0$, otherwise $\sin x = 0$. However, both sine and cosine cannot be zero (recall that $\sin^2 x + \cos^2 x = 1$). So, we can divide both sides by $\cos x$: $\tan x + 1 = 0 \implies \tan x = -1 \implies x_1 = -\dfrac{\pi}{4} + \pi n.$

(2) $1 + 2\cos x = 0. \implies \cos x = -\dfrac{1}{2} \implies x_2 = \pm\dfrac{2\pi}{3} + 2\pi n.$ ∎

Problem 17.25. Solve the equation $\sin^4 x + \cos^3 x = 1$.

Solution. We have

$\cos^3 x - (1 - \sin^4 x) = 0. \implies \cos^3 x - (1 - \sin^2 x)(1 + \sin^2 x) = 0.$
$\cos^3 x - \cos^2 x (1 + \sin^2 x) = 0. \implies \cos^2 x (\cos x - 1 - \sin^2 x) = 0 \implies$
$\cos^2 x (\cos x - 1 - 1 + 1 - \sin^2 x) = 0 \implies$
$\cos^2 x (\cos^2 x + \cos x - 2) \implies \cos^2 x (\cos x - 1)(\cos x + 2) = 0.$

Since the last factor $\cos x + 2$ cannot be zero (because $|\cos x| \leq 1$), we get two sets of solutions:

(1) $\cos x = 0 \implies x_1 = \dfrac{\pi}{2} + \pi n,$

(2) $\cos x = 1 \implies x_2 = 2\pi n.$ ∎

Solve the following problem on your own.

Problem 17.26. Solve the equation $\sin^5 x + \cos^8 x = 1$. ∎

Note. In a similar way, we can solve a general equation $\sin^n x + \cos^m x = 1$, in which n or m is a power of 2: $n = 2^k$ or $m = 2^k$.

Problem 17.27. Solve the equation $a \sin x + b \cos x = c$, where a, b, and c are arbitrary constants.

Solution. We can assume that coefficient $a \geq 0$. Otherwise, we multiply both sides of the equation by -1.

(1) **Case $c = 0$.** The equation takes the form: $a \sin x + b \cos x = 0$.
   (a) If $a = 0$ and $b \neq 0$, then the equation becomes $\cos x = 0$, so $x = \pi/2 + n\pi$.
   (b) If $a \neq 0$, then $\cos x \neq 0$. Indeed, if $\cos x = 0$, then we get that $\sin x = 0$. But sine and cosine cannot be equal to zero at the same time. Since $\cos x \neq 0$, we may divide both sides of the equation by $\cos x$, and get $a \sin x / \cos x + b = 0$, or $\tan x = -b/a$. By the odd property of arctangent, we can write the solution of the equation $a \sin x + b \cos x = 0$ in the form $x = -\arctan(b/a) + n\pi$.

(2) **Case $c \neq 0$.** In Problem 12.5 from Chapter 12, we represented an expression from the left side of the equation in the form of a single sinusoid. Here we represent this expression as a single cosine function. The procedure is similar to Problem 12.5.

$$a \sin x + b \cos x = \sqrt{a^2 + b^2}\left(\frac{a}{\sqrt{a^2 + b^2}} \sin x + \frac{b}{\sqrt{a^2 + b^2}} \cos x \right).$$

Since, $a/\sqrt{a^2 + b^2} < 1$ and $a^2/(a^2 + b^2) + b^2/(a^2 + b^2) = 1$, there exists an angle θ such that

$$\sin \theta = a / \sqrt{a^2 + b^2} \text{ and } \cos \theta = b / \sqrt{a^2 + b^2}.$$

Then $a \sin x + b \cos x = \sqrt{a^2 + b^2}(\sin x \sin \theta + \cos x \cos \theta) = \sqrt{a^2 + b^2} \cos(x - \theta) = c$. From here we obtain the simplest equation

$$\cos(x - \theta) = \frac{c}{\sqrt{a^2 + b^2}}.$$

We see that the given equation has solutions if and only if $|c| \le \sqrt{a^2 + b^2}$.
Next, because $a \ge 0$ and $\sin \theta = a/\sqrt{a^2 + b^2}$, we conclude that $\sin \theta \ge 0$.
Therefore, angle θ lies in the interval $[0, \pi]$. This interval is the range of the
inverse cosine function, and we can set $\theta = \arccos(b/\sqrt{a^2 + b^2})$.

Recall that the simplest equation $\cos \alpha = A$ has roots:
$\alpha = \pm \arcos A + 2\pi n$. In our case, $\alpha = x - \theta$, and $A = c/\sqrt{a^2 + b^2}$.
Thus, we get the following final result.

The equation

$$a \sin x + b \cos x = c, \ (a \ge 0)$$

has roots if and only if $|c| \le \sqrt{a^2 + b^2}$, and its general solution is described
by the formula

$$x = \pm\arccos(c/\sqrt{a^2 + b^2}) + \arccos(b/\sqrt{a^2 + b^2}) + 2\pi n. \ \blacksquare$$

Entertainment Problem

Problem E17.1. Traveling Professor Smartman returned from a trip to a
distant land where each inhabitant is either an honest person or a liar (an
honest person always tells the truth, and a liar always lies). Professor tells
the following story to his son: "I met two inhabitants and asked one of
them: 'Is one of you guys honest?' After getting the answer, I deducted who
was who." "So, who was who?" asked the son. "I've told you everything
you need to know to figure this out on your own." replied the Professor.
So, who was who?

Chapter 18

Applications to Geometry

The Moor has done his work, the Moor may go.

Friedrich Schiller (1759–1805),
a German poet, philosopher, physician, historian and playwright

In this chapter, we solve a few geometric problems using trigonometry as a tool. Here, trigonometry is used only in the intermediate steps of solving and "disappears" in the final result. We include some "classic" problems, such as finding the bisectors, heights, and medians in a triangle. Some of the problems are difficult to solve without the use of trigonometry. Such problems demonstrate the power of this science.

Problem 18.1. Find the bisector of a triangle given its three sides.

Solution. Depict the triangle with sides a, b, and c.

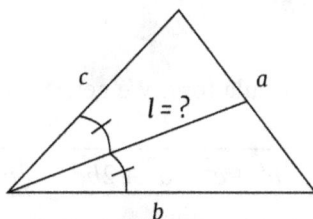

Let S be the area of the triangle, and θ be the angle between side b and bisector l (θ is half of the angle between b and c). Bisector l divides the triangle into two triangles with areas S_1 (between b and l) and S_2 (between

247

c and l). Using the formula (1) from Chapter 1 for the area of a triangle, we get

$$S = \frac{1}{2}\sin 2\theta \cdot cb = \sin \theta \cos \theta \cdot cb, \quad S_1 = \frac{1}{2}\sin \theta \cdot bl, \quad S_2 = \frac{1}{2}\sin \theta \cdot cl.$$

Substituting these expressions into equality $S = S_1 + S_2$, we have $\sin \theta \cos \theta \cdot bc = \frac{1}{2}\sin \theta (b+c)l$. Or $\cos \theta \cdot bc = \frac{1}{2}(b+c)l$. From here we can express the bisector l:

$$l = \frac{2bc \cos \theta}{b+c}.$$

What remains is to express $\cos \theta$ through the sides of the triangle. To do this, we use the Law of Cosines (see formulas (1) from Chapter 7):

$$\cos 2\theta = \frac{b^2 + c^2 - a^2}{2bc}.$$

Since $\cos 2\theta = 2\cos^2 \theta - 1$, then

$$\cos \theta = \sqrt{\frac{\cos 2\theta + 1}{2}} = \sqrt{\frac{1}{2}}\sqrt{\cos 2\theta + 1} = \sqrt{\frac{1}{2}\sqrt{\frac{b^2 + c^2 - a^2}{2bc} + 1}}$$

$$= \frac{1}{2}\sqrt{\frac{b^2 + c^2 - a^2 + 2bc}{bc}}.$$

Substitute into the formula for l. We get:

$$l = \frac{2bc \cos \theta}{b+c} = \frac{2bc}{(b+c)2}\sqrt{\frac{b^2 + c^2 - a^2 + 2bc}{bc}} = \frac{\sqrt{bc(b^2 + c^2 - a^2 + 2bc)}}{b+c}$$

$$= \frac{\sqrt{bc(a+b+c)(b+c-a)}}{b+c}. \quad \blacksquare$$

Problem 18.2. Prove that a bisector of an angle of a triangle divides the opposite side into two segments that are proportional to the other two sides of the triangle.

Solution. In the figure below, BD is a bisector of angle B, and h is the height to the base AC:

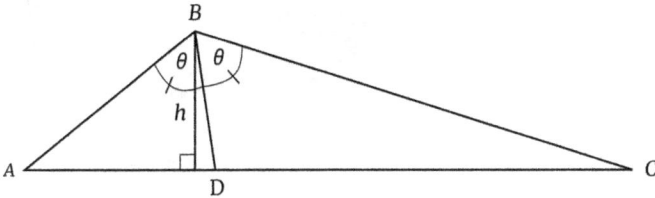

We need to prove that

$$\frac{AD}{DC} = \frac{AB}{BC}.$$

We have: $S_{ABD} = \frac{1}{2} AD \cdot h$, $S_{DBC} = \frac{1}{2} DC \cdot h$. From here,

$$\frac{S_{ABD}}{S_{DBC}} = \frac{AD}{DC}.$$

On the other hand, $S_{ABD} = \frac{1}{2} AB \cdot BD \cdot \sin\theta$, $S_{DBC} = \frac{1}{2} BC \cdot BD \cdot \sin\theta$. From here,

$$\frac{S_{ABD}}{S_{DBC}} = \frac{AB}{BC}.$$

Therefore,

$$\frac{AD}{DC} = \frac{AB}{BC}. \blacksquare$$

Problem 18.3. Prove that the sum of the squares of the diagonals of a parallelogram is equal to the sum of the squares of its sides.

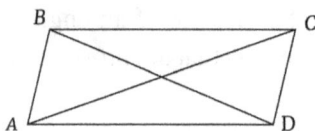

Solution. We need to prove that $AC^2 + BD^2 = BC^2 + DC^2 + AD^2 + AB^2$.
First, we apply the Law of Cosines to $\triangle ACD$:

$$AC^2 = AD^2 + DC^2 - 2AD \cdot DC \cdot \cos D.$$

Since $\angle D = 180° - \angle A$, then $\cos D = -\cos A$. Also $AD = BC$ and $DC = AB$.
Substituting values for $\cos D$, AD^2, and DC into the expression for AC^2, we get

$$AC^2 = BC^2 + DC^2 + 2AD \cdot AB \cdot \cos A.$$

Next, we apply the Law of Cosines to $\triangle ABD$:

$$BD^2 = AD^2 + AB^2 - 2AD \cdot AB \cdot \cos A.$$

Finally, we add AC^2 and BD^2. Then the expression $2AD \cdot AB \cdot \cos A$ is cancelled out, and we get $AC^2 + BD^2 = BC^2 + DC^2 + AD^2 + AB^2$. ∎

The next problem does not use trigonometry. However, its result directly follows from the previous problem.

Problem 18.4. Express the median of the triangle through three sides.

Solution. In triangle ABC with sides a, b, and c, we extend (double) the median AF in order to get the parallelogram $ABDC$.

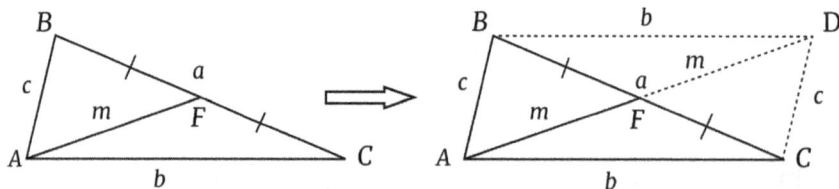

Using the result from the previous problem, we have, $(2m)^2 + a^2 = 2b^2 + 2c^2$. Or $4m^2 = 2(b^2 + c^2) - a^2$. From here

$$m = \frac{1}{2}\sqrt{2(b^2 + c^2) - a^2}. \blacksquare$$

Problem 18.5. Describe an idea of how to find a formula for the area of a triangle through three sides using a trigonometric approach.

Solution. Let's denote the area of the triangle as S, and the three given sides as a, b, and c. Also, denote by θ the angle between sides b and c.

First of all, note that it is not so easy to come up with a purely geometrical idea. In contrast, the trigonometric idea is straightforward. It can be described in three steps.

(1) Using the Law of Cosines, express cosine of the angle θ through sides a, b, and c: $\cos\theta = \dfrac{b^2 + c^2 - a^2}{2bc}$.

(2) Get $\sin\theta$ from $\cos\theta$: $\sin\theta = \sqrt{1 - \cos^2\theta}$.

(3) Get area S by the formula: $S = \dfrac{1}{2}bc\sin\theta$.

If you are able to go through these steps by doing some rather tedious but elementary algebraic operations, you will end up with the following wonderful formula, which is called the **Heron's formula**

$$S = \sqrt{p(p-a)(p-b)(p-c)},$$

where $p = (a + b + c)/2$. The number p is called the semi-perimeter. $\blacksquare$

The next result directly follows from the previous problem.

Problem 18.6. Express the height of a triangle through three sides.

Solution. Depict the triangle with sides a, b, and c:

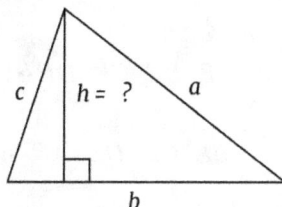

We need to find the height h. Denote the area of the triangle by S. According to the Heron's formula,

$$S = \sqrt{p(p-a)(p-b)(p-c)}, \text{ where } p = \frac{a+b+c}{2}.$$

On the other hand, S is expressed through base b and h:

$$S = \frac{1}{2}bh. \text{ From here } h = \frac{2S}{b} = \frac{2\sqrt{p(p-a)(p-b)(p-c)}}{b}. \blacksquare$$

Problem 18.7. The plot of land $ABCD$ is divided into four triangular parts according to the figure:

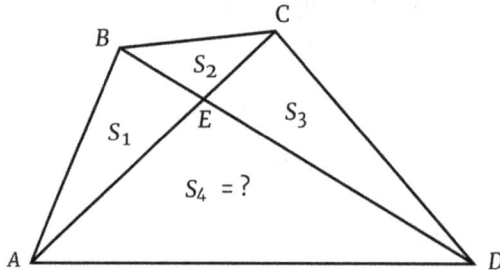

Workers were able to measure the area of three parts: $S_1 = 430$ ft², $S_2 = 300$ ft², and $S_3 = 645$ ft². However, because of swampy soil, they have difficulty measuring the area S_4 of the 4th plot. How do we calculate this area without measuring it?

Solution. Denote angle AED by θ. Since angle $BEA = 180° - \theta$, then $\sin BEA = \sin \theta$. Thus, all angles around point E have the same value of sine. We have,

$$S_1 = \frac{1}{2}AE \cdot BE \cdot \sin \theta, \quad S_2 = \frac{1}{2}BE \cdot CE \cdot \sin \theta,$$

$$S_3 = \frac{1}{2}CE \cdot DE \cdot \sin \theta, \quad S_4 = \frac{1}{2}DE \cdot AE \cdot \sin \theta,$$

From here,

$$S_1 \cdot S_3 = \frac{1}{4}AE \cdot BE \cdot CE \cdot DE \cdot \sin^2 \theta,$$

$$S_2 \cdot S_4 = \frac{1}{4}BE \cdot CE \cdot DE \cdot AE \cdot \sin^2 \theta.$$

Therefore, $S_1 \cdot S_3 = S_2 \cdot S_4$ and

$$S_4 = \frac{S_1 \cdot S_3}{S_2} = \frac{430 \cdot 645}{300} = 924.5 \text{ ft}^2.$$

Note that the problem on the Cover Page contains these data:

$S_1 = 20 \text{ ft}^2$, $S_2 = 10 \text{ ft}^2$, and $S_3 = 30 \text{ ft}^2$. Therefore, $S_4 = \frac{20 \cdot 30}{10} = 60 \text{ ft}^2$ ∎.

Problem 18.8. In triangle ABC all sides are given: $BC = a$, $AC = b$, $AB = c$. Point D divides side BC by the ratio $m{:}n$. Find the length of the segment AD.

Solution. Depict the problem in the figure below

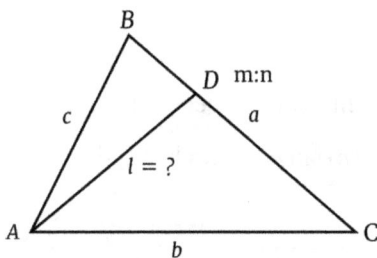

Since $\dfrac{BD}{DC} = \dfrac{m}{n}$ and $DC = a - BD$ then

$$\frac{BD}{a - BD} = \frac{m}{n} \Rightarrow BD = \frac{am}{m+n}.$$

According to the Law of Cosines for $\triangle ABC$, $\cos B = \dfrac{a^2 + c^2 - b^2}{2ac}$.

Again, using the Law of Cosine for $\triangle ABD$, we get

$$l^2 = c^2 + BD^2 - 2cBD \cos B = c^2 + \frac{a^2 m^2}{(m+n)^2} - 2c \frac{am}{m+n} \cdot \frac{a^2 + c^2 - b^2}{2ac}.$$

At this point, we have expressed l through known data. After some algebraic transformations (we skipped these steps), l takes the form:

$$l = \frac{\sqrt{(m+n)\left(b^2 m + c^2 n\right) - a^2 mn}}{m+n}. \blacksquare$$

Note. Problem 18.4 is a special case of problem 18.8.

Problem 18.9. Find the side of a triangle given two other sides and area of the triangle.

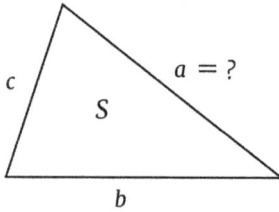

Solution. We need to find side a. Denote the angle between sides b and c by θ. We have $S = \frac{1}{2} b \cdot c \cdot \sin\theta$. From here, $\sin\theta = \frac{2S}{bc}$,

$$\cos\theta = \sqrt{1 - \sin^2\theta} = \sqrt{1 - \left(\frac{2S}{bc}\right)^2} = \frac{\sqrt{b^2 c^2 - 4S^2}}{bc}.$$

According to the Law of Cosines,

$$a = \sqrt{b^2 + c^2 - 2bc \cdot \cos\theta} = \sqrt{b^2 + c^2 - 2bc \cdot \frac{\sqrt{b^2 c^2 - 4S^2}}{bc}}$$

$$= \sqrt{b^2 + c^2 - 2\sqrt{b^2 c^2 - 4S^2}}. \blacksquare$$

Problem 18.10. Find the area S of the triangle given two sides and a bisector between them.

Solution. Depict the problem in the figure below

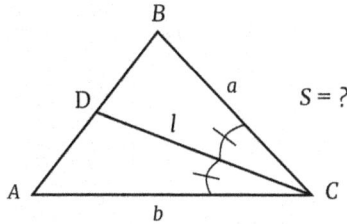

Let $BC = a$, $AC = b$, and bisector $CD = l$. Denote $\angle ACD = \theta$. Then $\angle ACB = 2\theta$.

Area S of the triangle ABC is defined by the formula:

$$S = \frac{1}{2}ab\sin 2\theta.$$

Bisector CD divides triangle ABC into two triangles with areas S_1 (between b and l) and S_2 (between l and a). We have

$$S_1 = \frac{1}{2}bl\sin\theta, \quad S_2 = \frac{1}{2}al\sin\theta. \quad \text{Since } S = S_1 + S_2,$$

$$\frac{1}{2}ab\sin 2\theta = \frac{1}{2}bl\sin\theta + \frac{1}{2}al\sin\theta. \quad \text{Or } ab\sin 2\theta = l\sin\theta(a+b).$$

Using formula $\sin 2\theta = 2\sin\theta\cos\theta$, we get $ab2\sin\theta\cos\theta = l\sin\theta\,(a+b)$. From here

$$\cos\theta = \frac{l(a+b)}{2ab}.$$

Express $\sin 2\theta$ through $\cos\theta$. $\sin 2\theta = 2\sin\theta\cos\theta = 2\cos\theta\sqrt{1-\cos^2\theta}$.

Substituting $\cos\theta = \dfrac{l(a+b)}{2ab}$ into the last expression, we get

$$\sin 2\theta = 2\frac{l(a+b)}{2ab}\sqrt{1 - \frac{l^2(a+b)^2}{4a^2b^2}} = \frac{l(a+b)\sqrt{4a^2b^2 - l^2(a+b)^2}}{2a^2b^2}.$$

Finally,

$$S = \frac{1}{2}ab\sin 2\theta = \frac{lab(a+b)\sqrt{4a^2b^2 - l^2(a+b)^2}}{4a^2b^2}. \blacksquare$$

Problem 18.11. Find x in the following figure. Write the answer in exact radical form.

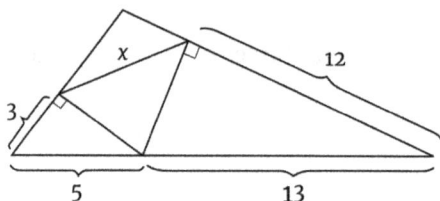

Solution. Let's mark the angles and the line segments:

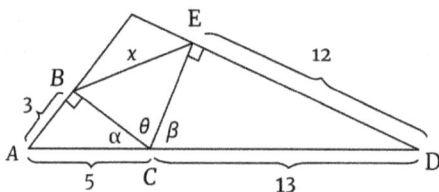

From the right $\triangle ABC$: $BC = \sqrt{5^2 - 3^2} = 4$, $\sin\alpha = 3/5$, $\cos\alpha = 4/5$.

From the right $\triangle CDE$: $CE = \sqrt{13^2 - 12^2} = 5$, $\sin\beta = 12/13$, $\cos\beta = 5/13$.

By the Law of Cosines,

$$x = \sqrt{BC^2 + CE^2 - 2BC \cdot CE \cdot \cos\theta} = \sqrt{41 - 40\cos\theta}.$$

Since $\theta = 180° - \alpha - \beta$, $\cos\theta = \cos(180° - \alpha - \beta) = -\cos(\alpha + \beta) = $

$$-(\cos\alpha\cos\beta - \sin\alpha\sin\beta) = -\frac{4}{5}\cdot\frac{5}{13} + \frac{3}{5}\cdot\frac{12}{13} = \frac{16}{65}.$$

Finally, we get $x = \sqrt{41 - 40\cdot(16/65)} = \sqrt{405/13} = 9\sqrt{5/13}. \blacksquare$

Problem 18.12. In the triangle ABC, two sides are given: $AB = c$, $AC = b$. Angle A is twice as large as angle C. Find a third side BC.

Solution. Depict triangle ABC:

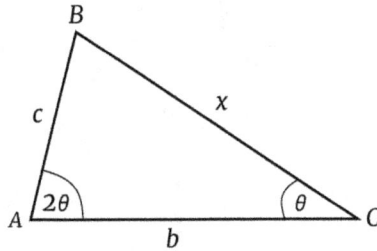

Using the Law of Cosines, we have $x^2 = b^2 + c^2 - 2bc \cos 2\theta$.
Since $\cos 2\theta = 2\cos^2 \theta - 1$, then

$$x^2 = b^2 + c^2 - 2bc(2\cos^2 \theta - 1) \qquad (*)$$

Using the Law of Sines, we have $\dfrac{c}{\sin \theta} = \dfrac{x}{\sin 2\theta}$.

Since $\sin 2\theta = 2 \sin \theta \cos \theta$, then $\dfrac{c}{\sin \theta} = \dfrac{x}{2 \sin \theta \cos \theta}$. From here,

$\cos \theta = \dfrac{x}{2c}$. $\cos^2 \theta = \dfrac{x^2}{4c^2}$. Substituting $\cos^2 \theta$ into (*), we get a lin-

ear equation for x^2:

$$x^2 = b^2 + c^2 - 2bc\left(2\frac{x^2}{4c^2} - 1\right).$$

From here, $x^2 + 2bc\left(2\dfrac{x^2}{4c^2} - 1\right) = b^2 + c^2$. $+c^2$. Or

$$x^2 + \frac{bx^2}{c} = b^2 + c^2 + 2bc = (b+c)^2, \quad x^2\left(1 + \frac{b}{c}\right) = (b+c)^2,$$

$$x^2\left(\frac{b+c}{c}\right) = (b+c)^2, \quad x^2 = c(b+c), \quad x = \sqrt{c(b+c)} . \blacksquare$$

Problem 18.13. A circle inscribed in triangle ABC touches sides AB, BC, and AC at points D, E, and F, respectively. FG is the radius of the circle. Find DE if $AF = 3$, $FC = 2$, and $FG = \dfrac{3}{2}$.

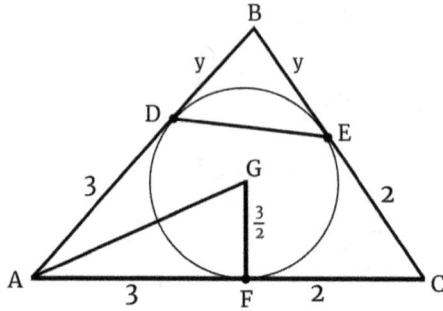

Solution. Find AG by the Pythagorean theorem:

$$AG = \sqrt{AF^2 + FG^2} = \sqrt{3^2 + \left(\frac{3}{2}\right)^2} = \frac{3\sqrt{5}}{2}.$$

Since AG is a bisector of $\angle A$, then $\angle GAF = \dfrac{A}{2}$.

Calculate $\cos \dfrac{A}{2}$ from $\triangle AGF$. We have, $\cos\dfrac{A}{2} = \dfrac{AF}{AG} = \dfrac{3 \cdot 2}{3\sqrt{5}} = \dfrac{2}{\sqrt{5}}$.

From $\cos \dfrac{A}{2}$ we can calculate $\cos A$, using the Half-Angle Formula:

$$\cos A = 2\cos^2 \frac{A}{2} - 1 = 2 \cdot \frac{4}{5} - 1 = \frac{3}{5}.$$

Using the Law of Cosines for $\triangle ABC$, obtain the equation for y:

$$BC^2 = AB^2 + AC^2 - 2AB \cdot AC \cdot \cos\theta.$$

Or

$$(2+y)^2 = (3+y)^2 + (3+2)^2 - 2(3+y)(3+2)\cos A.$$

$$4 + 4y + y^2 = 9 + 6y + y^2 + 25 - 2 \cdot (3+y) \cdot 5 \cdot \frac{3}{5} \quad \Rightarrow \quad y = 3.$$

Next, we will find $\cos B$, using the Law of Cosines for $\triangle ABC$ one more time:

$$AC^2 = AB^2 + BC^2 - 2AB \cdot BC \cdot \cos B.$$

From here

$$(3+2)^2 = (3+3)^2 + (2+3)^2 - 2(3+3)(2+3)\cos B \Rightarrow \cos B = \frac{3}{5}.$$

Finally, using the Law of Cosines for ΔDBE, we get

$$DE^2 = y^2 + y^2 - 2y \cdot y \cdot \cos B = 3^2 + 3^2 - 2 \cdot 3^2 \cdot \frac{3}{5} = \frac{36}{5}.$$

$$DE = \sqrt{\frac{36}{5}} = \frac{6}{\sqrt{5}} = \frac{6\sqrt{5}}{5}. \blacksquare$$

Problem 18.14. Three regular polygons are inscribed in a circle. The number of sides of each subsequent polygon is twice as large as that of the previous one. The areas of the first two are equal to A_1 and A_2. Find the area A_3 of the third.

Solution. First, let's figure out the area S_n of the regular n-gon. From the picture below, we see that $\angle\theta = 2\pi/n$.

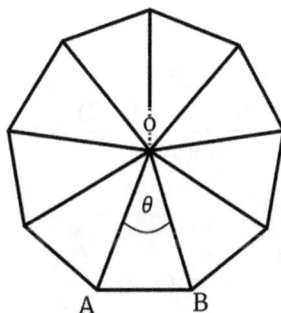

Let radius $AO = r$. Then $S_n = n \cdot S_{\Delta AOB} = \frac{1}{2}nr^2 \sin\theta = \frac{1}{2}nr^2 \sin\frac{2\pi}{n}$.

Since all three polygons are inscribed in one circle, they are related to the same radius. We have

$$A_1 = S_n = \frac{1}{2} n \cdot r^2 \cdot \sin \frac{2\pi}{n} = \frac{1}{2} n r^2 \sin \theta,$$

$$A_2 = S_{2n} = \frac{1}{2} 2n \cdot r^2 \cdot \sin \frac{2\pi}{2n} = n r^2 \sin \frac{\theta}{2},$$

$$A_3 = S_{4n} = \frac{1}{2} 4n \cdot r^2 \cdot \sin \frac{2\pi}{4n} = 2 n r^2 \sin \frac{\theta}{4}.$$

From here,

$$\frac{A_2}{A_3} = \frac{n r^2 \sin(\theta/2)}{2 n r^2 \sin(\theta/4)} = \frac{\sin(\theta/2)}{2 \sin(\theta/4)} = \left[\sin \frac{\theta}{2} = 2 \sin \frac{\theta}{4} \cos \frac{\theta}{4} \right]$$

$$= \frac{2 \sin(\theta/4) \cdot \cos(\theta/4)}{2 \sin(\theta/4)}.$$

Reducing by 2 sin(θ/4), we get

$$\frac{A_2}{A_3} = \cos(\theta/4) \ \Rightarrow \ A_3 = \frac{A_2}{\cos(\theta/4)}.$$

Next,

$$\frac{A_1}{A_2} = \frac{n r^2 \sin \theta}{2 n r^2 \sin(\theta/2)} = \frac{2 \sin(\theta/2) \cdot \cos(\theta/2)}{2 \sin(\theta/2)} = \cos(\theta/2) = 2 \cos^2(\theta/4) - 1.$$

Therefore,

$$\frac{A_1}{A_2} = 2 \cos^2(\theta/4) - 1 \ \Rightarrow \ \cos^2(\theta/4) = \left(\frac{A_1}{A_2} + 1 \right) / 2 = \frac{A_1 + A_2}{2 A_2}.$$

$$\cos(\theta/4) = \sqrt{\frac{A_1 + A_2}{2 A_2}} \cdot \text{ Substitute the last expression into}$$

$$A_3 = \frac{A_2}{\cos(\theta/4)} \cdot \ \Rightarrow \ A_3 = A_2 \cdot \sqrt{\frac{2 A_2}{A_1 + A_2}} \cdot \blacksquare$$

Problem 18.15. Consider a quadrilateral with two right angles, and a diagonal equal to 1:

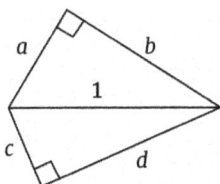

Prove that $ac + bd \leq 1$.

Solution. By the Pythagorean theorem, $a^2 + b^2 = 1$ and $c^2 + d^2 = 1$.
There exist angles α and β, such that
$a = \sin \alpha$, $b = \cos \alpha$, $c = \sin \beta$, and $d = \cos \beta$.
Then $ac + bd = \sin \alpha \sin \beta + \cos \alpha \cos \beta = \cos(\alpha - \beta) \leq 1$. ■

Problem 18.16. A farmer has a triangular plot of land with sides $AB = 50$ feet and $AC = 90$ feet. He plans to divide his land into two plots for planting potatoes and cabbage. The area $DEBC$ for potatoes should be twice as large as AED for cabbage. Our farmer would like to start making a border at a distance of 40 feet from point A on side AC. In order to find the location of the second point E for the border, he needs to figure out the distance between points A and E. Calculate this distance.

Solution. Depict the problem on the figure below:

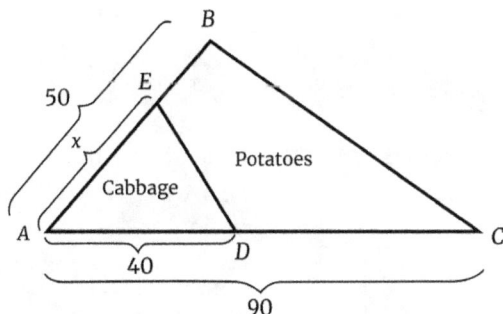

Area $S_{ABC} = \dfrac{1}{2} AB \cdot AC \sin A$.

By the condition of the problem, area $S_{AED} = \dfrac{1}{3} S_{ABC} = \dfrac{1}{6} AB \cdot AC \sin A$.

On the other hand, $S_{AED} = \dfrac{1}{2} AE \cdot AD \sin A$.

From here we get an equation for AE:

$$\frac{1}{2} AE \cdot AD \sin A = \frac{1}{6} AB \cdot AC \sin A.$$

Solving it, we get $AE = \dfrac{AB \cdot AC}{3AD} = \dfrac{50 \cdot 90}{3 \cdot 40} = 37.5.$ ∎

Exercises for Solving on Your Own

Exercise 18.1. Two friends Anna (A) and Beatrice (B) return to their homes after a meeting. Each of them moves along a straight road in different directions. After driving 60 miles, A stops for the night at a hotel. To reach home, A needs to drive 110 miles more. B also stops at a hotel for the night after driving 30 miles, and she will need to drive home 90 miles more. The distance between the hotels is 40 miles. How far apart do the friends live?

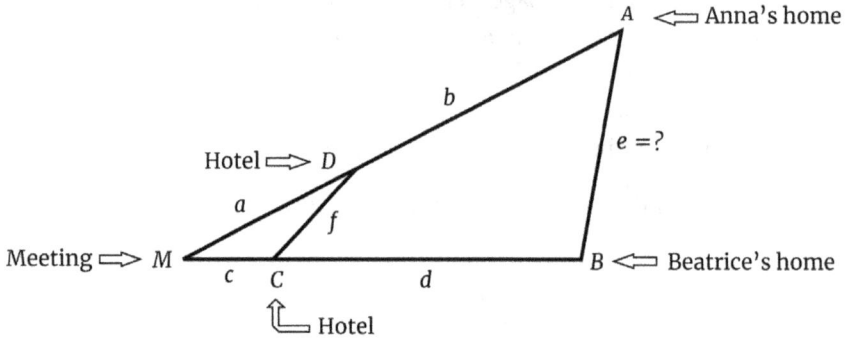

Exercise 18.2. Geodesists presented the land area measurement results in the following scheme:

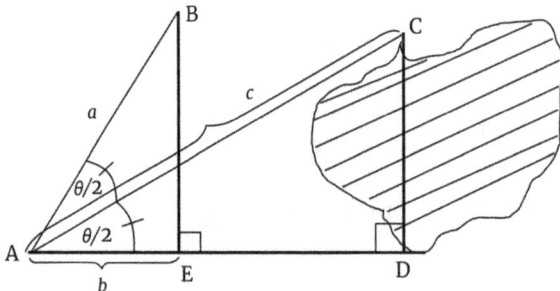

The shaded area indicates a lake. Here are the results of their measurements (we will use letters instead of numbers to solve the problem in general form):

$AB = a$, $AE = b$, $AC = c$. The segment AC is the bisector of the angle BAD, and BE and CD are perpendicular to AD.

From these data we need to determine the width CD of the lake.

Exercise 18.3. Describe an idea of how to solve the following problem. The final answer should not contain any trig functions. You don't need to derive a formula for the answer.

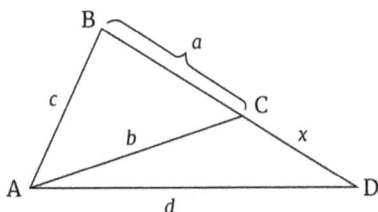

Given: a, b, c, and d. Find CD.

Exercise 18.4. Among all the triangles with two given sides a and b, find the triangle with the largest area.

Entertainment Problems

Problem E18.1. A plant appeared on the lake. It reproduces at such a rate that its area doubles every day. The lake was completely overgrown after 60 days. In how many days was it half overgrown?

Problem E18.2. There are coins on the table. Fifteen of them lie heads-up. The rest are heads-down. While blindfolded, you need to arrange these coins into two piles such that in each pile the number of coins lying heads-up will be the same. The number of coins in piles can be different. Coins can be flipped, but it is impossible to feel which way the coin is facing.

Chapter 19

Historical Background of Complex Numbers. Cardano's Formula

In science, the credit goes to the man who convinces the world, not to whom the idea first occurs.

Francis Darwin (1848–1925), a British botanist

In this chapter, we observe some historical events related to the appearance in math of so-called imaginary or complex numbers. The theory of complex numbers or, more generally complex analyses, is a comprehensive area of mathematics that deserves separate consideration. Although this is an independent topic, it is closely related to trigonometry. Many important results in the theory of complex numbers were obtained precisely due to trigonometry.

Roughly speaking, imaginary numbers are the square roots of negative numbers. The question immediately arises: how can this be? We cannot take the square root of a negative number, because the square of any number cannot be negative. Or, in other words, the equation $x^2 = a$ has no solutions for $a < 0$. Yes, this is true if we are dealing with real numbers. However, as we will see, it is possible to extend the set of real numbers in such a way that the mentioned equation will have solutions.

It might seem that imaginary numbers arose precisely from an attempt to solve quadratic equations such as $x^2 = -1$. However, it is not so. Although the formula for solving quadratic equations was known for a very long time (about 2000 years BC in Babylon), ancient mathematicians did not find anything strange in the fact that sometimes this formula led to the operation of taking the square root of negative numbers. In such cases, they simply looked at the quadratic equation as having no solutions.

The efforts of mathematicians were aimed at obtaining formulas for solving equations of degrees higher than 2, particularly for equations of the 3rd degree (cubic equations). However, this task was beyond their ability. Only after more than 3500 years, in about 1500, an Italian mathematician del Ferro (1465–1526), through tremendous effort, was the first to discover the formula for the cubic equation. After attaining it, as we will see below, a suspicion began to arise that there are new numbers that are different from the real ones. We can say that this formula specifically had a motivating effect on the introduction of complex numbers.

There is an interesting story related to Ferro's discovery. Having found the formula, Ferro made a secret of it. The concealment of scientific discoveries in those days was of particular importance for the career of their authors. In Italy, mathematical contests were widely practiced: in crowded meetings, opponents challenged each other with problems. The winner was awarded not only fame and a monetary prize, but also the opportunity to hold a university position. On the contrary, a person who was defeated at such an event often lost his place. Since Ferro's student Fiore participated in these competitions, Ferro (shortly before his death) informed him of the secret formula. Armed with this knowledge, Fiore called for a public debate with one of the most powerful mathematicians of the time, Nicolo Tartaglia (1499–1557). Having received the challenge, Tartaglia somehow learned that most of the problems would be related to finding roots of cubic equations. Rumors reached Tartaglia that Fiore had a formula. He began to intensively search for this formula. Here is what Tartaglia wrote about it: "I worked days and nights and applied all my zeal and mathematical skill to find this algorithm, and thanks to a favorable fate, I managed to do this eight days before the start of the fight." It is amazing how Tartaglia, in such a short amount of time was able to find the formula for which mathematicians unsuccessfully searched for several thousand years. Of course, a tremendous incentive was that he learned about the existence of such a formula, and he had a great desire to win. It is also possible that somehow, he found out about some of the details unknown to us that helped him derive the formula. The debate took place in 1535, and Tartaglia within 2 h solved all 30 problems proposed by the opponent. Fiore, however, was unable to solve any of the 30 problems chosen by Tartaglia from various areas of mathematics. After the debate Tartaglia became famous throughout Italy. However, he also kept secret the formula he found.

At the same time, another Italian scientist Girolamo Cardano (1501–1576) was in the process of writing a book in which he wanted to include all the existing achievements in algebra, and he also tried (unsuccessfully) to derive this formula.

Hearing of Tartaglia's secret, Cardano reached out to him in hopes of convincing Tartaglia to share the formula. After numerous attempts, Tartaglia was finally persuaded when Cardano promised to introduce him to the Spanish governor, who could help his future career. Cardano vowed never to publish the formula and keep it a secret, which he did for several years. But in 1543 he managed to get acquainted with late Ferro's papers and became convinced that Ferro already knew this formula. Cardano thought that he is not obligated to keep the secret in this case, because it belonged not only to Tartaglia. Cardano published the formula in his book *Ars Magna*. In the preface, he wrote that "Scipione del Ferro discovered this formula. And, competing with him, Nicolo Tartaglia, our friend, also solved the same problem, and after long requests passed it to me." The publication of this formula provoked Tartaglia's great indignation. He wrote: "Cardano treacherously stole the best achievement of my work in algebra." Over time, the role of Tartaglia and Ferro was forgotten. This result came to be associated only with the book *Ars Magna*, and began to be called, unfairly, the Cardano's formula, although Cardano himself honestly indicated in his book who really discovered it. Interestingly, all his life, Cardano dreamed of gaining recognition and fame. He published a large number of works (over 120) in various fields of science, technology, and medicine. His description of a mechanical device for transmitting torque and rotation (Cardan shaft) is well known. However, this device was previously depicted by Leonardo da Vinci. Cardano himself believed that he received his main achievements in medicine. In his autobiography *The Book of My Life*, Cardano tried to perpetuate his name. He succeeded. However, in history, he remains known, in fact, mainly for two achievements: the Cardan shaft (driveshaft) and the formula for solving cubic equations. Both of these achievements were not his.

Let us now reveal this famous formula. It is fascinating that its derivation in modern notation isn't very complicated and is actually quite elementary. It seems that the problem of obtaining such a formula could even be solved by students at a mathematical Olympiad.

We proceed to the derivation of Cardano's formula. Any cubic equation can be written in the form:

$$x^3 + ax^2 + bx + c = 0.$$

Our first goal is to get rid of the term ax^2. This is done in the next problem, which is proposed to be solved on your own.

Problem 19.1. Make the following substitution in the above cubic equation: $x = y - \dfrac{a}{3}$. Prove that the result is a cubic equation for y that can be written in the form:

$$y^3 + py + q = 0,$$

where p and q are the new coefficients. ∎

Problem 19.2. Find the formula for the roots of the equation $y^3 + py + q = 0$.

Solution. The decisive idea is to present the solution as the sum of two new variables u and v: $y = u + v$. Replacing y with $u + v$ in our equation, we get:

$$(u + v)^3 + p(u + v) + q = 0$$

$$u^3 + 3u^2v + 3uv^2 + v^3 + p(u + v) + q = 0$$

$$u^3 + v^3 + 3uv(u + v) + p(u + v) + q = 0$$

$$u^3 + v^3 + (3uv + p)(u + v) + q = 0. \tag{1}$$

Since we introduced two variables u and v, we can impose one more condition on them so that the third term in the last equation vanishes. The condition is $3uv + p = 0$. It allows us to replace the equation with the following system of two equations for u and v:

$$\begin{cases} u^3 + v^3 + q = 0 \\ 3uv + p = 0. \end{cases}$$

We solve the second equation for uv: $uv = -\dfrac{p}{3}$. Then we raise this equation to the 3rd power: $u^3v^3 = -p^3/27$. Our system of equations becomes a system for u^3 and v^3:

$$\begin{cases} u^3 + v^3 = -q \\ u^3v^3 = -p^3 / 27. \end{cases}$$

Now we are given the sum and product of the unknown variables u^3 and v^3. Then by the Vieta's theorem[1], u^3 and v^3 are roots of the quadratic equation with coefficients q and $-\dfrac{p^3}{27}$ (we use variable t instead of x):

$$t^2 + qt - \frac{p^3}{27} = 0.$$

Solving this equation by the quadratic formula, we get u^3 and v^3. Then take cube roots and find u and v:

$$u = \sqrt[3]{-\frac{q}{2} + \sqrt{\frac{q^2}{4} + \frac{p^3}{27}}}, \quad v = \sqrt[3]{-\frac{q}{2} - \sqrt{\frac{q^2}{4} + \frac{p^3}{27}}}.$$

Since $y = u + v$, we arrive at **Cardano's** famous formula for the equation $y^3 + pv + q = 0$:

$$y = \sqrt[3]{-\frac{q}{2} + \sqrt{\frac{q^2}{4} + \frac{p^3}{27}}} + \sqrt[3]{-\frac{q}{2} - \sqrt{\frac{q^2}{4} + \frac{p^3}{27}}}. \quad \blacksquare$$

---

[1] Vieta's theorem says that if $u + v = a$ and $uv = b$, then u and v are roots of the equation $x^2 - ax + b = 0$. Indeed, replace a with $u + v$ and b with uv: $x^2 - (u + v)x + uv = 0$. The last equation can be factored: $(x - u)(x - v) = 0$, so u and v are its roots.

Having received the formula for the cubic equation, scientists proceeded to the equation of the 4th degree. Almost immediately, the desired formula was found. This was done by Cardano's assistant, an Italian mathematician Lodovico de Ferrari (1522–1565). The formula was obtained by reducing a 4th degree equation to cubic.

After attaining formulas in radicals for equations of 3rd and 4th degrees, scientists enthusiastically started searching for similar result for equations of 5th degree. The best mathematicians in the world tried to do this. But all their sophisticated attempts for almost three centuries turned out to be unsuccessful. Eventually, a suspicion began to creep in that such a formula does not exist at all. A breakthrough in this direction was made by a young Norwegian mathematician Niels Henrik Abel (1802–1829). He was able to prove that in the general case, such a formula does not exist. It was a great discovery made by him at the age of 22. Subsequently, Abel received many other important results in various fields of mathematics. However, despite his outstanding scientific achievements, he was in poverty his entire life. Sometimes he did not even have enough money for food. Abel died of tuberculosis at the age of 26. After about 50 years, when his name was recognized worldwide, Abel's portrait was placed on a Norwegian banknote. In Norway, a monument was erected to him. In 2002, in honor of Abel's 200th birthday, the Norwegian government instituted the Abel Prize in Mathematics — analogous to the Nobel Prize (there is no Nobel Prize for math).

After Abel's results, it became clear that Italian mathematicians, having solved the equations for 3rd and 4th degrees in the 16th century, reached the limit in this area. However, one question remained open. The fact is that in some special cases, it was possible to obtain formulas for the roots of equations of higher degrees. The question arose of how to establish whether the given equation can be solved in radicals or not.

This very difficult problem was solved by a French mathematician Evarist Galois (1811–1832) at the age of 19. His solution was so original and unusual that even the greatest mathematicians of that time did not understand him. To solve the problem, Galois introduced a completely new mathematical object called a group. Sadly, his life and career were prematurely interrupted. Galois was killed in a duel in 1832. There are many speculations about what caused the duel: his political activities, a woman, or something else.

Today the theory of groups is a powerful mathematical apparatus for studying various types of symmetry in nature. It became one of the main tools in modern physics.

Let us come back to Cardano's formula. Using it, we can obtain some interesting numerical identities. The following problem illustrates one of them.

Problem 19.3. Using the Cardano's formula, prove that $\sqrt[3]{2+\sqrt{5}} + \sqrt[3]{2-\sqrt{5}} = 1$.

Solution. Consider the equation: $x^3 + 3x - 4 = 0$. It is factored $(x - 1)(x^2 + x + 4) = 0$. One can verify that the discriminant of the second factor is negative. It means that this expression cannot be equal to zero for any real number, and the cubic equation has a unique real solution equal to 1. On the other hand, this solution can also be obtained by Cardano's formula: $x = \sqrt[3]{2+\sqrt{5}} + \sqrt[3]{2-\sqrt{5}}$. Therefore, $\sqrt[3]{2+\sqrt{5}} + \sqrt[3]{2-\sqrt{5}} = 1$. ■

Note that direct proof of this equality is complicated, since it is unclear how to get rid of cubic roots.

Even though Cardano's formula was a triumph of the human mind, one alarming fact remaining: this formula contains a **square root**. What happens if a negative number appears inside the root, but the equation contains only real solutions? Here is an example. Solve the following problem on your own.

Problem 19.4. Using Cardano's formula, prove that one of the roots of the equation

$$(x - 1)(x - 2)(x + 3) = 0$$

is expressed by the formula:

$$x = \sqrt[3]{-3+\frac{10}{9}\sqrt{-3}} + \sqrt[3]{-3-\frac{10}{9}\sqrt{-3}}. ■$$

This result looks rather strange. On the one hand, the initial form of the equation clearly shows that it has three real roots 1, 2, and −3 because they make each factor within the parentheses equal to zero. On the other hand, the negative number −3 appears under the square root in the resulting formula. The first thought might be that something is wrong with the formula. However, a thorough analysis showed that there was no mistake. Cardano himself was discouraged by this result. Nevertheless, he tried to perform actions similar to ordinary numbers on the roots of negative numbers. Eventually, he could get the real roots of the equations. However, he did not dare to develop this direction further and referred to such numbers as "sophisticated."

We can assume that an obstacle to the perception of complex numbers was the view of numbers as something that is used only to count or measure various quantities (sizes, speeds, etc.). Then, of course, the roots of negative numbers are not suitable for this. Note that for the same reason even negative numbers were rejected in those times. Mathematicians in Babylon (approximately 2000 BC) believed that if the result of subtraction of two numbers is negative, then it should be discarded as impossible, since "how can you count or measure less than nothing of something?" They believed that negative numbers are not related to anything in the real world. Moreover, even zero was considered nonsense because it expressed "nothing."

For the first time negative numbers were used in a Chinese mathematical book *The Nine Chapters on the Mathematical Art* in the 2nd century BC, and then from about the 7th century in India, where positive numbers were interpreted as profit (or property), and negative numbers as debts (or shortages). In Europe, for a long time negative numbers were called "false" or "absurd." Their first description in the European literature appeared in 1202 in a book entitled *Liber Abaci* by an Italian mathematician Leonardo of Pisa (c. 1170– c. 1240–50; best known by the nickname Fibonacci). He interpreted negative numbers as debt ("loss"). And, of course, square roots of negative numbers seemed even more absurd.

A decisive step in the application of imaginary numbers was made by another Italian mathematician, Rafael Bombelli (1526–1572). In his book *Algebra*, he outlined the theory of complex numbers. Bombelli's big guess was that the square roots of negative one in the formula for the solutions of cubic equations should be canceled out when they are summed up. Bombelli considered an equation similar to $(x - 1)(x - 2)(x + 3) = 0$. He called his approach a "wild thought." Bombelli assumed that the formulas for the roots could be represented as the sum of terms $u = a + b\sqrt{-1}$ and $v = a - b\sqrt{-1}$ with some coefficients a and b. Then, when adding, the imaginary parts (with $\sqrt{-1}$) should be canceled out. After quite laborious manipulations with $\sqrt{-1}$ as an ordinary number, he was able to calculate a and b: $a = -\frac{3}{2}$ and $b = 6$. So, he got:

$$u = \sqrt[3]{-3 + \frac{10}{9}\sqrt{-3}} = -\frac{3}{2} + \frac{\sqrt{-3}}{6}, \quad v = \sqrt[3]{-3 - \frac{10}{9}\sqrt{-3}} = -\frac{3}{2} - \frac{\sqrt{-3}}{6}.$$

These expressions for u and v could be verified by raising the right sides to the 3rd power. In doing so, we should use the usual rules of arithmetic operations with a single exception: we assume that the square of $\sqrt{-1}$ is -1. For instance, you can verify that:

$$u^3 = \left(-\frac{3}{2} + \sqrt{\frac{-3}{6}}\right)^3 = -3 + \frac{10}{9}\sqrt{-3}. \text{ Similarly, } v^3 = -3 - \frac{10}{9}\sqrt{-3}.$$

As we see, the expressions for u and v make sense, despite the fact that they contain the square root of a negative number. One root of the equation $(x - 1)(x - 2)(x + 3) = 0$ is

$$x = u + v = -\frac{3}{2} + \frac{\sqrt{-3}}{6} - \frac{3}{2} - \frac{\sqrt{-3}}{6} = -3.$$

The idea that ordinary numbers could be generated by imaginary numbers according to the usual rules for arithmetic operations became the basis for the introduction of one of the most important mathematical objects — complex numbers. Subsequently, many mathematicians made a significant contribution to the development of complex numbers. We will mention only two of them: Leonhard Euler and Carl Gauss.

Swiss mathematician Leonhard Euler (1707–1783), who worked for many years in The Academy of Sciences at Saint Petersburg in Russia, significantly developed the theory of complex numbers, and also made a fundamental contribution to the development of many other branches of mathematics, physics, astronomy, and a number of applied sciences. Among other things, he introduced the modern notation for the trigonometric functions, as well as the letter i to denote the imaginary unit, taking the first letter of the Latin word *imaginarius*. The following fact emphasizes the extraordinary skills of Euler. In 1735 the Russian Academy of Sciences was given the task of performing an urgent and very cumbersome mathematical calculation. A group of academics asked for 3 months for this work, but Euler was able to do the work alone in 3 days. However, probably due to overdoing, he lost sight in his right eye. However, Euler himself attributed the loss of eyesight to his mapping work in the Academy's geographic department. Surprisingly, after vision loss Euler's productivity not only did not decrease but even increased. He explained this by the fact that he began to be less distracted from math. Euler owns the famous formula, which establishes the fundamental relationship between the trigonometric functions and the complex exponential function:

$$e^{-ix} = \cos x + i \sin x.$$

Here, e is a constant number, the so-called base of natural logarithms (it is related to exponential growth). This number is denoted by the letter e in Euler's honor. The approximate value of e is 2.718. In the entire field of mathematics, perhaps there are two most important real numbers: π and e. If we substitute π for x in the above equation, it becomes:

$$e^{-i\pi} = -1.$$

This equation connects four remarkable numbers: π, -1, i, and e. In 1988 a mathematics journal, *The Mathematical Intelligencer*, conducted a survey asking about the most beautiful theorem in mathematics. The winner was the above equation. Another survey conducted by "Physics World" magazine in 2004 also asked a similar question of physicists. Euler's equation came second after Maxwell's equations.

German mathematician Carl Friedrich Gauss (1777–1855) proved the fundamental theorem of algebra, which says that the number of roots (including complex numbers) of any algebraic equation is equal to its degree. For his numerous achievements, Gauss became known as the Prince of Mathematics. His phenomenal talent was evident in early childhood. As one story is often told, his school teacher needed to leave the classroom.

To keep the children busy, he gave them a time-consuming problem: calculate the sum of numbers from one to one hundred. Before the teacher could leave the class, the young Gauss had already given him the answer.

Pretend to be little Gauss and solve the following problem on your own.

Problem 19.5. Mentally calculate the sum of all (natural) numbers from 1 to 100. ■

In the history of mathematics, there were cases when scientists were afraid to express their revolutionary ideas. Thus, for example, Carl Gauss came to the discovery of non-Euclidean geometry, which denied the axiom that only one straight line parallel to a given line could be drawn through a point on the plane. But he did not publish his discovery. Gauss believed that if he published this idea, he would look like a madman, and thus his high authority could be undermined. However, there were young mathematicians who were not afraid of this: a Hungarian mathematician Janos Bolyai (1802–1860) and a Russian Nikolai Lobachevsky (1792–1856), as they had nothing to lose. And the glory of this remarkable discovery went to them.

In the next chapter, we will discuss in detail the appearance of complex numbers due to adding of the square root of a negative unit to the set of real numbers. Below we consider an equation that reduces to an equation of degree 4. The solution does not use Ferrari formulas, but it is so beautiful that we decided to present it here. The idea is to treat a particular constant number as a variable. It sounds unusual, doesn't it?

Problem 19.6. Solve the equation: $x^2 - 5 = \sqrt{x+5}$.

Solution. The first few steps are standard. Square both sides: $(x^2 - 5)^2 = x + 5$.

Next, expand the parentheses and move all the terms from the right side to the left. We get the equation:

$$x^4 - 10x^2 + 25 - x - 5 = 0.$$

Such a 4th degree equation is not so easy to solve. We look at it as a quadratic equation for the number 5. Such a view seems, of course, unusual. To make this clearer, we denote number 5 by some symbol, for instance, s. In other words, we replace 5 with s, and rewrite the equation, replacing 25 with s^2, and 10 with $2s$. We can get the same equation by replacing 5 with s in the initial equation.

$$x^4 - 2sx^2 + s^2 - x - s = 0.$$

Now, look at this equation as a quadratic for s. Combine like terms:

$$s^2 - (2x^2 + 1)s + (x^4 - x) = 0.$$

Solve this equation by the quadratic formula:

$s_{1,2} = \dfrac{2x^2 + 1 \pm \sqrt{D}}{2}$, where $D = (2x^2 + 1)^2 - 4(x^4 - x) = 4x^2 + 4x + 1 = (2x + 1)^2$.

From here $s_{1,2} = \frac{1}{2}[2x^2 + 1 \pm (2x + 1)]$. Therefore, $s_1 = x^2 + x + 1$ and $s_2 = x^2 - x$.

Now recall that $s = 5$. We replace s_1 and s_2 with 5 and consider two quadratic equations: $x^2 + x + 1 = 5$ and $x^2 - x = 5$. These equations are easily solved by quadratic formulas. We leave the rest of the solution to the reader. ∎

Entertainment Problem

We believe that young Gauss would have been able to solve the problem below. You should also try solving it without a calculator (in your head).

Problem E19.1. Below is a copy of the famous painting by a Russian artist Nikolay Bogdanov-Belsky (1868–1945). The painting is called "Mental Calculation. In Public School of S. A. Rachinsky." It is on display at the Tretyakov Gallery in Moscow, Russia.

The problem presented on the blackboard requires to mentally compute the following expression:

$$\frac{10^2 + 11^2 + 12^2 + 13^2 + 14^2}{365}.$$

Hint: You can use the formula $(a \pm b)^2 = a^2 \pm 2ab + b^2$, and take $a = 12$, $b = \pm1, \pm2$.

Chapter 20

Definition and Properties of Complex Numbers

In the previous chapter, we described some stories about a new element (number) i, called the **imaginary unit**, with the property $i^2 = -1$. It is also often written as $i = \sqrt{-1}$. A simple addition of this single element to the set of real numbers expands this set, as well as our ability to solve math problems. It opens up a new and wonderful "complex world." As you will see below, in this way we transfer from a line where we can only move left-right to a plane where we have much more freedom to move.

Standard (Rectangular) Form of Complex Numbers

Since we want to interpret this new element i as a number, we need to define the basic arithmetic operations: addition, subtraction, multiplication, and division, which can be performed together with it and real numbers.

Let's start with raising number i to a positive integer power. We already know how to multiply i by itself: by definition, $i \cdot i = i^2 = -1$.

Problem 20.1. Calculate

(1) i^3; (2) i^4

Solution.

(1) $i^3 = i^2 \, i = (-1) \, i = -i$.
(2) $i^4 = i^3 \, i = (-i) \, i = -(i^2) = -(-1) = 1$. ∎

Let's see how we can calculate i^n for an arbitrary natural number n. Note that $(-1)^k = 1$ if k is even, and $(-1)^k = -1$ if k is odd. To calculate i^n, separately consider cases of even and odd n.

(1) If n is even, it can be written as $n = 2k$, where k is an integer. Then $i^n = i^{2k} = (i^2)^k = (-1)^k$. From here, i^n equals to 1 or -1, depending on whether k is even or odd. For even k, $i^n = 1$, and for odd k, $i^n = -1$.

(2) If n is odd, it can be written as $n = 2k + 1$, where k is an integer. Then $i^n = i^{2k+1} = i^{2k}{\cdot}i = (i^2)^k{\cdot}i = (-1)^k{\cdot}i$. From here, i^n equals to either i or $-i$, depending on whether k is even or odd. For even k, $i^n = i$, and for odd k, $i^n = -i$.

Solve the following problem on your own.

Problem 20.2. Calculate

(a) i^{100}; (b) i^{150}; (c) i^{27}; (d) i^{37}. ■

Now we determine how to multiply and add a real number a to an imaginary i. To multiply a and i, we simply write the result in the form $a{\cdot}i$ (or ai), and nothing more. To add a and i, we act in the same way: we just write the result as $a + i$. Continuing these operations, we write the sum of a and bi in the form $a + bi$, where a and b are real numbers. Expression of the form $a + bi$ we call a **complex number**. We also say that this number is written in **standard form** (or **rectangular form**). The number a is called the **real part**, and b is the **imaginary part** of the complex number $a + bi$. The following are examples of complex numbers:

$$i, 5i, 1 + i, 1 - 3i, 2 + 4i, 3 - 2i.$$

If the imaginary part $b = 0$, we have $b \cdot i = 0 \cdot i = 0$. Therefore, we can treat the complex number $a + 0i$ as a real number a. In other words, any real number can be interpreted as a special case of a complex number, in which the imaginary part is equal to zero. So, the set of complex numbers can be considered as an extension of the set of real numbers. If the real part $a = 0$, then the complex number takes the form bi, which is called an **imaginary number** (or purely imaginary number). For example, $5i$ is the imaginary number. If $a = b = 0$, then the complex number is our usual 0. Also, two complex numbers are equal if their real and imaginary parts are equal, respectively.

Complex numbers are added or subtracted in an obvious way, which corresponds to our intuition: to perform these operations, we simply add or subtract the real and imaginary parts separately.

Problem 20.3. (a) Add $(3 + 2i)$ and $(1 + 5i)$; (b) Subtract $(4 - 7i)$ from $(6 + 3i)$.

Solution: (a) $(3 + 2i) + (1 + 5i) = (3 + 1) + (2 + 5)i = 4 + 7i$.
     (b) $(6 + 3i) - (4 - 7i) = (6 - 4) + [(3 - (-7)]i = 2 + 10i$. ∎

Now consider multiplication of complex numbers. They can be multiplied in the same way as regular binomials (distribute and combine like terms) with the only additional property $i^2 = -1$.

Problem 20.4. Multiply complex numbers $z_1 = 3 + 2i$ and $z_2 = 1 - 5i$.

Solution.

$$z_1 z_2 = (3 + 2i)(1 - 5i) = 3 - 15i + 2i - 10i^2$$
$$= 3 - 10i^2 - 13i = 3 - 10(-1) - 13i$$
$$= 3 + 10 - 13i = 13 - 13i. ∎$$

Solve the following problem on your own.

Problem 20.5. Derive a general formula for the product of two complex numbers

$$z_1 = a + bi \text{ and } z_2 = c + di. ∎$$

In some cases, for the complex number $z = a + bi$ it is useful to consider its "relative" — complex number $\overline{z} = a - bi$. Numbers z and $\overline{z}$ are called **complex conjugate** to each other. The important property is that their product is a real number.

Solve the following problem on your own.

Problem 20.6. Let $z = a + bi$ and $\overline{z} = a - bi$. Prove that $z\overline{z} = a^2 + b^2$. ∎

Conjugate numbers help to divide complex numbers.

Problem 20.7. Divide $(5 + 4i)$ by $(3 - 2i)$.

Solution. At first glance, the quotient $\dfrac{5+4i}{3-2i}$ does not look like a single complex number in the standard form $a + bi$. Here is a way to get it: multiply both sides of the fraction by a complex number which is conjugate to the denominator, so multiply by $3 + 2i$.

$$\frac{5+4i}{3-2i} = \frac{(5+4i)(3+2i)}{(3-2i)(3+2i)} = \frac{15+10i+12i+8i^2}{9+4} = \frac{(15-8)+(10+12)i}{13}$$

$$= \frac{7+22i}{13} = \frac{7}{13} + \frac{22}{13}i. \blacksquare$$

Solve the following problem on your own.

Problem 20.8. Derive a general formula for division of two complex numbers

$$z_1 = a + bi \quad \text{and} \quad z_2 = c + di. \blacksquare$$

Geometric Interpretation of Complex Numbers

As we know, real numbers can be "materialized" (represented geometrically) as points on a number line. Since the complex number $z = a + bi$ is determined by a pair of real numbers (a, b), complex numbers can also be materialized. Namely, they can be represented as points on a plane in the Cartesian coordinate system. In it, the real part a is the abscissa (x-coordinate), and the imaginary part b is the ordinate (y-coordinate). Passing from real numbers to complex numbers, we are freed from the framework of the straight line to the freedom of the plane. For example, complex numbers

$$z_1 = 4i, z_2 = 4 + 3i, z_3 = -5 + 2i, z_4 = 3 - i$$

are these points:

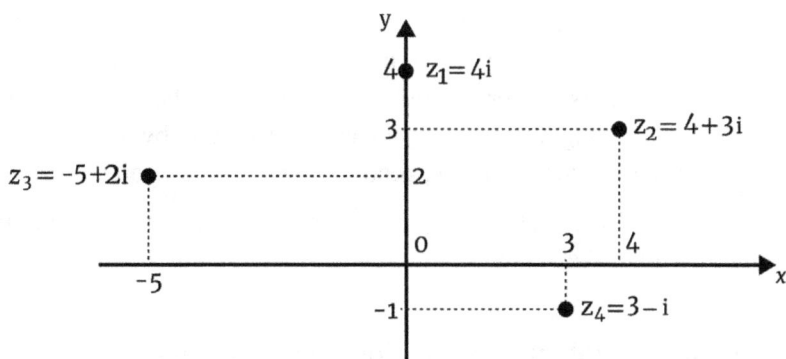

The horizontal x-axis is called the real axis, and the vertical y-axis is called the imaginary axis. Accordingly, real numbers lie on the real axis, and imaginary numbers lie on the imaginary axis. We call the entire plane a **complex plane**. The complex number $z = a + bi$ can also be interpreted on the complex plane not only as a point, but also as a vector:

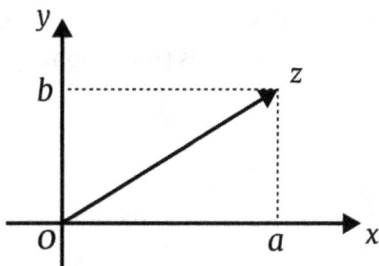

With this interpretation, the addition and subtraction of the complex numbers z_1 and z_2 are represented as operations with vectors. This is illustrated in the following figures:

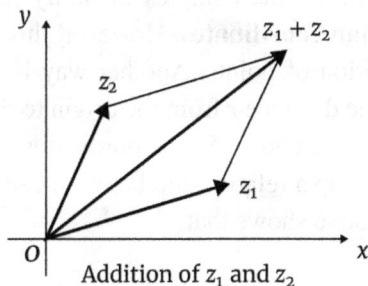

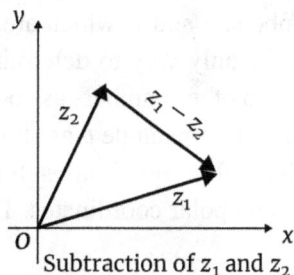

Addition of z_1 and z_2

Subtraction of z_1 and z_2

Multiplying a vector by a complex number causes the vector to rotate about the origin (and possibly change its length). For example, if $z = a + bi$, then $iz = -b + ai$. As vectors, z and iz are perpendicular to each other. Therefore, multiplying a vector z by i causes z to rotate by a right angle. The length of the vector $z = a + bi$, according to the Pythagorean theorem is $\sqrt{a^2 + b^2}$ and it is denoted by $|z|$ or $\|z\|$. This number is called the **magnitude** or **norm** of the vector.

Trigonometric (Polar) Form of Complex Numbers

We have seen that if complex numbers are written in a standard form, it is very easy to add or subtract them. Multiplication and division require a bit more work, but not too much. But what if we need to raise a complex number to some power? For example, how do you calculate $(1 + i)^{10}$? Here we develop a method of how to solve these kinds of problems.

Let's reconsider the representation of complex numbers as points on the complex plane. The complex number $z = x + yi$ can be represented as a point in the coordinate system like this (for simplicity, we consider a point in the 1st quadrant):

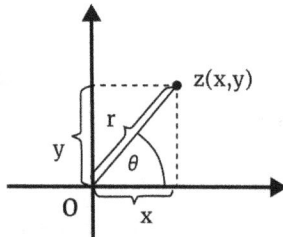

We described the position of a point on the complex plane by two numbers: x and y, which are the **cartesian coordinates**. However, this is not the only way to determine the position of points. Another way is to use two other numbers as coordinates: the distance r from the origin to the point, and the angle θ as shown in the figure above. Such coordinates are called **polar coordinates**. It is easy to set up a relationship between cartesian and polar coordinates. The figure above shows that

$$x = r \cos \theta$$
$$y = r \sin \theta. \tag{1}$$

Thus, the complex number $z = x + yi$ takes the form:

$$z = r(\cos \theta + i \sin \theta).$$

It is called the **trigonometric** or **polar form**. As we already mentioned, the distance r is defined by the formula $r = \sqrt{x^2 + y^2}$. It is called the **modulus** or **absolute value** of the complex number and is denoted by $|z|$.

Angle θ is called the **argument** of a complex number. It is defined for all complex numbers except zero. Angle θ satisfies the equation: $\tan \theta = \frac{y}{x}$. To get this equation, just divide the second expression (1) by the first.

Warning: Complex number $z = r(\sin \theta + i \cos \theta)$ is **not** in trigonometric form.

Problem 20.9. Present $z = \sin \theta + i \cos \theta$ in the trigonometric form.

Solution. We need to convert sine to cosine, and vice versa. To do this, we use the reduction formulas: $z = \sin \theta + i \cos \theta = \cos\left(\dfrac{\pi}{2} - \theta\right) + i \sin\left(\dfrac{\pi}{2} - \theta\right)$. ■

The argument θ is not uniquely defined, since the equation $\tan \theta = \frac{y}{x}$ has infinitely many solutions. Not all solutions to this equation are valid for the argument θ. In some cases, we can take $\theta = \tan^{-1}(y/x)$, in some other cases, $\theta = \tan^{-1}(y/x) + \pi$. To get the correct value, we need to identify the quadrant in which point (x, y), corresponding to the complex number $z = x + yi$, is located.

Problem 20.10. Convert to polar form complex numbers shown below. In parts (a) and (b), use radians and round the argument to the nearest hundredth. In parts (c) – (f), use degrees.
(a) $z_1 = 2 + 3i$, (b) $z_2 = -3 + 4i$, (c) $z_3 = -\sqrt{3} - i$, (d) $z_4 = 2 - 2i$, (e) $z_5 = -3i$, (f) $z_6 = -5$.

Solution.
(a) $z_1 = 2 + 3i$. $r = \sqrt{2^2 + 3^2} = \sqrt{13}$. Point $(2, 3)$ is in the 1st quadrant, and we can use the inverse tangent, since its values are in the 1st and 4th quadrants. $\theta = \tan^{-1}3/2 \approx 0.98$.
Answer: $z_1 = \sqrt{13}(\cos 0.98 + i \sin 0.98)$.

(b) $z_2 = -3 + 4i$. $r = \sqrt{(-3)^2 + 4^2} = \sqrt{25} = 5$. Point $(-3, 4)$ is in the 2nd quadrant, which is not in the range of inverse tangent. Therefore, we take $\theta = \tan^{-1}(4/-3) + \pi \approx -0.93 + 3.14 = 2.21$.

Answer: $z_2 = 5(\cos 2.21 + i \sin 2.21)$.

(c) $z_3 = -\sqrt{3} - i$. $r = \sqrt{(-\sqrt{3})^2 + (-1)^2} = \sqrt{4} = 2$. Point $(-\sqrt{3}, -1)$ is in the 3rd quadrant, which is not in the range of inverse tangent. Therefore, we take $\theta = \tan^{-1}(-1/-\sqrt{3}) + 180° = 30° + 180° = 210°$.

Answer: $z_3 = 2(\cos 210° + i \sin 210°)$.

(d) $z_4 = 2 - 2i$. $r = \sqrt{2^2 + (-2)^2} = \sqrt{8} = 2\sqrt{2}$. Point $(2, -2)$ is in the 4th quadrant, and we can use the inverse tangent, since its values are in the 1st and 4th quadrants. $\theta = \tan^{-1}(-2/2) = -45°$.

Answer: $z_4 = 2\sqrt{2}[\cos(-45°) + i \sin(-45°)]$.

(e) $z_5 = -3i$. $r = \sqrt{0^2 + (-3)^2} = \sqrt{9} = 3$. Point $(0, -3)$ lies on the vertical axis below the horizontal axis. Such a point corresponds to angles $270°$ or $-90°$. Tangent for these angles does not exist. However, for polar form, we need sine and cosine. We can take any of these angles. Let's take the angle of $-90°$.

Answer: $z_5 = 3[\cos(-90°) + i \sin(-90°)]$.

(f) $z_6 = -5$. $r = \sqrt{(-5)^2 + 0^2} = \sqrt{25} = 5$. Point $(-5, 0)$ lies on the horizontal axis to the left of the vertical axis. Corresponding angles are $180°$ or $-180°$. We take angle $180°$.

Answer: $z_6 = 5(\cos 180° + i \sin 180°)$. ∎

It may seem that the trigonometric form does not give anything new, and only complicates the notation of a complex number. However, it is not so. The first thing we will see is a surprisingly significant simplification of the multiplication of complex numbers.

Problem 20.11. Let $z_1 = r_1 (\cos \theta_1 + i \sin \theta_1)$ and $z_2 = r_2 (\cos \theta_2 + i \sin \theta_2)$. Prove that

$$z_1 z_2 = r_1 r_2 [\cos(\theta_1 + \theta_2) + i \sin(\theta_1 + \theta_2)].$$

Solution. We have,

$$z_1 z_2 = r_1 (\cos \theta_1 + i \sin \theta_1) r_2 (\cos \theta_2 + i \sin \theta_2)$$

$$= r_1 r_2 (\cos\theta_1 \cos\theta_2 + i \cos\theta_1 \sin\theta_2 + i \sin\theta_1 \cos\theta_2 - \sin\theta_1 \sin\theta_2)$$

$$= r_1 r_2 [(\cos\theta_1 \cos\theta_2 - \sin\theta_1 \sin\theta_2) + i (\sin\theta_1 \cos\theta_2 + \sin\theta_1 \cos\theta_2)]$$

$$= [\text{use addition formulas}] = r_1\, r_2\, [\cos(\theta_1 + \theta_2) + i \sin(\theta_1 + \theta_2)]. \blacksquare$$

The resulting formula gives a simple recipe for multiplying complex numbers in polar form. Namely, to multiply two complex numbers, we just multiply their moduli and add up the arguments. As we can see, multiplying complex numbers becomes much easier if the numbers are written in polar form rather than the standard form.

Problem 20.12. Multiply

$$z_1 = 2\left(\cos\frac{2\pi}{5} + i \sin\frac{2\pi}{5}\right) \text{ and } z_2 = 3\left(\cos\frac{3\pi}{5} + i \sin\frac{3\pi}{5}\right).$$

Solution.

$$z_1 z_2 = 2\left(\cos\frac{2\pi}{5} + i \sin\frac{2\pi}{5}\right) 3\left(\cos\frac{3\pi}{5} + i \sin\frac{3\pi}{5}\right)$$

$$= 6\left[\cos\left(\frac{2\pi}{5} + \frac{3\pi}{5}\right) + i \sin\left(\frac{2\pi}{5} + \frac{3\pi}{5}\right)\right] = 6(\cos\pi + \sin\pi) = -6. \blacksquare$$

Solve the problem below on your own.

Problem 20.13. Prove that

$$\boxed{(\cos\theta + i \sin\theta)^n = \cos n\theta + i \sin n\theta,}$$

where n is any natural number. $\blacksquare$

The above formula is called **de Moivre's formula**. It was discovered by an English mathematician of French origin Abraham de Moivre (1667–1754). Here is a short story about him. At a young age, he moved from France to London, where he lived all his life. Moivre soon became known as a talented mathematician. Unfortunately, in those days, as a foreigner, he had no chance of getting a position at the university and was forced to earn his whole life as a private tutor of mathematics. In London, Moivre met Isaac Newton (1642–1726/27) and became his assistant. Newton extremely appreciated Moivre. According to gossip of the time, Newton told visitors who distracted him with petty mathematical things, using the following phrase: "Go to Moivre, he knows this better than I."

Moivre, as Newton's closest friend, participated on his behalf in a Commission dealing with a fierce debate between Newton and a German mathematician Gottfried Leibniz (1646–1716) over who had first authored the fundamental theorem of calculus. This theorem establishes a relationship between differentiation and integration. The Commission concluded that the authorship should belong to Newton. However, over time, objectivity triumphed, and it is believed that Newton and Leibniz came to their discoveries independently of each other. Therefore, their famous theorem is written through a conciliation dash as the Newton–Leibniz theorem.

Moivre himself made a significant contribution to the development of trigonometry and probability theory. There is a legend according to which Moivre was able to accurately predict the day of his own death. He found that the duration of his sleep began to increase in arithmetic progression. He concluded that he would die when the dream reached 24 hours. Based on this, he calculated the date of death, and really died on that day, November 27, 1754.

Let's get back to math and apply de Moivre's formula to the following problem.

Problem 20.14. Calculate $(1 + i)^{10}$.

Solution. It would be tedious to calculate it directly using the standard form. Let's convert the number in parentheses to polar form, and then use de Moivre's formula. For complex number $1+i$, $r = \sqrt{1^2 + 1^2} = \sqrt{2}$. Point (1,1) is in the 1st quadrant, and $\theta = \tan^{-1}(1/1) = 45°$. Therefore, $1+i = \sqrt{2}(\cos 45° + i \sin 45°)$. Using de Moivre's formula,

$$(1+i)^{10} = (\sqrt{2})^{10}[\cos(10 \cdot 45°) + i \sin(10 \cdot 45°)]$$
$$= 2^5(\cos 450° + i \sin 450°) = 32[\cos(90° + 360°) + i \sin(90° + 360°)]$$
$$= 32(\cos 90° + i \sin 90°) = 32i. \quad \blacksquare$$

As you can see, the original problem and the answer to it do not contain any trig functions. They participate here as a tool that is used in intermediate calculations, and then disappear.

de Moivre's formula also allows us to express $\sin n\theta$ and $\cos n\theta$ through $\sin \theta$ and $\cos \theta$. To do this, it is enough to clear the parenthesis on the left part of the formula and combine like terms.

Problem 20.15. Express $\sin 3\theta$ and $\cos 3\theta$ through $\sin \theta$ and $\cos \theta$.

Solution. We will use the following cube of the sum formula:

$$(a + b)^3 = a^3 + 3a^2b + 3ab^2 + b^3.$$

Let $z = \cos \theta + i \sin \theta$. Then

(1) By de Moivre's formula: $z^3 = (\cos \theta + i \sin \theta)^3 = \cos 3\theta + i \sin 3\theta$.
(2) By the cube of sum formula: $z^3 = (\cos \theta + i \sin \theta)^3$

$$\begin{aligned}
&= \cos^3 \theta + i\, 3 \cos^2 \theta \sin \theta + i^2\, 3 \cos \theta \sin^2 \theta + i^3 \sin^3 \theta \\
&= \cos^3 \theta + i\, 3 \cos^2 \theta \sin \theta - 3 \cos \theta \sin^2 \theta - i \sin^3 \theta \\
&= (\cos^3 \theta - 3 \cos \theta \sin^2 \theta) + i(3 \cos^2 \theta \sin \theta - \sin^3 \theta).
\end{aligned}$$

Using expressions (1) and (2) for z^3, separately equate real and imaginary parts:

$$\begin{aligned}
\cos 3\theta &= \cos^3 \theta - 3 \cos \theta \sin^2 \theta = \cos^3 \theta - 3 \cos \theta\, (1 - \cos^2 \theta) \\
&= 4 \cos^3 \theta - 3 \cos \theta. \\
\sin 3\theta &= 3 \cos^2 \theta \sin \theta - \sin^3 \theta = 3(1 - \sin^2 \theta) \sin \theta - \sin^3 \theta \\
&= 3 \sin \theta - 4 \sin^3 \theta.
\end{aligned}$$

So,

$$\begin{aligned}
\cos 3\theta &= 4 \cos^3 \theta - 3 \cos \theta, \\
\sin 3\theta &= 3 \sin \theta - 4 \sin^3 \theta. \quad \blacksquare
\end{aligned}$$

Complex numbers participated here as tools: they are used to derive the formulas and are not present in the final results.

Solve problems below on your own.

Problem 20.16. Prove that the number z^{-1}, the inverse to $z = r (\cos \theta + i \sin \theta)$, is determined by the formula:

$$z^{-1} = r^{-1}[\cos(-\theta) + i \sin(-\theta)]. \blacksquare$$

Problem 20.17. Let $z_1 = r_1 (\cos \theta_1 + i \sin \theta_1)$ and $z_2 = r_2(\cos \theta_2 + i \sin \theta_2)$, $z_2 \neq 0$. Prove that

$$\frac{z_1}{z_2} = \frac{r_1}{r_2}[\cos(\theta_1 - \theta_2) + i \sin(\theta_1 - \theta_2)]. \blacksquare$$

This formula, which is similar to that in Problem 20.11, shows a simple way to divide complex numbers in polar form: just divide their moduli and subtract the arguments.

Roots of Unity

In this section we discuss the expression $\sqrt[n]{1}$. You may be wondering "what's the big deal: the answer is 1." However, this is true if we only use real numbers. For complex numbers, the situation is different and more interesting. First of all, let's formulate the problem more accurately. We are interested in solving the equation $z^n = 1$ in the set of complex numbers and natural n. For real numbers, we may have one or two solutions. For example, the equation $z^2 = 1$ has two solutions $z = \pm 1$, while the equation $z^3 = 1$ has only one $z = 1$. To solve the equation $z^n = 1$ for an arbitrary natural n in the set of complex numbers, we will use the polar form and de Moivre's formula. Solutions of this equation are called **de Moivre numbers**.

Problem 20.18. Solve the equation $z^n = 1$ for complex numbers.

Solution. We represent the unknown z in the polar form: $z = r (\cos \theta + i \sin \theta)$. Then by de Moivre's formula, $z^n = r^n(\cos n\theta + i \sin n\theta)$. Since $z^n = 1$, we have $r^n(\cos n\theta + i \sin n\theta) = 1 = 1 + 0i$. Equating the real and imaginary parts, we get $r^n \cos n\theta = 1$ and $r^n \sin n\theta = 0$. From the last equation we obtain $n\theta = m\pi$, where m is an arbitrary integer. From the equation $r^n \cos n\theta = 1$ it follows that m must be even number (otherwise, $\cos n\theta = \cos m\pi = -1$). So, m = 2k, $n\theta = 2k\pi$ and $\theta = \dfrac{2\pi k}{n}$. For each k, we can represent the solution z_k of the equation $z^n = 1$ in the form

$$z_k = \cos\theta + i\sin\theta = \cos\frac{2\pi k}{n} + i\sin\frac{2\pi k}{n}.$$

It might seem that there are infinitely many solutions z_k, since k is an arbitrary integer. However, this is not so. Let's take a closer look at the number of different complex numbers z_k that are in the above formula. To see this, let's list angles θ for different values of parameter k. We will denote them as

$$\theta_k = \frac{2\pi k}{n}, \ k = 0, 1, 2, \dots.$$

We have

$$k = 0: \ \theta_0 = 0.$$

$$k = 1: \ \theta_1 = \frac{2\pi}{n}.$$

$$k = 2: \ \theta_2 = \frac{4\pi}{n}.$$

$$\dots\dots\dots\dots\dots\dots\dots\dots$$

$$k = n-1: \ \theta_{n-1} = \frac{2\pi(n-1)}{n}.$$

$$k = n: \ \theta_n = \frac{2\pi n}{n} = 2\pi = \theta_0 + 2\pi.$$

$$k = n+1: \ \theta_{n+1} = \frac{2\pi(n+1)}{n} = \frac{2\pi}{n} + 2\pi = \theta_1 + 2\pi.$$

As we can see, starting with $k = n$, angles $\theta_n = \theta_0 + 2\pi$, $\theta_{n+1} = \theta_1 + 2\pi$, and so on. Since angles in pairs (θ_n, θ_0), (θ_{n+1}, θ_1), ... differ by 2π, then sines of angles in each pair (as well as cosines) are the same. Therefore, starting from $k = n$, solutions z_k are repeated. Finally, we get the following result:

Equation $z^n = 1$ has only n roots z_k $(k = 0, 1, 2, \ldots, k - 1)$, which are defined by the formula:

$$z_k = \cos\frac{2\pi k}{n} + i\sin\frac{2\pi k}{n}. \ \blacksquare$$

Roots z_k have an interesting geometric interpretation. Since the modulus of z_k is equal to 1, all z_k numbers are represented as points on the unit circle in the complex plane. Here is a more detailed description.

Problem 20.19. Show that on the unit circle in the complex plane, roots to the equation $z^n = 1$ are located at the vertices of the regular n-gon, starting from the vertex $(1, 0)$.

Solution. The first root z_0 is equal to 1, and the corresponding point is $(1, 0)$. The next root z_1 is defined by the formula: $z_1 = \cos\frac{2\pi}{n} + \sin\frac{2\pi}{n}$. This root, called the primitive n-th root, is located on the unit circle at angle $\frac{2\pi}{n}$ to the horizontal x-axis. To understand the location of other roots, let's write together roots z_k and z_{k+1}:

$$z_k = \cos\frac{2\pi k}{n} + i\sin\frac{2\pi k}{n},$$

$$z_{k+1} = \cos\frac{2\pi(k+1)}{n} + \sin\frac{2\pi(k+1)}{n} = \cos\left(\frac{2\pi k}{n} + \frac{2\pi}{n}\right) + i\sin\left(\frac{2\pi k}{n} + \frac{2\pi}{n}\right).$$

Comparing both, we can conclude that their angles differ by $\frac{2\pi}{n}$. Therefore, in order to move from the root z_k to the next z_{k+1}, we rotate the point for z_k along the unit circle counterclockwise by an angle $\frac{2\pi}{n}$. It means that all points corresponding to root z_k are at the vertices of the regular n-gon. $\blacksquare$

For illustration, here are the figures for the cases $n = 3$ and $n = 4$, i.e., for roots to the equations $z^3 = 1$ and $z^4 = 1$:

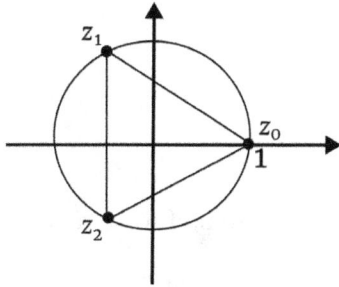

roots of $z^3 = 1$

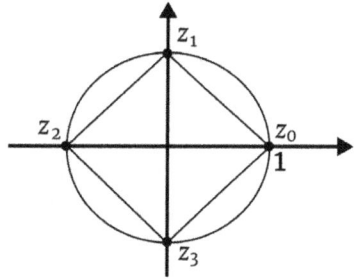

roots of $z^4 = 1$

For $n = 3$, the roots are $z_0 = 1$, $z_1 = -\frac{1}{2} + \frac{\sqrt{3}}{2}i$, $z_2 = -\frac{1}{2} - \frac{\sqrt{3}}{2}i$.

For $n = 4$, the roots are $z_0 = 1$, $z_1 = i$, $z_2 = -1$, $z_3 = -i$.

Problem 20.20. Prove that the sum of all roots to the equation $z^n = 1$ for $n > 1$ is equal to zero.

Solution. This result follows from one of the formulas of the Vieta's theorem. This theorem, in particular, asserts that if $z_1, z_2, \dots, z_n$ are roots (not necessarily distinct) of the polynomial $P(z) = z^n + a_1 z^{n-1} + \cdots + a_n$, then $z_1 + z_2 + \cdots + z_n = -a_1$.[1] Let's write the equation $z^n = 1$ in the form $z^n - 1 = 0$, and consider the polynomial $P(z) = z^n - 1$. Then the coefficient for z^{n-1} is zero, and this is the sum of all the roots of the given equation. ∎

The above problem can be used to prove the following trig identities.

Problem 20.21. Let n be any natural number greater than 1. Prove that

(a) $\sin\dfrac{2\pi}{n} + \sin\dfrac{4\pi}{n} + \sin\dfrac{6\pi}{n} + \cdots + \sin\dfrac{2(n-1)\pi}{n} = 0.$

(b) $\cos\dfrac{2\pi}{n} + \cos\dfrac{4\pi}{n} + \cos\dfrac{6\pi}{n} + \cdots + \cos\dfrac{2(n-1)\pi}{n} = -1.$

---

[1] Here is the proof: polynomial $P(z)$ can be factored like this: $P(z) = (z - z_1)(z - z_2)\cdots(z - z_n)$. By distributing and combining like terms, we get that the coefficient for z^{n-1} is $-(z_1 + z_2 + \cdots + z_n)$, which is a_1.

Solution. According to the results of Problem 20.19, the roots of the equation $z^n = 1$ are

$$1, \ \cos\frac{2\pi}{n} + i\sin\frac{2\pi}{n}, \ \cos\frac{4\pi}{n} + i\sin\frac{4\pi}{n}, \ \cos\frac{6\pi}{n} + i\sin\frac{6\pi}{n}, \ldots,$$

$$\cos\frac{2(n-1)\pi}{n} + i\sin\frac{2(n-1)\pi}{n}.$$

According to the results of Problem 20.20, the sum of these roots is zero. Separating the real and imaginary parts, we have

$$1 + \cos\frac{2\pi}{n} + \cos\frac{4\pi}{n} + \cos\frac{6\pi}{n} + \cdots + \cos\frac{2(n-1)\pi}{n} = 0, \quad \text{and}$$

$$\sin\frac{2\pi}{n} + \sin\frac{4\pi}{n} + \sin\frac{6\pi}{n} + \cdots + \sin\frac{2(n-1)\pi}{n} = 0. \ \blacksquare$$

In conclusion, we'd like to mention a few advanced topics related to the application of complex numbers. They are used in many fields of science and technology, such as mechanics, quantum physics, electrical engineering, aerodynamics, the flow of fluid around objects and much more. Here we give just one illustrative example. To study the lift of an airplane wing, a function of a complex variable with complex values is used, named after a Russian scientist Nikolay Zhukovsky (1847–1921). Zhukovsky's function (also called the Joukowsky transformation) is defined by the formula:

$$f(z) = \frac{1}{2}\left(z + \frac{1}{z}\right).$$

Its use in aerodynamics is based on the fact that Zhukovsky's function transforms a circle into a curve, which corresponds to the shape of an airplane wing:

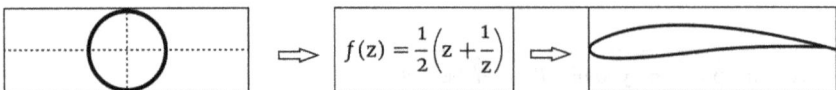

Zhukovsky's function transforms a circle on a complex plane into a shape of an airplane wing

By changing the radius and position of the circle relative to its center, it is possible to determine from this function the angle of bending and the thickness of the wing. Based on these data, conclusions can be drawn about the lifting force of the wing. Interestingly, Zhukovsky himself never liked to fly. There is a joke that he was afraid to fly in an airplane because he did not trust his own function. Nevertheless, the Russians call him the Father of Russian Aviation.

Exercises for Solving on Your Own

Exercise 20.1. Calculate $\dfrac{1}{z}$, where $z = a + bi$.

Exercise 20.2. Convert the given complex numbers into Polar form.
   (a) $z_1 = -3 - \sqrt{3}i$; (b) $z_2 = 4 - 4i$; (c) $z_3 = -\sqrt{3} + 3i$.

Exercise 20.3. Calculate
(a) $z_1 = [1/2 - (\sqrt{3}/2)i])^{30}$;   (b) $z_2 = [-\sqrt{2}/2 + (\sqrt{2}/2)i]^{20}$;
(c) $z_3 = [\sqrt{3}/2 + (1/2)i]^{40}$.

Exercise 20.4. Solve the equation $z^6 = 1$ for complex numbers and describe solutions geometrically.

Entertainment Problem

Problem E20.1. Three bandit brothers were captured and sentenced by a mighty king: the first brother was given 10 years in jail, the second 20 years in jail, and the third life in jail. The king soon died, and his son ascended the throne. On this occasion, the newly crowned king announced an amnesty and reduced every prisoner's sentence by half. Thus, the first brother would only need to sit for 5 years, and the second for 10 years. What should be done about the third brother, as to properly fulfill the amnesty requirements? If the jailers do not precisely follow the order of the king, they will be executed. What can you suggest?

Appendix

Two Remarkable Triangles

In Chapter 2, we observed the Egyptian triangle. Here, we consider the two other remarkable triangles: Golden triangle and Kepler triangle.

Golden Triangle

The term "Golden triangle" has become so popular that it is used in numerous areas: there are communities with this name, restaurants, shopping malls and much more. In mathematics, the golden triangle is a triangle in a five-pointed star (pentagram):

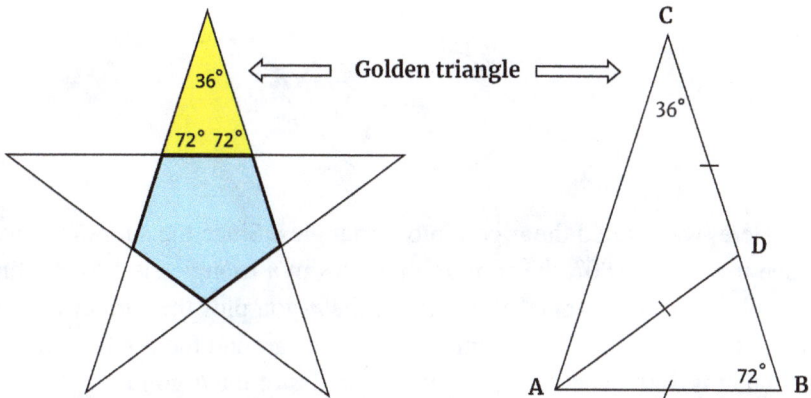

Figure A.1.

The golden triangle is an isosceles triangle with the top angle of 36° and the base angles of 72° (we will prove that in Problem A.1). It got its name because the ratio of its slant side to the base is equal to the so-called **golden ratio**, a name that was given by the Greeks. By definition, two numbers are in the golden ratio if the ratio of the larger number to the smaller is equal to the ratio of their sum to the larger number. The golden ratio is widely used in architecture and art due to the mysterious fact that it creates proportions that are aesthetically pleasing to the human eye.

One of the first to use the golden ratio in construction was the Greek sculptor and mathematician Phidias (500 BC – 432 BC). He applied the golden ratio to some constructions in the temple Parthenon located on the hill Acropolis in Athens, Greece. His work included the famous statue of Zeus at Olympia that also reflects the golden ratio. This huge statue (seated Zeus is about 12.4 m high), like the Great Pyramid of Giza, is also listed in the Seven Wonders of the Ancient World. In 1900 American mathematician Mark Barr used the Greek letter φ (phi) to designate golden ratio, named after Phidias.

Problem A.1. Prove that the angles in the golden triangle are 36° and 72°.

Solution. Consider a convex polygon with n sides (n-gon). We will show that the sum of all its interior angles has a measurement of $180°(n - 2)$. Look at this figure:

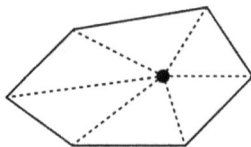

Here, we divided the n-gon into n triangles. Since the sum of angles in one triangle is 180°, the sum of all angles in n triangles is $180°n$. This sum consists of the sum of all angles in the n-gon plus the sum of angles around the dot shown in the figure. The angles around the dot form a full circle that is 360°. Therefore, the sum of angles of the n-gon is

$$180°n - 360° = 180°(n - 2).$$

In particular, the sum of angles in any pentagon (5-gon) is $180° (5 - 2) = 540°$. In a regular pentagon (in which all angles are equal), the measure of one angle is $540°/5 = 108°$. Now take a look at the above Figure A.1 of the golden triangle shown in the star. You can see that the bottom angles of this triangle are supplementary to the angles of the pentagon (together they form straight angles measuring 180°). Therefore, the bottom angles in the golden triangle are $180° - 108° = 72°$. The top angle is $180° - 72° \cdot 2 = 36°$. ∎

Note that in Chapter 9, we found the exact values for the trig functions of angles 36° and 72°.

Problem A.2. Show that sides of the golden triangle are in the golden ratio φ, and calculate the number φ.

Solution. Take a look at the golden triangle on the right-hand side of Figure A.1. In it, we drew a bisector AD. We have:

$$\angle CAD = \angle DAB = 72°/2 = 36°$$

and

$$\angle ADB = 180° - \angle DAB - \angle ABD = 180° - 36° - 72° = 72°.$$

Therefore, $\triangle ACD$ and $\triangle ADB$ are isosceles, and $CD = AD = AB$. Also, $\triangle ADB$ is a golden triangle that is similar to $\triangle ABC$. We can set up this proportion:

$$\frac{AD}{DB} = \frac{CB}{AB}.$$

Let's denote $x = AD$ and $y = DB$. We have $CB = CD + DB = x + y$ and $AB = x$. The above proportion becomes:

$$\frac{x}{y} = \frac{x+y}{x}.$$

This proportion shows that in golden triangle ADB, sides AD and DB are in the golden ratio φ. To calculate this ratio, we divide the numerator and denominator of the last fraction by y:

$$\varphi = \frac{x}{y} = \frac{x+y}{x} = \frac{x/y+1}{x/y} = \frac{\varphi+1}{\varphi}.$$

From here we get the equation $\varphi = \dfrac{\varphi+1}{\varphi}$. This equation can be reduced to a quadratic equation. Let's solve it to find a positive value of φ (the golden ratio is positive):

$$\varphi = \frac{\varphi+1}{\varphi} \;\Rightarrow\; \varphi^2 = \varphi+1 \;\Rightarrow\; \varphi^2 - \varphi - 1 = 0 \;\Rightarrow\; \varphi = \frac{1+\sqrt{5}}{2}. \;\blacksquare$$

As we see, the golden ratio φ is an irrational number: $\varphi = \dfrac{1+\sqrt{5}}{2} \approx 1.618.$[1] Also, we see that this ratio can be defined as the positive root of the quadratic equation

$$x^2 - x - 1 = 0.$$

Kepler Triangle

This is a right triangle in which the sides are in geometric progression.[2] It turns out (see Problem A.3 below) that the Kepler triangle is related to the golden ratio φ:

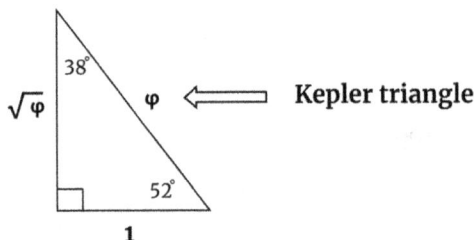

Kepler triangle

---

[1] A reader familiar with Fibonacci numbers may notice their relationship to the golden ratio.
[2] Numbers $a_1, a_2, a_3, \dots$ are in geometric progression if $a_2 = a_1 \cdot r$, $a_3 = a_2 \cdot r$, $\dots$. The number r is called the common ratio.

This triangle is named after a German mathematician and astronomer Johannes Kepler (1571–1630). He was fascinated by the golden ratio and considered it as one of the two great treasures in geometry. The other one is the Pythagorean theorem. Kepler is best known for his laws of planetary motion. One of the important conclusions of his laws is the realization that the planets in the solar system, when moving around the Sun, form not circular orbits, but elliptical ones, such that the Sun is located in one of the elliptical foci. These laws were not accepted during Kepler's life. Only after his death, his discoveries were recognized, mainly when Isaac Newton (1642–1726/27) derived Kepler's laws from his theory of gravitation.

Problem A.3. Let r be the common ratio in a Kepler triangle (i.e., the ratio of sides is $1{:}r{:}r^2$). Show that $r = \sqrt{\varphi} \approx 1.272$, where φ is the golden ratio.

Solution. By the similarity property of triangles, we may assume that the sides of the triangle are 1, r and r^2. Since this is a right triangle, its sides satisfy the Pythagorean theorem:

$$1^2 + r^2 = \left(r^2\right)^2 \Rightarrow 1 + r^2 = r^4 \Rightarrow r^4 - r^2 - 1 = 0.$$

If we denote $x = r^2$, then x satisfies the equation $x^2 - x - 1 = 0$. This is the same equation shown above for the golden ration φ, so $x = r^2 = \varphi$ and $r = \sqrt{\varphi}$. ∎

Solve the following problems on your own.

Problem A.4. Prove that angles in the Kepler triangle are $51.83°$ and $38.17°$. ∎

Problem A.5. A triangle (not necessarily a right one) has sides in the geometric progression. Let r be the common ratio of the sides (i.e., the ratios of sides are $1{:}r{:}r^2$). Prove that $1/\varphi < r < \varphi$, where φ is the golden ratio. ∎

According to definition, both Egyptian and Kepler triangles are based on arithmetic and geometric progressions respectively. It turned out that triangles with such properties are unique.

Problem A.6. Prove that Egyptian and Kepler triangles are the only right triangles whose sides are in arithmetic and geometric progressions (up to similarity).

Solution. As it is shown in Problem A.3, a geometric progression $1, r, r^2$ for sides in a right triangle has common ratio equals to the square root of the gold ratio. It proves the uniqueness of the Kepler triangle. Now look at a triangle with an arithmetic progression. By the similarity property, we can assume that the smallest side in a triangle is 3. Let $3, 3 + d, 3 + 2d$ be sides in a right triangle. We will prove that $d = 1$. Since a triangle is a right one, its sides satisfy the Pythagorean theorem:

$$3^2 + (3 + d)^2 = (3 + 2d)^2 \quad \Rightarrow \quad d^2 + 2d - 3 = 0 \quad \Rightarrow \quad (d - 1)(d + 3) = 0.$$

Since $d > 0$, then we conclude that $d = 1$. Therefore, if the sides of a right triangle are in the arithmetic progression and the smallest side is equal to 3, then it is the Egyptian 3:4:5 triangle. ∎

Note. The "top" angles in the Egyptian, Golden, and Kepler triangles are $37°$, $36°$, and $38°$. These angles are so close to each other that it is unclear which triangle (or maybe all of them or none), was used by the Egyptians to build the pyramids.

Summary of Results

1. Definition of trigonometric functions for right triangles

Let α be an acute angle in a right triangle with legs a, b, and hypotenuse c:

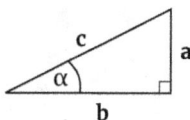

Then $\sin \alpha = \dfrac{a}{c}$, $\cos \alpha = \dfrac{b}{c}$, $\tan \alpha = \dfrac{a}{b}$, $\cot \alpha = \dfrac{b}{a}$, $\sec \alpha = \dfrac{c}{b}$, $\csc \alpha = \dfrac{c}{a}$.

Or

$$\sin \alpha = \frac{\text{opposite side}}{\text{hypotenuse}}, \quad \cos \alpha = \frac{\text{adjucent side}}{\text{hypotenuse}}, \quad \tan \alpha = \frac{\text{opposite side}}{\text{adjucent side}}.$$

2. Positive and negative angles

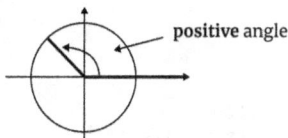

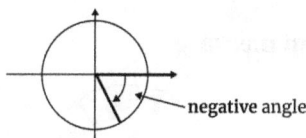

positive angle

negative angle

3. General definition of trigonometric functions

Consider a unit circle in a Cartesian coordinate system. Let α be an angle with sides Ox and OA:

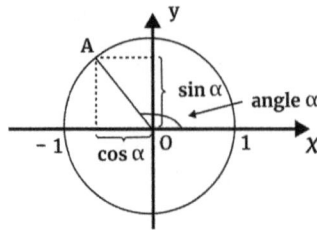

Then $\sin \alpha$ and $\cos \alpha$ of angle α are defined as the y and x coordinates of the point A correspondingly: $\sin \alpha = y$, $\cos \alpha = x$.

4. Relations between trigonometric functions

$$\tan \alpha = \frac{\sin \alpha}{\cos \alpha}, \quad \cot \alpha = \frac{1}{\tan \alpha} = \frac{\cos \alpha}{\sin \alpha}, \quad \sec \alpha = \frac{1}{\cos \alpha}, \quad \csc \alpha = \frac{1}{\sin \alpha}.$$

5. Formulas for triangles
Let a triangle have sides a, b, c, and angles α, β, γ:

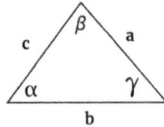

Law of sines:

$$\frac{a}{\sin \alpha} = \frac{b}{\sin \beta} = \frac{c}{\sin \gamma}.$$

Law of cosines:

$$a^2 = b^2 + c^2 - 2bc \cos \alpha.$$

6. Radian measure

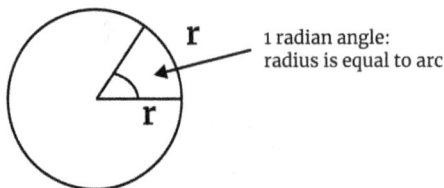

1 radian angle:
radius is equal to arc

π rad $= 180°$, 1 rad $= 180°/\pi \approx 57°$, $1° = \pi/180$ rad ≈ 0.017 rad.

7. Special values

Degrees	Radians	sin	cos	tan
0°	0	0	1	0
30°	$\pi/6$	1/2	$\sqrt{3}/2$	$\sqrt{3}/3$
45°	$\pi/4$	$\sqrt{2}/2$	$\sqrt{2}/2$	1
60°	$\pi/3$	$\sqrt{3}/2$	1/2	$\sqrt{3}$
90°	$\pi/2$	1	0	Undefined
180°	π	0	−1	0
270°	$3\pi/2$	−1	0	Undefined

8. Special angles on the unit circle

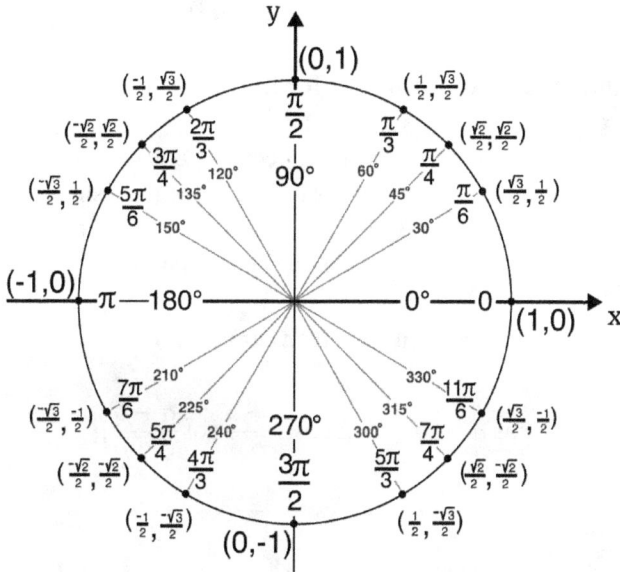

9. Main identity

$$\sin^2 \theta + \cos^2 \theta = 1.$$

10. Reduction formulas

$$\sin(\pi/2 + \alpha) = \cos \alpha, \qquad \cos(\pi/2 + \alpha) = -\sin \alpha,$$
$$\sin(\pi/2 - \alpha) = \cos \alpha, \qquad \cos(\pi/2 - \alpha) = \sin \alpha,$$

$$\sin(\pi + \alpha) = -\sin \alpha, \qquad \cos(\pi + \alpha) = -\cos \alpha,$$
$$\sin(\pi - \alpha) = \sin \alpha, \qquad \cos(\pi - \alpha) = -\cos \alpha,$$
$$\sin(3\pi/2 + \alpha) = -\cos \alpha, \qquad \cos(3\pi/2 + \alpha) = \sin \alpha,$$
$$\sin(3\pi/2 - \alpha) = -\cos \alpha, \qquad \cos(3\pi/2 - \alpha) = -\sin \alpha.$$

11. Periodic properties

$$\sin(\alpha + 2\pi) = \sin \alpha \text{ (period is } 2\pi),$$
$$\cos(\alpha + 2\pi) = \cos \alpha \text{ (period is } 2\pi),$$
$$\tan(\alpha + \pi) = \tan \alpha \text{ (period is } \pi).$$

12. Even–odd properties

$$\sin(-\alpha) = -\sin \alpha \text{ (odd)},$$
$$\cos(-\alpha) = \cos \alpha \text{ (even)},$$
$$\tan(-\alpha) = -\tan \alpha \text{ (odd)}.$$

13. Sum and difference formulas (for angles)

$$\sin(\alpha + \beta) = \sin \alpha \cdot \cos \beta + \cos \alpha \cdot \sin \beta,$$
$$\cos(\alpha + \beta) = \cos \alpha \cdot \cos \beta - \sin \alpha \cdot \sin \beta,$$
$$\sin(\alpha - \beta) = \sin \alpha \cdot \cos \beta - \cos \alpha \cdot \sin \beta,$$
$$\cos(\alpha - \beta) = \cos \alpha \cdot \cos \beta + \sin \alpha \cdot \sin \beta.$$

14. Sum and difference formulas (for functions)

$$\cos \alpha + \cos \beta = 2 \cos \frac{\alpha + \beta}{2} \cos \frac{\alpha - \beta}{2},$$

$$\sin \alpha + \sin \beta = 2 \sin \frac{\alpha + \beta}{2} \cos \frac{\alpha - \beta}{2},$$

$$\cos \alpha - \cos \beta = 2 \sin \frac{\alpha + \beta}{2} \sin \frac{\beta - \alpha}{2},$$

$$\sin \alpha - \sin \beta = 2 \sin \frac{\alpha - \beta}{2} \cos \frac{\alpha + \beta}{2}.$$

15. Product formulas

$$2 \sin \alpha \sin \beta = \cos(\alpha - \beta) - \cos(\alpha + \beta),$$
$$2 \cos \alpha \cos \beta = \cos(\alpha + \beta) + \cos(\alpha - \beta),$$
$$2 \sin \alpha \cos \beta = \sin(\alpha + \beta) + \sin(\alpha - \beta).$$

16. Half-angle formulas

$$\sin^2(\alpha/2) = (1 - \cos \alpha)/2,$$
$$\cos^2(\alpha/2) = (1 + \cos \alpha)/2.$$

17. Double-angle formulas

$$\sin 2\alpha = 2 \sin \alpha \cos \alpha,$$
$$\cos 2\alpha = \cos^2 \alpha - \sin^2 \alpha,$$
$$\cos 2\alpha = 1 - 2\sin^2 \alpha,$$
$$\cos 2\alpha = 2\cos^2 \alpha - 1.$$

18. Squared formulas

$$\sin^2 \alpha = (1 - \cos 2\alpha)/2,$$
$$\cos^2 \alpha = (1 + \cos 2\alpha)/2.$$

19. Range of trigonometric functions

$$-1 \leq \sin \alpha \leq 1,$$
$$-1 \leq \cos \alpha \leq 1,$$
$$-\infty < \tan \alpha < \infty.$$

20. Graphs of trigonometric functions

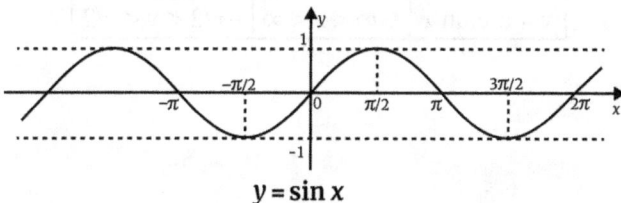

$$y = \sin x$$

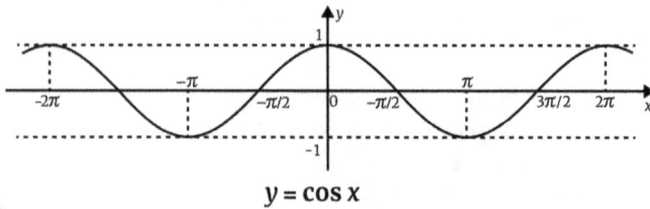

$$y = \cos x$$

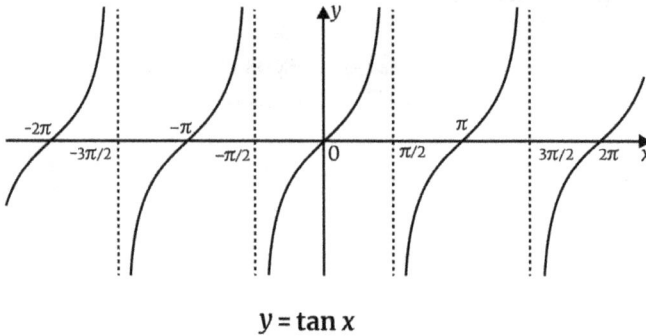

$$y = \tan x$$

21. Definition of inverse trigonometric functions

arcsin x is an angle from the interval $[-\pi/2, \pi/2]$, whose sine is equal to x.

arccos x is an angle from the interval $[0, \pi]$, whose cosine is equal to x.

arctan x is an angle from the interval $(-\pi/2, \pi/2)$, whose tangent is equal to x.

22. Domain and range of inverse trigonometric functions

Function	Domain	Range
$y = \arcsin x$	$-1 \le x \le 1$	$-\pi/2 \le y \le \pi/2$
$y = \arccos x$	$-1 \le x \le 1$	$0 \le y \le \pi$
$y = \arctan x$	$-\infty < x < \infty$	$-\pi/2 < y < \pi/2$

23. Graphs of inverse trigonometric functions

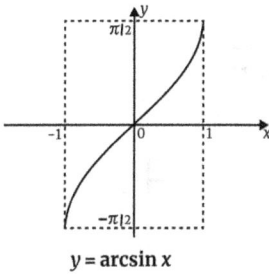

$y = \arcsin x$

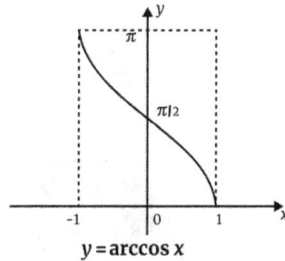

$y = \arccos x$

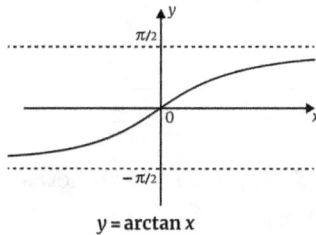

$y = \arctan x$

24. Properties of inverse trigonometric functions

$\sin(\arcsin x) = x \ (-1 \le x \le 1)$

$\cos(\arccos x) = x \ (-1 \le x \le 1)$

$\arcsin(\sin \alpha) = \alpha$, if and only if $\alpha \in [-\pi/2, \pi/2]$

$\arccos(\cos \alpha) = \alpha$, if and only if $\alpha \in [0, \pi]$

$\arcsin x + \arccos x = \pi/2$

$\arcsin(-x) = -\arcsin x \ \text{(odd property)}$

$\arctan(-x) = -\arctan x \ \text{(odd property)}$

$\arccos(-x) = \pi - \arccos x$

$\sin(\arccos x) = \sqrt{1 - x^2}$

$\cos(\arcsin x) = \sqrt{1 - x^2}$

$\sin(\arctan x) = x/\sqrt{1 - x^2}$

$\cos(\arctan x) = 1/\sqrt{1 - x^2}$

25. Special trigonometric equations

(*n* is any integer number)

Equation	Solution
$\sin \alpha = -1$	$\alpha = -\pi/2 + 2\pi n$
$\sin \alpha = 1$	$\alpha = \pi/2 + 2\pi n$
$\sin \alpha = 0$	$\alpha = \pi n$
$\sin \alpha = 1/2$	$\alpha = (-1)^n \pi/6 + \pi n$
$\cos \alpha = -1$	$\alpha = \pi + 2\pi n$
$\cos \alpha = 1$	$\alpha = 2\pi n$
$\cos \alpha = 0$	$\alpha = \pi/2 + \pi n$
$\cos \alpha = 1/2$	$\alpha = \pm \pi/3 + 2\pi n$

26. General solutions of simplest trigonometric equations

(*n* is any integer number)

Equation	Solution
$\sin \alpha = A \ (\,|A| \leq 1\,)$	$\alpha = (-1)^n \arcsin A + \pi n$
$\cos \alpha = A \ (\,|A| \leq 1\,)$	$\alpha = \pm \arccos A + 2\pi n$
$\tan \alpha = A$ (*A* is any number)	$\alpha = \arctan A + \pi n$

27. Equation $a \cdot \sin \alpha + b \cdot \cos \alpha = c \ (a > 0)$

(1) Left side of this equation can be presented in the form:

$$a \cdot \sin \alpha + b \cdot \cos \alpha = \sqrt{a^2 + b^2} \, \cos(\alpha - \beta),$$

where $\beta = \arccos\left(b / \sqrt{a^2 + b^2}\right)$.

(2) The equation has solutions if and only if $|c| \leq \sqrt{a^2 + b^2}$.

(3) General solution is described by the formula:

$$\alpha = \pm \arccos\left(c / \sqrt{a^2 + b^2}\right) + \arccos\left(b / \sqrt{a^2 + b^2}\right) + 2\pi n,$$

where *n* is any integer number.

28. Complex numbers definition

Complex number z is defined by the expression:

$$z = a + bi,$$

where real number a is called the real part, real number b is called the imaginary part, and i is called an imaginary unit with the property: $i^2 = -1$.

29. Complex conjugate

The complex conjugate of a complex number $z = a + bi$ is the complex number

$$\bar{z} = a - bi.$$

The product of z and $\bar{z}$ is a real number, which is defined by the formula:

$$z\bar{z} = a^2 + b^2.$$

30. Trigonometric (Polar) form

Trigonometric form of the complex number $z = a + bi$ is defined by the expression

$$z = r(\cos \theta + i \sin \theta).$$

$r = \sqrt{a^2 + b^2}$ and it is called the modulus or absolute value of the complex number.

θ satisfies the equation $\tan \theta = \dfrac{b}{a}$ and it is called the argument of z.

a and b are expressed through r and θ by the formulas: $a = r \cos \theta$, $b = r \sin \theta$.

31. De Moivre's formula

$$(\cos \theta + i \sin \theta)^n = \cos n\theta + i \sin n\theta,$$

where n is an integer number.

32. Roots of unity

Equation $z^n = 1$ has n complex roots z_k that are defined by the formula:

$$z_k = \cos\frac{2\pi k}{n} + i\sin\frac{2\pi k}{n}, \quad \text{where } k = 0,1,2,\ldots,n-1.$$

Geometrically, the roots are the vertices of a regular n-sided polygon inscribed in the unit circle. For example, roots of the equation $z^{10} = 1$ look like this:

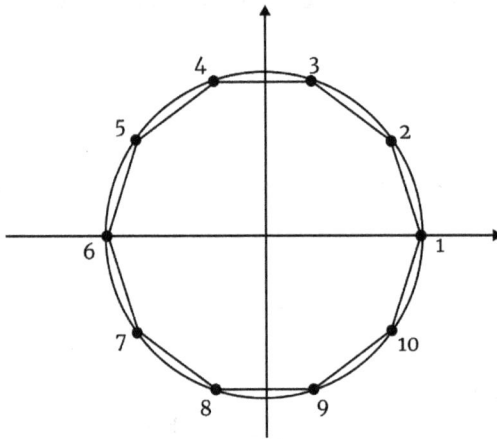

Answers and Solutions to Problems and Exercises

Preface Problem. Since the books stand as shown in the picture, the first page of book 1 is located on its right side. The last page of book 2 adjoins it closely. Therefore, the bookworm chewed only these two pages: the first from book 1 and the last from book 2.

Abbreviations: P X.X. — Problems to solve on your own, **E X.X.** — Exercises, **EP X.X.** — Entertainment problems, **A X.X.** — Solutions from Appendix.

Chapter 1

E 1.0. Red and blue triangles are not similar: the ratio of the legs in the red triangle is 2/5, while the ratio in the blue one is 3/8. Therefore, they have different angles. As a result, the "hypotenuses" of the given figures are not really straight lines. In the top figure, the "hypotenuse" is slightly concave, and in the bottom figure, it is slightly convex. Therefore, these figures are not triangles (they are quadrilaterals), and their areas are different. The area of the top figure is 32, and the area of the bottom one is 33. So, we have difference of one square unit.

E 1.1.

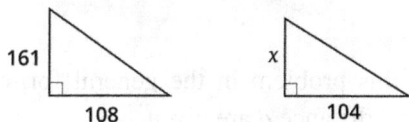

These two triangles are similar, and we can set up the proportion:

$$\frac{161}{108} = \frac{x}{104} \quad \Rightarrow \quad x = \frac{161 \cdot 104}{108} = 155 (\text{cm}).$$

E 1.2. (a) $\sin K = 5/13$, (b) $\cos K = 12/13$, (c) $\tan L = 12/5$.

E 1.3. (a) $\sin A = a/c \quad \Rightarrow \quad a = c \sin A = 6.8 \cdot 0.35 = 2.38$.

(b) $\tan B = b/a \quad \Rightarrow \quad b = a \tan B = 5.2 \cdot 6.25 = 32.5$.

(c) $\cos A = b/c \quad \Rightarrow \quad c = b/\cos A = 3.8/0.65 = 6$.

E 1.4. Diagonals $d_1 = AC$ and $d_2 = BD$. These diagonals split the quadrilateral into four triangles: AEB, BEC, CED, and DEA. Let's denote their areas as S_1, S_2, S_3 and S_4 respectively.

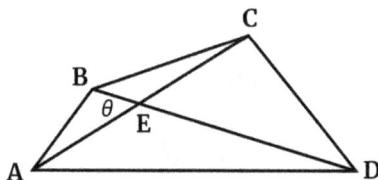

Using formula (1) from Problem 1.5 and the note below it, we have

$$S_1 = \frac{1}{2} AE \cdot BE \cdot \sin\theta, \quad S_2 = \frac{1}{2} BE \cdot CE \cdot \sin\theta,$$

$$S_3 = \frac{1}{2} CE \cdot DE \cdot \sin\theta, \quad S_4 = \frac{1}{2} DE \cdot AE \cdot \sin\theta.$$

The area of the entire quadrilateral is the sum:

$$S_1 + S_2 + S_3 + S_4 = \frac{1}{2}\sin\theta (AE \cdot BE + BE \cdot CE + CE \cdot DE + DE \cdot AE) =$$

$$\frac{1}{2}\sin\theta \left[BE(AE + CE) + DE(CE + AE)\right] = \frac{1}{2}\sin\theta (AE + CE)(BE + DE) =$$

$$\frac{1}{2}\sin\theta d_1 \cdot d_2.$$

E 1.5. Let's solve this problem in the general form: In the following figure, angles α, β and distance d are given.

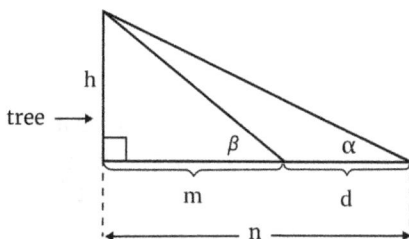

The problem is to calculate m (the width of the river). We have

$$\cot \alpha = n/h \quad \Rightarrow \quad n = h \cot \alpha, \; \cot \beta = m/h \quad \Rightarrow \quad m = h \cot \beta.$$
$$d = n - m = h \cot \alpha - h \cot \beta = h(\cot \alpha - \cot \beta) \quad \Rightarrow \quad h = d/(\cot \alpha - \cot \beta).$$

We get $m = h \cot \beta = (d \cot \beta)/(\cot \alpha - \cot \beta)$.

According to the problem data, $\alpha = 32°$, $\beta = 43°$, $d = 10$. We can find $\cot \alpha = \cot 32° = 1/\tan 32° = 1.6$ and $\cot \beta = \cot 43° = 1/\tan 43° = 1.07$. Then $m = (10 \cot 43°)/(\cot 32° - \cot 43°) = 20.3$ m. Since the rope's length is 30 m, the professor will be able to swim across the river.

EP 1.1. It is enough to make just one check with two shortest and one longest stick. If Eli can get a triangle, then the sum of the lengths of the two shortest sticks is greater than the length of the longest. In this case, the sum of the lengths of any two sticks of the set is longer than any other. This means that any stick can be used to make a triangle. The statement on the box is proved.

Chapter 2

E 2.1.

(a) $\tan A = a/b = 4/9 \quad \Rightarrow \quad A = \tan^{-1}(4/9) = 23.96°$.
(b) $\sin B = b/c = 7/8 \quad \Rightarrow \quad B = \sin^{-1}(7/8) = 61.04°$.
(c) $\cos A = b/c = 6/11 \quad \Rightarrow \quad A = \cos^{-1}(6/11) = 56.94°$.

E 2.2. We need to check that $a^2 + b^2 = c^2$. We have $a^2 + b^2 = (m^2 - n^2)^2 + (2mn)^2 = m^4 - 2m^2n^2 + n^4 + 4m^2n^2 = m^4 + 2m^2n^2 + n^4 = (m^2 + n^2)^2 = c^2$.

E 2.3. Denote as a, b, and c two legs and the hypotenuse of the triangle such that m, n and h are the corresponding heights. Let S be the area of the triangle. Then $2S = am = bn = ch$. From here, $a = 2S/m$, $b = 2S/n$, $c = 2S/h$.

Substitute these expressions into the Pythagorean $a^2 + b^2 = c^2$, and reduce by $4S^2$.

E 2.4. Let three sides be a, b and c, such that $a = 5$ and $b = 1$. In any triangle, $c < a + b$, so $c < 5 + 1 = 6$. On the other hand, $c + b > a$, so $c + 1 > 5 \Rightarrow c > 4$. Since c is an integer, c = 5.

E 2.5. $\sin 17° = x/25 \Rightarrow x = 25 \sin 17° = 7.3$ miles.

E 2.6. $\cos 55° = x/20 \Rightarrow x = 20 \cos 55° = 11.5$ feet.

E 2.7. Let h be the height of the building. Then $h = m + 5$. We have $\tan 75° = m/15 \Rightarrow m = 15 \tan 75° = 55.98 \Rightarrow h = 55.98 + 5 = 60.98$.

E 2.8. Let x be the length of the ladder. Then $\cos 56° = 2/x \Rightarrow x = 2/\cos 56° = 3.58$ m

E 2.9. $\sin 35° = 5.3/x \Rightarrow x = 5.3/\sin 35° = 9.24$ km.

E 2.10. Angle $A = 70° \Rightarrow \tan 70° = 80/x \Rightarrow x = 80/\tan 70° = 29$ ft.

E 2.11. $\cos \theta = 22/50 \Rightarrow \theta = \cos^{-1}(22/50) = 63.9°$.

E 2.12. $\sin \theta = 2.5/8.6 \Rightarrow \theta = \sin^{-1}(2.5/8.6) = 16.9°$.

E 2.13. $\tan \theta = 40/15 = 2.667 \Rightarrow \theta = \tan^{-1}(2.667) = 69.4°$.

E 2.14. Let's $\alpha = \measuredangle EBA = \measuredangle NBC$. We draw the horizontal line MN. Then $EM = AE - AM = 1$ and $MN = \sqrt{2^2 - 1^2} = \sqrt{3}$.

From $\triangle AEB$, $\tan \alpha = \dfrac{2}{AB} \Rightarrow AB = \dfrac{2}{\tan \alpha}$.

From $\triangle BNC$, $\tan \alpha = \dfrac{1}{BC} \Rightarrow BC = \dfrac{1}{\tan \alpha}$.

$AB + BC = AC = MN = \sqrt{3} \Rightarrow \dfrac{2}{\tan \alpha} + \dfrac{1}{\tan \alpha} = \sqrt{3}$.

$\dfrac{3}{\tan \alpha} = \sqrt{3} \Rightarrow \tan \alpha = \dfrac{3}{\sqrt{3}} = \sqrt{3} \Rightarrow \alpha = \tan^{-1}\left(\sqrt{3}\right) = 60°$.

E 2.15. The altitude to the hypotenuse cannot be greater than its half. It is because half of the hypotenuse is equal to the radius of the circumscribed circle and the radius is longer than (or equal to) the altitude:

height ≤ 5   radius = 5

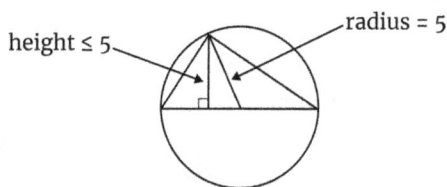

E 2.16. Here is the idea:

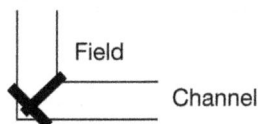

Field

Channel

Let's make calculations to see whether the lengths of the boards are sufficient.

Consider the picture in more detail.

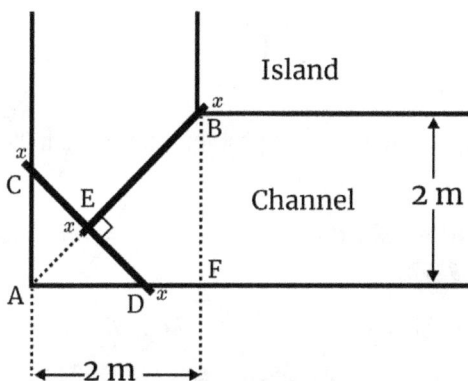

Island

x
B

x
C   E

Channel   2 m

F

A   D x

←—2 m —→

Let x be the protrusion beyond the edges. We will find what the value of x could be. As a measure, we will use centimeters. We have $AB = 200\sqrt{2}$, and $EB = 200 - 2x$ (length of the board minus twice the protrusion).

$$AE = AB - EB = 200\sqrt{2} - (200 - 2x) = 200\sqrt{2} - 200 + 2x.$$

$$CD = 200 - 2x \quad \Rightarrow \quad AE = CE = CD/2 = 100 - x.$$

$$200\sqrt{2} - 200 + 2x = 100 - x \quad \Rightarrow \quad x = \frac{300 - 200\sqrt{2}}{3} \approx 5.7.$$

According to conditions, the acceptable protrusion is at least 4 cm. Since $x = 5.7$ cm, the professor will be able to get to the island.

E 2.17. Professor Smartman took advantage of the Egyptian triangle 3:4:5. He made a box in the form of a right-angled parallelepiped with sides 6 feet (length) and 4.5 feet (width). Then the diagonal is 7.5 feet. The triangle $4.5 \times 6 \times 7.5$ is similar to the Egyptian (multiply sides 3, 4, and 5 by 1.5). The professor placed the fishing rod diagonally into the box.

EP 2.1. Since the doors of the bus are located on the opposite side of the picture, the bus moves to the left (assuming right-hand traffic).

EP 2.2. The player who first begins the game wins. They should place a coin on the center of the table as the first move and make all subsequent moves such that the arrangement of coins on the table is centrally symmetrical. Obviously, if the second player has a move, so does the first. This problem can be generalized to any table surface that is centrally symmetrical. Also, coins (or other items) of various sizes and shapes can be used.

Chapter 3

P 3.3. According to the previous problem, $\tan \theta = \sin \theta / \cos \theta$. Therefore, $\cot \theta = 1/\tan \theta = \cos \theta / \sin \theta$.

P 3.4. By definition, $\sin \theta = a/c$ and $\cos \theta = b/c$. Since $c = 1$, then $\sin \theta = a$ and $\cos \theta = b$.

P 3.7. $\tan \theta = \sin \theta / \cos \theta = \sqrt{1 - \cos^2 \theta} / \cos \theta$.

P 3.8. $b/c = \left(a\sqrt{3}\right)/2a = \sqrt{3}/2$. From here, $b = c\sqrt{3}/2$.

P 3.9. $c/b = 2a/a\sqrt{3} = 2\sqrt{3} = 2\sqrt{3}/3 \implies c = 2b\sqrt{3}/3$.

P 3.10. $a/c = a/a\sqrt{2} = \sqrt{2}/2 \implies a = c\sqrt{2}/2$.

P 3.14. According to the previous problem, $\sin \alpha = \cos \beta$ and $\cos \alpha = \sin \beta$, if $\alpha + \beta = 90°$. From here, $\alpha = 90° - \beta$. Then, $\sin (90° - \beta) = \cos \beta$ and $\cos (90° - \beta) = \sin \beta$. Just rename β to θ.

E 3.1. Since $\tan \theta = \sin \theta / \cos \theta$, divide numbers from the 2nd row by the numbers from the 3rd row. We get: tan   0.62   1.33   2.76

E 3.2. For any angle θ, $\sin^2 \theta + \cos^2 \theta = 1$, so the answer is 1.

E 3.3.   Special angles are 30°, 45°, and 60°. Since $\cot\theta = 1/\tan\theta$,

$$\cot 30° = 1/(\sqrt{3}/3) = \sqrt{3}, \quad \cot 45° = 1, \quad \cot 60° = 1/\sqrt{3} = \sqrt{3}/3.$$

E 3.4.   Since $\cos\theta = \sqrt{1 - \sin^2\theta}$, $\cot\theta = \cos\theta / \sin\theta = \sqrt{1 - \sin^2\theta} / \sin\theta$.

E 3.5.   Consider a right triangle with the angle θ and the adjacent side equal to 1.

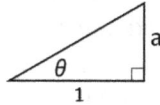

Denote the opposite side as a. By the Pythagorean theorem, the hypotenuse is equal to $\sqrt{1 + a^2}$. Then $\sin\theta = a / \sqrt{1 + a^2}$, and $\tan\theta = a$. Therefore, $\sin\theta = \tan\theta / \sqrt{1 + \tan^2\theta}$.

E 3.6.   Since $\cos\theta < 1$, $1/\cos\theta > 1$. Multiply both sides of the last inequality by $\sin\theta$: $\sin\theta / \cos\theta > \sin\theta \implies \tan\theta > \sin\theta$.

E 3.7.

(1) Since $79° + 11° = 90°$, by the reduction formula, $\sin 79° = \cos 11° = 0.98$.
(2) Using the main identity, $\sin 8° = \sqrt{1 - \cos^2 8°} = 0.14$, and $\cos 82° = \sin 8° = 0.14$.
(3) $\sin 62° = \cos 28° = \sqrt{1 - \sin^2 28°} = \sqrt{1 - 0.47^2} = 0.883$, $\cos 62° = \sin 28° = 0.47$, $\tan 62° = \sin 62°/\cos 62° = 0.883/0.47 = 1.88$.

E 3.8.   Using the reduction formulas for sine and cosine, $\tan(90° - \theta) = \sin(90° - \theta)/\cos(90° - \theta) = \cos\theta/\sin\theta = \cot\theta$.

E 3.9. $f(30°) = \sin 30° + \cos 60° = \dfrac{1}{2} + \dfrac{1}{2} = 1.$

E 3.10. The result contradicts with $\cot\alpha = 1/\tan\alpha$.

E 3.11.

(a) $c = 2a = 12$. $b = \sqrt{c^2 - a^2} = \sqrt{4a^2 - a^2} = a\sqrt{3} = 6\sqrt{3}$.

(b) $c = 2a \implies a^2 + b^2 = 4a^2 \implies a^2 + 9 = 4a^2 \implies 3a^2 = 9$
$\implies a^2 = 3 \implies a = \sqrt{3}, c = 2\sqrt{3}$.

(c) $a = c/2 = 8/2 = 4$. According to part (a), $b = a\sqrt{3} = 4\sqrt{3}$.

E 3.12.

(a) $b = a = 4$. $c = \sqrt{a^2 + b^2} = \sqrt{4^2 + 4^2} = 4\sqrt{2}$.

(b) $a = b = 7$. $c = \sqrt{a^2 + b^2} = \sqrt{7^2 + 7^2} = 7\sqrt{2}$.

(c) $a = b \;\Rightarrow\; a^2 + a^2 = c^2 \;\Rightarrow\; 2a^2 = 100 \;\Rightarrow\; a^2 = 50$

$\Rightarrow\; a = \sqrt{50} = 5\sqrt{2},\; b = 5\sqrt{2}$.

EP 3.1. First, we numerate all workers. Then we take one coin from the first worker, two from the second, three from the third, and so on until the last. It is not difficult to calculate the total weight if all the coins were genuine (let's call this "genuine weight"). The result of the test weighing must be subtracted from the "genuine weight," and the difference in grams will show the number assigned to the worker. That would be our counterfeiter!

Chapter 4

P 4.2.   The angle is $-360° \cdot 3 + 10° = -1070°$.

P 4.3.   From $\triangle ABO$ we have: $\cot \measuredangle ABO = \dfrac{AB}{AO} = \dfrac{AB}{1} = AB$.

Since line l is parallel to the x-axis, $\measuredangle\theta = \measuredangle ABO$ and $\cot\theta = AB$.

E 4.1.   Angle $1 = 360° + 35° = 395°$. Angle $2 = -35°$. Angle $3 = 360° - 35° = 325°$.

E 4.2. $500° = 360° + 140°$. Angle of $140°$ is between $90°$ and $180°$. So angle of $500°$ lies in the 2nd quadrant.

E 4.3. $-430° = -360° - 70°$, so angle of $-430°$ is coterminal with $-70°$. Here is the figure:

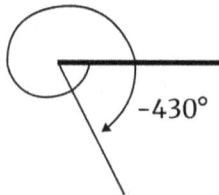

$-430°$

E 4.4. Sine.

E 4.5. Angle of $-180°$ is coterminal to $180°$. Using a table for quadrantal angles, we have

(1) $\sin(-180°) = 0$; (2) $\cos(-180°) = -1$; (3) $\tan(-180°) = 0$.

Angle of $-90°$ is coterminal to the quadrantal angle of $270°$. Therefore,

(4) $\sin(-90°) = -1$; (5) $\cos(-90°) = 0$; (6) $\tan(-90°)$ is undefined.

Since $\cot \theta = \cos \theta / \sin \theta$, we get:

(7) $\cot 0°$ is undefined; (8) $\cot 180°$ is undefined; (9) $\cot(-90°) = 0$; (10) $\cot(-270°) = 0$.

E 4.6. $\tan \theta$ is undefined for $\theta = 90°$ and $\theta = 270°$, so $x + 30° = 90°$ or $x + 30° = 270°$. From here, $x = 60°$ or $x = 240°$.

E 4.7. Tangent is undefined for angles at which cosine is zero. So the answer is (a) $90°$ and (b) $270°$.

E 4.8. The radius r of the circle is $r = 2.3$ and $x = -1.2$. In the 2nd quadrant, the y-coordinate is positive, so $y = \sqrt{r^2 - x^2} = \sqrt{2.3^2 - 1.2^2} = 1.962$. If α is the corresponding angle, then

$$\sin \alpha = y/r = 1.962/2.3 = 0.85, \cos \alpha = x/r = (-1.2)/2.3 = -0.52,$$

$$\tan \alpha = y/x = 1.962/(-1.2) = -1.64.$$

E 4.9. Let the coordinates of point A be (x, y) and the corresponding angle be α. Then $\tan \alpha = y/x = -3.3$. From here, $y = -3.3x$. On the unit circle, $x^2 + y^2 = 1 \Rightarrow x^2 + (-3.3x)^2 = 1 \Rightarrow 11.89x^2 = 1 \Rightarrow x^2 = 0.0841$. In the 2nd quadrant x is negative, therefore $x = -\sqrt{0.0841} = -0.29$, $y = -3.3 \cdot (-0.29) = 0.96$. So, A has coordinates $(-0.29, 0.96)$.

E 4.10. We have $x = 2$, $y = -3$ $r = \sqrt{x^2 + y^2} = \sqrt{2^2 + (-3)^2} = \sqrt{13}$. By definition,

$$\sin \theta = y/r = -3/\sqrt{13}, \quad \cos \theta = x/r = 2/\sqrt{13}, \quad \tan \theta = y/x = -3/2.$$

E 4.11. (1) Sine is negative in the 3rd quadrant. (2) Cosine is positive in the 4th quadrant.

E 4.12. Angle of $-50°$ is in the 4th quadrant. Therefore, sine and tangent are negative, and cosine is positive. Angle of $200°$ is in the 3rd quadrant, so sine and cosine are negative, and tangent is positive.

E 4.13. Let $0 < \theta < 180°$. In this interval, $\sin \theta$ is positive. We have $0 < \theta/2 < 90°$. Here, $\tan (\theta/2)$ is also positive. Let $180° < \theta < 360°$. $\sin \theta$ is negative in this interval. We have $90° < \theta/2 < 180°$. Here $\tan(\theta/2)$ is also negative.

E 4.14. Signs of cotangent are the same as signs of tangent: positive in the 1st and 3rd quadrants, negative in the 2nd and 4th.

E 4.15. (1) 4th, (2) 2nd, (3) 2nd, (4) 3rd, (5) 3rd, (6) 4th.

E 4.16. Sine and cosine are y- and x-coordinates, so $x = y$. Also, $x^2 + y^2 = 1$. From here, $2x^2 = 1$, and the coordinates are $(\sqrt{2}/2, \sqrt{2}/2)$ and $(-\sqrt{2}/2, -\sqrt{2}/2)$. The quadrants are 1st and 3rd.

EP 4.1. The professor should tie the rope to the tree, which is on the shore. Then go around the lake with the other end of the rope in his hands and again tie it to the same tree. As a result, the rope is stretched between the two trees.

Chapter 5

P 5.1. By definition, sine is the y-coordinate and cosine is the x-coordinate of the point on the unit circle. If we draw the sine and cosine as line segments, we will see that the expression $x^2 + y^2$, according to the Pythagorean theorem, is equal to the square of the radius of the unit circle. Thus, $\sin^2 \theta + \cos^2 \theta = 1$.

P 5.2. Divide both sides of the main identity $\sin^2 \theta + \cos^2 \theta = 1$ by $\sin^2 \theta$:

$$\frac{\sin^2 \theta}{\sin^2 \theta} + \frac{\cos^2 \theta}{\sin^2 \theta} = \frac{1}{\sin^2 \theta} \Rightarrow 1 + \cot^2 \theta = \csc^2 \theta.$$

P 5.4. If the angle θ is in the 2nd quadrant, then angle $-\theta$ is in the 3rd quadrant, and vice versa.

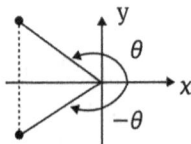

Points on the unit circle, which correspond to angles θ and $-\theta$ have the same x-coordinate, which is cosine, so $\cos \theta = \cos(-\theta)$. It means that cosine is an even function. As for y-coordinates, which is sine, they are equal in absolute value and differ in sign, so $\sin \theta = -\sin (-\theta)$. It means that sine is an odd function.

P 5.5. $\quad \tan(-\theta) = \dfrac{\sin(-\theta)}{\cos(-\theta)} = \dfrac{-\sin \theta}{\cos \theta} = -\tan \theta.$

P 5.11.

For $\sin(90° + \theta)$:

(1) Angle $90° + \theta$ is in the 2nd quadrant. Here the sine is positive.
(2) The quadrantal angle $90°$ points to the vertical axis, so we change sine to cosine. The answer is $\sin(90° + \theta) = \cos \theta$.

For $\cos(180° + \theta)$:

(1) Angle $180° + \theta$ is in the 3rd quadrant. Here the cosine is negative.
(2) The quadrantal angle $180°$ points to the horizontal axis, so we do not change cosine to sine. The answer is $\cos(180° + \theta) = -\cos \theta$.

P 5.13. $\sin 925° = \sin(2 \cdot 360° + 205°) = \sin 205° = \sin(180° + 25°)$
$= -\sin 25°.$

$\qquad \cos 430° = \cos(360° + 70°) = \cos 70° = \cos(90° - 20°) = \sin 20°.$

E 5.1. According to the Main Identity, $\sin^2 \theta + \cos^2 \theta = 1$. However, $0.4^2 + 0.6^2 \neq 1$.

E 5.2. $\quad \cot \theta = \dfrac{\cos \theta}{\sin \theta} = \dfrac{-\sin(\theta + 90°)}{\cos(\theta + 90°)} = -\tan(\theta + 90°).$

E 5.3.

(a) Angle $130°$ is in the 2nd quadrant. Ref. angle $= 180° - 130° = 50°$.
(b) Angle $320°$ is in the 4th quadrant. Ref. angle $= 360° - 320° = 40°$.
(c) Angle $250°$ is in the 3rd quadrant. Ref. angle $= 250° - 180° = 70°$.
(d) $760° = 2 \cdot 360° + 40°$. Angle $40°$ is in the 1st quadrant. Ref. angle $= 40°$.
(e) Angle $-210°$ is in the 2nd quadrant. Ref. angle $= 210° - 180° = 30°$.

E 5.4.

(a) $\sin 210° = \sin(180° + 30°) = -\sin 30° = -1/2.$
(b) $\cos 330° = \cos(270° + 60°) = \sin 60° = \sqrt{3} / 2.$
(c) $\tan 135° = \tan(90° + 45°) = -\cot 45° = -1.$

E 5.5.

(a) Angle $315°$ is in the 4th quadrant and ref. angle $= 360° - 315° = 45°$.
Therefore, $\sin 315° = -\sin 45° = -\sqrt{2} / 2$.

(b) Angle $150°$ is in the 2nd quadrant and ref. angle $= 180° - 150° = 30°$.
Therefore, $\cos 150° = -\cos 30° = -\sqrt{3} / 2$.

(c) Angle $240°$ is in the 3rd quadrant and ref. angle $= 240° - 180° = 60°$.
Therefore, $\tan 240° = \tan 60° = \sqrt{3}$.

E 5.6. Let θ_r be a reference angle. We have

(a)

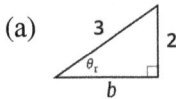

$b = \sqrt{3^2 - 2^2} = \sqrt{5}$, $\sin \theta_r = 2 / 3$. Since both sine and tangent are negative, angle θ is in the 4th quadrant, in which cosine is positive. Therefore,
$\cos \theta = \sqrt{5} / 3$, $\sec \theta = 3 / \sqrt{5}$, $\tan \theta = -2 / \sqrt{5}$, $\cot \theta = -\sqrt{5} / 2$,
$\csc \theta = -3 / 2$.

(b)

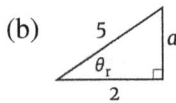

$a = \sqrt{5^2 - 2^2} = \sqrt{21}$, $\cos \theta_r = 2 / 5$. Since cosine is negative and sine is positive, angle θ is in the 2nd quadrant, in which tangent is negative. Therefore,

$\sin \theta = \sqrt{21} / 5$, $\csc \theta = 5 / \sqrt{21}$, $\tan \theta = -\sqrt{21} / 2$, $\cot \theta = -2 / \sqrt{21}$,
$\sec \theta = -5 / 2$.

c)

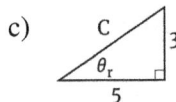

$c = \sqrt{3^2 + 5^2} = \sqrt{34}$, $\tan \theta_r = 3 / 5$. Since tangent is positive and cosine is negative, angle θ is in the 3rd quadrant, in which sine is negative. Therefore,
$\sin \theta = -3 / \sqrt{34}$, $\csc \theta = -\sqrt{34} / 3$, $\cos \theta = -5 / \sqrt{34}$, $\sec \theta = -\sqrt{34} / 5$,
$\cot \theta = 5 / 3$.

E 5.7. By definition, $f(x + P) = f(x)$ for any x. Plug in $x = 0$. Then $f(P) = f(0) = c$, or $aP^2 + bP + c = c \implies aP^2 + bP = 0 \implies aP + b = 0$. Next, plug in $x = 2P$. Then $f(2P) = f(P) = c$, or $4aP^2 + 2bP + c = c \implies 4aP^2 + 2bP = 0 \implies 2aP + b = 0$. From the two equations, $aP + b = 0$ and $2aP + b = 0$, we conclude that $a = b = 0$ (just subtract one equation from the other).

E 5.8. $f(x + 1) = f(x + 1 + 3 \cdot 3 - 2 \cdot 5) = f(x)$.

E 5.9. Our goal is to prove that $f(x + 2P) = f(x)$. We have

$$f(x + 2P) = f[(x + P) + P] = -f(x + P) = f(x).$$

As an example, we can take $\sin x$ or $\cos x$. Number $P = \pi$. According to the reduction formulas, $\sin(x + \pi) = -\sin x$ and $\cos(x + \pi) = -\cos x$.

EP 5.1. You need to light one candle at both ends, and the other at one end. When the first candle burns down, half an hour will pass (since it's burning from both directions at the same time). At that moment, you need to light the second candle at its other end. It will burn down in 15 min. Thus, the total burning time will be 45 min.

EP 5.2. Eli needs to lock the container with his padlock. Then he sends the container to Ben. Ben also locks the container with his padlock and sends it back to Eli. Eli removes his padlock and sends the container to Ben once again. Ben opens the container, unlocking his own padlock with his key.

Chapter 6

P 6.2. From $\triangle ABD$: $\sin A = h/AB \implies h = AB \sin A = 12 \sin 65° \approx 10.9$ feet.

E 6.1.

T 1. $BC/\sin 20° = 25/\sin 110° \implies BC = 25 \sin 20°/\sin 110° = 9.1$ m.

T 2. $B = 180° - 50° - 110° = 20°$, $AB/\sin 110° = 12/\sin 20°$. $AB = 12 \sin 110°/\sin 20° = 33$ m.

T 3. $18/\sin A = 23/\sin 55° \implies \sin A = 18 \sin 55°/23 = 0.6411 \implies A = \sin^{-1} 0.6411 = 39.9°$.

T 4. $17/\sin C = 25/\sin 50° \Rightarrow \sin C = 17 \sin 50°/25 = 0.5209 \Rightarrow C = \sin^{-1} 0.5209 = 31.4° \Rightarrow B = 180° - 50° - 31.4° = 98.6°$.

T 5. $12/\sin B = 19/\sin 110° \Rightarrow \sin B = 12 \sin 110°/19 = 0.5935$.

$B = \sin^{-1} 0.5935 = 36.4° \Rightarrow A = 180° - 110° - 36.4° = 33.6°$.

$BC/\sin 33.6° = 19/\sin 110° \Rightarrow BC = 19 \sin 33.6°/\sin 110° = 11.2$ m.

E 6.2.

R 1. Side b is greater than side a, therefore $B > A$. Since A is obtuse, B also obtuse. A triangle cannot have two obtuse angles.

R 2. $A + B = 71° + 115° = 186° > 180°$. The sum of angles in any triangle is equal to $180°$.

R 3. $A + B + C = 176° < 180°$. The sum of angles in any triangle is equal to $180°$.

R 4. $B = 180° - A - C = 180° - 71° - 42° = 67°$. In any triangle, the larger the side, the larger the opposite angle. However, $b > a$, but $B < A$.

R 5. In a triangle, the sum of any two sides is greater than the third side, but $a + c = 370 + 120 = 490 < 500 = b$.

R 6. Calculate the altitude h to the side AB: $h = b \sin A = 400 \sin 71° = 378 > a = 370$. Such a triangle does not exist: a must be greater than h.

R 7. $B = 180° - A - C = 180° - 71° - 54° = 55°$. According to the Law of Sines, $a/\sin A = b/\sin B$. However, $a/\sin A = 370/\sin 71° = 391$, but $b/\sin B = 350/\sin 55° = 427$.

R 8. Calculate the altitude h to the side AB: $h = b \sin A = 400 \sin 41° = 262$. We have $h < a < b$. According to Proposition 6.2, part 4, there are two triangles, and one of them is obtuse. Calculate $b \tan A = 400 \tan 41° = 348$. We have $a > b \tan A$. According to Problem 6.7, part 3, the second triangle is also obtuse. Thus, it is not certain which land plot she should take into account.

E 6.3. Calculate the altitude h to the side AB: $h = b \sin A = 400 \sin 61° = 350$. We have $h < a < b$. According to Proposition 6.2, part 4, there are two triangles. One of them is obtuse. Calculate $b \tan A = 400 \tan 61° = 722$. We have $a < b \tan A$. According to Problem 6.7, part 2, the second triangle is acute. The additional information can be the type of triangle in question: acute or obtuse.

E 6.4.

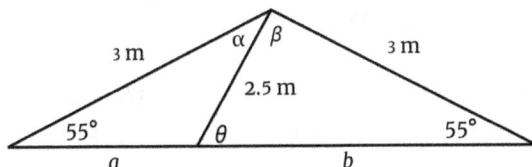

From the triangle on the right:

$$\frac{2.5}{\sin 55°} = \frac{3}{\sin \theta} \Rightarrow \sin \theta = \frac{3\sin 55°}{2.5} = 0.983 \Rightarrow \theta = \sin^{-1} 0.983 = 79.4°.$$

$$\beta = 180° - 55° - \theta = 180° - 55° - 79.4° = 45.6°.$$

$$\frac{b}{\sin 45.6°} = \frac{2.5}{\sin 55°} \Rightarrow b = \frac{2.5\sin 45.6°}{\sin 55°} = 2.18.$$

$$\alpha + \beta = 180° - 2 \cdot 55° = 70° \quad \Rightarrow \quad \alpha = 70° - \beta = 70° - 45.6° = 24.4°.$$

From the triangle on the left:

$$\frac{a}{\sin 24.4°} = \frac{2.5}{\sin 55°} \Rightarrow a = \frac{2.5\sin 24.4°}{\sin 55°} = 1.26.$$

Answer: $a = 1.26$ m, $b = 2.18$ m.

EP 6.1. Flip two switches to ON and, after some time, flip one of them back to OFF. Enter the room and physically touch the light bulbs which are OFF. The warm one will match the second switch, which was just flipped OFF. With the rest of the bulbs, everything is obvious.

Chapter 7

E 7.1. The distance is $\sqrt{750^2 + 830^2 - 2 \cdot 750 \cdot 830 \cdot \cos 70°} = 909$ m.

E 7.2. $\angle ACB = 180° - 145° = 35°$. Using the Law of Cosines, we get

$$AB = \sqrt{240^2 + 180^2 - 2 \cdot 240 \cdot 180 \cdot \cos 35°} = 139 \text{ feet.}$$

E 7.3. $\cos\theta = \dfrac{65^2 + 60^2 - 16^2}{2\cdot 65\cdot 60} = 0.970 \Rightarrow \theta = \cos^{-1} 0.970 = 14°.$

E 7.4. $\cos F = \dfrac{500^2 + 800^2 - 700^2}{2\cdot 500\cdot 800} = 0.5 \Rightarrow F = 60° > 40°.$ The answer is "No."

E 7.5. Let a, b, and c be the sides of a triangle ABC, which correspond to angles A, B, and C. By the Law of Cosines,

$$a = \sqrt{b^2 + c^2 - 2bc\cos A} = \sqrt{4^2 + 9^2 - 2\cdot 4\cdot 9\cdot\cos 30°} = 5.886.$$

Using the Law of Cosines again,

$$\cos C = \frac{a^2 + b^2 - c^2}{2ab} = \frac{5.886^2 + 4^2 - 9^2}{2\cdot 5.886\cdot 4} = -0.645,$$

so $C = \cos^{-1}(-0.645) = 130°.$

Note: if we use the Law of Sines $\dfrac{a}{\sin A} = \dfrac{c}{\sin C}$, then $C = 50°$, which is wrong.

E 7.6. The problem is to find the angle θ. First, we find $\angle ACB$ by the Law of Cosines:

$$\cos\angle ACB = \frac{11^2 + 10^2 - 19^2}{2\cdot 11\cdot 10} = -0.6364. \quad \angle ACB = \cos^{-1}(-0.6364) = 130°.$$

$\theta = 180° - \angle ACB = 180° - 130° = 50°.$

E 7.7. According to Problem 7.5, calculate $E = 5^2 + 9^2 - 13^2 = -63$. Since $E < 0$, the triangle is obtuse.

E 7.8. First, we find the angles that both hands make from the position of 12:00 pm (vertical position) to the given time 1:25 pm. We convert this time to minutes: 1 h 25 min = 85 min.

(1) The hour hand makes a full rotation (360°) in 12 h = 720 min. So, in 1 min it makes a 360°/720 = 0.5° angle. Therefore, in 85 min, the hour hand will make a 0.5° · 85 = 42.5° angle.

(2) The minute hand makes a full rotation in 1 h = 60 min, so in one min it makes a 360°/60 = 6° angle. During the given time 1 h 25 min, in 1 h

the minute hand returns to the vertical position, and for the remaining 25 min, it makes a $6° \cdot 25 = 156°$ angle.

(3) The angle between both hands is $156° - 42.5° = 113.5°$. Now, we can use the Law of Cosines to find the distance d between the hands:

$$d = \sqrt{4^2 + 6^2 - 2 \cdot 4 \cdot 6 \cdot \cos 113.5°} = 8.43 \, \text{cm}.$$

EP 7.1. Before leaving home, he wound up the wall clock and made a note of what time he left.

Having arrived at his friend's house, he noted his arrival time and when leaving departure time back home.

After arriving back home, he noted the time indicated on his wall clock. At this point he has the following information:

- Exact time of his departure from his friend's house: T_d
- Total time he was absent (looking at his wall clock): T_a
- Time spent at a friend's house: T_f

The total time he was absent is the sum of the time he spent at a friend's house and double time T_w he walked to/from a friend's house: $T_a = T_f + 2T_w$. From here, $T_w = (T_a - T_f)/2$.

Finally, the correct time T on his clock is defined by the formula:
$T = T_d + T_w$.

Chapter 8

P 8.5.

$$\cos 75° = \cos(30° + 45°) = \cos 30° \cos 45° - \sin 30° \sin 45°$$

$$= \frac{\sqrt{3}}{2} \cdot \frac{\sqrt{2}}{2} - \frac{1}{2} \cdot \frac{\sqrt{2}}{2} = \frac{\sqrt{6} - \sqrt{2}}{4}.$$

P 8.7. Using odd and even properties of sine and cosine,

$\sin(\alpha - \beta) = \sin[\alpha + (-\beta)] = \sin \alpha \cos(-\beta) + \cos \alpha \sin(-\beta) = \sin \alpha \cos \beta - \cos \alpha \sin \beta$.

P 8.9. Set $\beta = \alpha$ in the Sine Sum formula:

$$\sin(\alpha + \alpha) = \sin \alpha \cdot \cos \alpha + \cos \alpha \cdot \sin \alpha = 2 \sin \alpha \cos \alpha.$$

E 8.1. Let (x, y) be the coordinates of a point on the circle. The radius $r = 3$ is equal to the distance from this point to the origin. By the distance formula, $x^2 + y^2 = r^2 = 9$. So, the equation has the form: $x^2 + y^2 = 9$.

E 8.2.

$$\cos 330° = \cos(360° - 30°) = \cos 30° = \sqrt{3}\,/\,2, \sin 330° = -\sin 30° = -1\,/\,2.$$

From here,

$$\cos(330° + \theta) = \cos 330° \cos \theta - \sin 330° \sin \theta = \sqrt{3}\,/\,2 \cos \theta + 1\,/\,2 \sin \theta.$$

E 8.3.   Using the result of Problem 8.9, we have $\sin 8\theta = 2 \sin 4\theta \cos 4\theta = 4 \sin 2\theta \cos 2\theta \cos 4\theta = 8 \sin \theta \cos \theta \cos 2\theta \cos 4\theta$. Now, divide both sides by $8 \sin \theta$.

E 8.4. $\tan(\alpha + \beta) = \dfrac{\sin(\alpha + \beta)}{\cos(\alpha + \beta)} = \dfrac{\sin \alpha \cos \beta + \cos \alpha \sin \beta}{\cos \alpha \cos \beta - \sin \alpha \sin \beta}$.   Divide both sides of the last fraction by $\cos \alpha \cos \beta$, and get the answer.

E 8.5.   Just set $\beta = \alpha$ in the Tangent of the Sum formula.

E 8.6.
(a) $\sin 10° \cos 20° + \cos 10° \sin 20° = \sin(10°+20°) = \sin 30° = 1/2$.
(b) $\sin 55° \sin 65° - \sin 25° \sin 35° = \sin(90° - 25°) \sin(90° - 35°) - \sin 25° \sin 35° = \cos 25° \cos 35° - \sin 25° \sin 35° = \cos(35° + 25°) = \cos 60° = 1/2$.

E 8.7.

$$\cos 36° \cos 72° = \frac{2\sin 36° \cos 36° \cos 72°}{2\sin 36°} = \frac{\sin 72° \cos 72°}{2\sin 36°}$$

$$= \frac{2\sin 72° \cos 72°}{2 \cdot 2\sin 36°} = \frac{\sin 144°}{4\sin 36°} = \frac{\sin\left(180° - 36°\right)}{4\sin 36°} = \frac{\sin 36°}{4\sin 36°} = \frac{1}{4}.$$

E 8.8.   Since $\triangle ABC$ is an isosceles, $\angle BAC = C \Rightarrow 2C + B = 180° \Rightarrow 2C = 180° - B \Rightarrow C = 90° - B/2 \Rightarrow \sin C = \sin(90° - B/2) = \cos(B/2)$.
From $\triangle ADC$, $\sin C = h/b \Rightarrow \cos(B/2) = h/b$.
Using the Double Angle Cosine formula,

$$\cos B = 2\cos^2 \frac{B}{2} - 1 = \frac{2h^2}{b^2} - 1 = \frac{2h^2 - b^2}{b^2} \Rightarrow B = \cos^{-1}\left(\frac{2h^2 - b^2}{b^2}\right).$$

E 8.9. See the solution of Exercise 8.10 below for the general case.

Answer: 3.8 mi.

E 8.10. $\tan A = \dfrac{h}{m} \Rightarrow m = \dfrac{h}{\tan A}$, $\tan B = \dfrac{h}{n} \Rightarrow n = \dfrac{h}{\tan B}$.

$$d = m + n = \frac{h}{\tan A} + \frac{h}{\tan B} = h(\cot A + \cot B) = h\left(\frac{\cos A}{\sin A} + \frac{\cos B}{\sin B}\right)$$

$$= h\frac{\cos A \sin B + \sin A \cos B}{\sin A \sin B} = h\frac{\sin(A + B)}{\sin A \sin B} \Rightarrow$$

$$h = \frac{d \sin A \sin B}{\sin(A + B)}.$$

E 8.11. See the solution of Exercise 8.12 below for the general case.

Answer: 6.3 mi.

E 8.12. $\tan A = \dfrac{h}{m} \Rightarrow m = \dfrac{h}{\tan A}$, $\tan B = \dfrac{h}{n} \Rightarrow n = \dfrac{h}{\tan B}$.

$$d = m - n = \frac{h}{\tan A} - \frac{h}{\tan B} = h(\cot A - \cot B) = h\left(\frac{\cos A}{\sin A} - \frac{\cos B}{\sin B}\right)$$

$$= h\frac{\cos A \sin B - \sin A \cos B}{\sin A \sin B} = h\frac{\sin(B - A)}{\sin A \sin B} \Rightarrow$$

$$h = \frac{d \sin A \sin B}{\sin(B - A)}.$$

If the aircraft is on the left side of both radar stations, in the above formula we switch A and B. We can combine both cases by the formula

$$h = \frac{d \sin A \sin B}{\sin|A - B|}.$$

EP 8.1. Denote 3-L and 5-L container by C3 and C5. Here are the steps:

(1) Fill C5 with water.
(2) Pour from C5 to C3. Two liters will remain in C5.
(3) Empty C3 and pour the two liters from C5.
(4) Fill C5 and pour into C3. Since C3 had two liters, there is only room for one liter. As a result, four liters remain in C5.

EP 8.2. After two transfers, both glasses will contain the original volume of liquid. From the glass with milk, some amount of milk went out, and because the initial volume remains unchanged, the same amount of coffee came in. The answer is: the same. The number of transfers does not matter.

Chapter 9

P 9.2. To get the first identity, add two expressions:

$\cos(\alpha + \beta) = \cos \alpha \cdot \cos \beta - \sin \alpha \cdot \sin \beta$ and
$\cos (\alpha - \beta) = \cos \alpha \cdot \cos \beta + \sin \alpha \cdot \sin \beta$.

Then divide by 2. To get the second identity, replace α with $90° - \alpha$ in the first identity. Then use the reduction formula $\cos(90° - \alpha) = \sin \alpha$.

P 9.4. $\cos^2 \alpha = 1 - \sin^2 \alpha = 1 - \dfrac{1 - \cos 2\alpha}{2} = \dfrac{2 - (1 - \cos 2\alpha)}{2} = \dfrac{1 + \cos 2\alpha}{2}$.

P 9.8.

(a) Use the Difference of Sines formula and odd property of sine:

$$\sin \alpha + \sin \beta = \sin \alpha - \sin(-\beta) = 2\sin \frac{\alpha - (-\beta)}{2} \cos \frac{\alpha + (-\beta)}{2}$$

$$= 2\sin \frac{\alpha + \beta}{2} \cos \frac{\alpha - \beta}{2}.$$

(b) Use the reduction formula $\cos \theta = \sin(90° - \theta)$ and the Difference of Sines formula:

$$\cos \alpha - \cos \beta = \sin(90° - \alpha) - \sin(90° - \beta) =$$

$$2\sin\frac{\left[(90° - \alpha) - (90° - \beta)\right]}{2}\cos\frac{\left[(90° - \alpha) + (90° - \beta)\right]}{2} =$$

$$2\sin\frac{-\alpha + \beta}{2}\cos\frac{180° - \alpha - \beta}{2} = 2\sin\frac{\beta - \alpha}{2}\cos\left(90° - \frac{\alpha + \beta}{2}\right)$$

$$= 2\sin\frac{\beta - \alpha}{2}\sin\frac{\alpha + \beta}{2}.$$

E 9.1.

(a) $\cos 15° \cos 75° = \frac{1}{2}\left[\cos(15° + 75°) + \cos(75° - 15°)\right]$

$$= \frac{1}{2}(\cos 90° + \cos 60°) = \frac{1}{2}\left(0 + \frac{1}{2}\right) = \frac{1}{4}.$$

(b) $\sin 15° + \sin 75° = 2\sin\frac{15° + 75°}{2}\cos\frac{75° - 15°}{2} = 2\sin 45° \cos 30°$

$$= 2 \cdot \frac{\sqrt{2}}{2} \cdot \frac{\sqrt{3}}{2} = \frac{\sqrt{6}}{2}.$$

E 9.2.

(a) $\sin 20° + \sin 40° = 2\sin\frac{20° + 40°}{2}\cos\frac{20° - 40°}{2} = 2\sin 30° \cos(-10°)$

$$= 2 \cdot \frac{1}{2} \cdot \cos 10° = \cos(90° - 80°) = \sin 80°.$$

(b) In part a, use the reduction formula $\sin(90° - \alpha) = \cos \alpha$ to change sine to cosine. Thus, $\sin 20° = \cos 70°$, $\sin 40° = \cos 50°$, $\sin 80° = \cos 10°$.

E 9.3.

$$\sin(60° + \theta) - \sin(60° - \theta) = 2\cos\frac{60° + \theta + 60° - \theta}{2}\sin\frac{60° + \theta - 60° + \theta}{2} =$$

$$2\cos 60° \sin \theta = 2 \cdot (1/2)\sin \theta = \sin \theta.$$

E 9.4.

$$\sin(\alpha+\beta)\sin(\alpha-\beta)=\frac{1}{2}(\cos 2\beta-\cos 2\alpha)$$

$$=\frac{1}{2}\left[\cos^2\beta-\sin^2\beta-\left(\cos^2\alpha-\sin^2\alpha\right)\right]$$

$$=\frac{1}{2}\left(\cos^2\beta-\sin^2\beta-\cos^2\alpha+\sin^2\alpha\right)$$

$$=\frac{1}{2}\left[\left(1-\sin^2\beta\right)-\sin^2\beta-\left(1-\sin^2\alpha\right)+\sin^2\alpha\right]$$

$$=\frac{1}{2}\left(2\sin^2\alpha-2\sin^2\beta\right)=\sin^2\alpha-\sin^2\beta.$$

E 9.5.

$$\tan\frac{\theta}{2}=\frac{\sin(\theta/2)}{\cos(\theta/2)}=\frac{2\sin(\theta/2)\cos(\theta/2)}{2\cos(\theta/2)\cos(\theta/2)}=\frac{\sin\theta}{2\cos^2(\theta/2)}=\frac{\sin\theta}{1+\cos\theta}.$$

The identity $\dfrac{\sin\theta}{1+\cos\theta}=\dfrac{1-\cos\theta}{\sin\theta}$ can be proved by cross-multiplication, and using the main identity $\sin^2\theta+\cos^2\theta=1$. Finally, $\dfrac{1-\cos\theta}{\sin\theta}=\dfrac{1}{\sin\theta}$

$$-\frac{\cos\theta}{\sin\theta}=\csc\theta-\cot\theta.$$

EP 9.1. Professor Smartman cut the rope into two pieces of 25 and 50 m. Then he attached one end of a piece of 25 m to the top and made a loop at the end of the second end. Then he put in one end of a piece of 50 m long in a loop and tied it to the second end. In total, he got a rope 50 meters long. This is enough to go down to the ledge, on which he pulled out a piece of rope 50 meters long from the loop. Finally, he attached the rope to the ledge and went down.

EP 9.2. Professor Smartman should add a fourth pill to these three pills, then cut each of them in half and put these halves in two different heaps. As the result, in each heap there will be two halves of a cobra pill and two halves of a viper pill, which together make up two whole pills: one from the cobra and one from the viper. So, today he takes pills from one heap (these are four halves), and tomorrow from another.

Chapter 10

P 10.2. For gradian measure, one full cycle corresponds to 400 grad. We can set up the proportion

$$\frac{\theta_{grad}}{s} = \frac{400}{2\pi r} \implies s = \frac{2\pi r\theta_{grad}}{400} = \frac{\pi}{200}\theta_{grad}r,$$

so $k = \pi/200 \approx 0.0157$.

P 10.4. Substitute $\theta_r = 1/2$ into the formula $s = \theta_r \cdot r$ and get $s = r/2$.

P 10.11. Replace π with $180°$ and reduce the fraction.

P 10.14. It is easier to convert radians to degrees by replacing π with $180°$ and reducing each fraction. The values of trig functions for special angles are already calculated in Chapter 6.

E 10.1.

(a) Using the basic relation $180° = \pi_{rad}$, we can set up the proportion

$$\frac{180°}{\pi} = \frac{50°}{\theta_r} \implies \theta_r = \frac{50°\pi}{180°} \approx 0.87.$$

(b) Using the same basic relation as in part a, we can set up the proportion

$$\frac{180°}{\pi} = \frac{\theta°}{1.7} \implies \theta° = \frac{180° \cdot 1.7}{\pi} \approx 97.4°.$$

E 10.2. Sector area S is defined by the equation: $S = r^2\theta_r/2$. Solve it for r: $r^2 = 2S/\theta_r \implies r = \sqrt{2S/\theta_r}$. In radians, the given angle of $30°$ is $\theta_r = \pi/6$. Substitute this value and $S = 3.6$ into the above formula and get the radius:

$$r = \sqrt{\frac{2 \cdot 3.6}{\pi/6}} \approx 3.71\text{feet}.$$

E 10.3. According to given ratios, we can denote the measures of the angles as $3x$, $4x$ and $5x$, where x is an unknown number. To find x, we can use the property that the sum of all angles in any triangle is 180°, which is π radians. Therefore, we can set up the equation $3x + 4x + 5x = \pi$. From here, $x = \pi/12$, and the angles are $3\pi/12 = \pi/4$, $4\pi/12 = \pi/3$, and $5\pi/12$.

E 10.4. It is given that the central angle $\theta_r = \pi/8$. The entire circle contains 2π radians. Therefore, the number of cabs is $2\pi \div \pi/8 = 16$. The radius $r = 16/2 = 8$ m. Hence the arc length $s = \theta_r\, r = (\pi/8) \cdot 8 \approx 3.14$ m.

E 10.5. Given: $r = 30$ m, $\theta_r = 6$. The problem is to find arc s: $s = \theta_r\, r = 6 \cdot 30 = 180$ m per min.

E 10.6. The second hand makes a full revolution of 2π in 1 min, or 60 s. Therefore, in 1 s it makes $\alpha = 2\pi/60 = \pi/30$ radians. It is given that $r = 1.5$ m $= 150$ cm and $s = 48$ cm. We have $s = \theta_r\, r \;\Rightarrow\; \theta_r = s/r = 48/150 = 8/25$. The fly will spend

$$\theta_r / \alpha = (8/25) \div (\pi/30) = 48/5\pi \approx 3 \ s.$$

E 10.7. If $\alpha = \pi/3$, the triangle ABC is equilateral and the chord $AB = 1$. The arc $\overset{\frown}{AB} = \alpha = \pi/3 \approx 1.05$. If $\alpha = 2\pi/3$, then $A = B = (\pi - 2\pi/3) \div 2 = \pi/6$. By the Law of Sines,

$$\frac{1}{\sin(\pi/6)} = \frac{AB}{\sin(2\pi/3)} \;\Rightarrow\; AB = \frac{\sin(2\pi/3)}{\sin(\pi/6)} = \frac{\sqrt{3}/2}{1/2} = \sqrt{3} \approx 1.73.$$

The arc $\overset{\frown}{AB} = \alpha = 2\pi/3 \approx 2.1$. Here are the results:
For $\alpha = \pi/3$, chord $AB = 1$ and arc $\overset{\frown}{AB} = \pi/3$.
For $\alpha = 2\pi/3$, chord $AB \approx 1.73$ and arc $\overset{\frown}{AB} = 2\pi/3$.
Thus, the chord is not proportional to the central angle.

E 10.8. We have, arc $s = \theta r \;\Rightarrow\; \theta = s/r$. To find the chord c, we can use either the Law of Sines or the Law of Cosines. Let's use both.

(1) Usage of the Law of Sines.
    Since $2A + \theta = \pi$, then $A = (\pi - \theta)/2 = \pi/2 - \theta/2$. By the Law of Sines,

$$\frac{c}{\sin\theta} = \frac{r}{\sin A} \;\Rightarrow\; c = \frac{r\sin\theta}{\sin(\pi/2 - \theta/2)} = \frac{r2\sin(\theta/2)\cos(\theta/2)}{\cos(\theta/2)} \Rightarrow$$

$c = 2r\sin(\theta/2) = 2r\sin(s/2r) = d\sin(s/d)$, where d is the diameter.

(2) Usage of the Law of Cosine.

$$c^2 = r^2 + r^2 - 2r \cdot r \cos \theta = 2r^2 - 2r^2 \cos \theta = 4r^2 \left(\frac{1 - \cos \theta}{2} \right) =$$

$d^2 \sin^2 \theta/2 = d^2 \sin^2 (s/d)$. From here, $c = d \sin (s/d)$.

E 10.9. From Problem 10.8, $\sin x < x$, if $0 < x < \pi/2$. Since $\sin 0 = 0$ and $\sin (\pi/2) = 1$, the value $x = 0$ is the only one for which $\sin x = x$ is in the given interval.

E 10.10. The number 100 can be presented in the form: $100 = 15(2\pi) + \theta$, where $\theta \approx 5.8$.

Since 100-radian angle and θ differ by $15(2\pi)$, both of them lie in the same quadrant. Angle θ belongs to the interval $\left(\dfrac{3\pi}{2} \approx 4.71, 2\pi \approx 6.28 \right)$ Therefore, both angle θ and 100-radian angle lie in the 4th quadrant.

E 10.11.

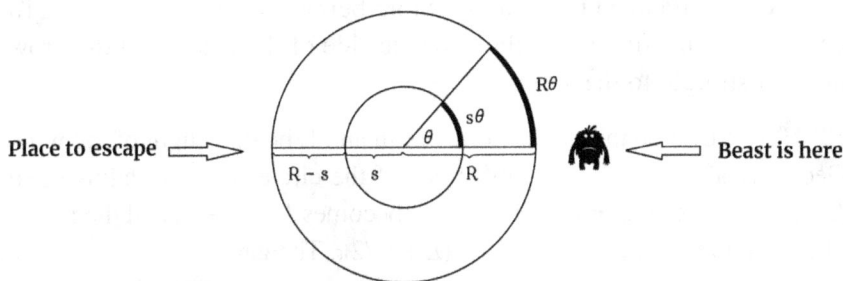

Place to escape ⇨ $R\theta$ $s\theta$ θ $R - s$ s R ⇦ Beast is here

Obviously, when the boat is near the center of the lake, the professor can move around the center faster than the beast along the shore. This means that at a certain distance from the center, it is easy for the professor to be on the opposite side of the beast.

Let's determine the acceptable distance between the boat and the center at which the boat will move around the lake faster than the beast. The largest distance from the center is that, at which the angular speed of the boat around the circle will be equal to the angular speed of the beast along the shore. Denote this distance by r. Suppose that at a certain time, both the boat and the beast would both shift by the same central angle θ radians. Then the boat will cover a distance of θr, and the beast a distance of $4\theta r = \theta R$, where R is the radius of the lake. From here $r = \dfrac{R}{4}$.

This means that if the professor is at a distance $s < \dfrac{R}{4}$ from the center, then his boat will move around the circle faster than the beast. Therefore, being at such a distance, the professor can position his boat on the opposite side of the beast.

Now, let's find the range for the distance s, at which the professor could reach the shore before the beast. After his boat is on the opposite side from the beast, it would need to travel the remaining distance of $R - s$. This distance should be 4 times less than the distance for the beast. The distance for the beast is half of the circle, which is πR, so $R - s < \pi R/4$. Solving this inequality for s, we get $s > R\left(1 - \dfrac{\pi}{4}\right)$. Taking into account that $s < \dfrac{R}{4}$, we obtain the following range for s:

$R(1 - \pi/4) < s < R/4$, or $0.21R < s < 0.25R$. Thus, if the professor will be at a distance s (in the range $0.21R < s < 0.25R$) away from the center, then he is guaranteed to reach the shore before the beast gets there. To do this, he must first be on the opposite side of the beast, and then row his boat straight to the shore.

EP 10.1. Let R be the radius of the Earth and L be the length of the wire. Geometrically, L is the circumference of the circle with the radius R, so $L = 2\pi R$. By extending L by 1 meter, it becomes $L_1 = L + 1$, and the radius of the circle becomes $R_1 = L_1/2\pi = (L + 1)/2\pi$. The gap is

$$R_1 - R = (L + 1)/2\pi - L/2\pi = 1/2\pi \approx 0.16 \text{ m} = 16 \text{ cm}.$$

EP 10.2. The smart boy did the following. Since the denomination of money was not specified, he presented \$4 as 400 cents. According to the promise, the professor must square this number. The result is $400^2 = 160{,}000$ cents, which is \$1,600.

Chapter 11

E 11.1.

(a) $g(x) = f(x - 7) = 5\sin(x - 7) + (x - 7)^2$;

(b) $g(x) = f(x) - 4 = 5\sin x + x^2 - 4$;

(c) $g(x) = f(-x) = 5\sin(-x) + (-x)^2 = -5\sin x + x^2$;

(d) $g(x) = -f(-x) = -5\sin(-x) - (-x)^2 = 5 \sin x - x^2$.

E 11.2. Any point (x, y) on the graph of $f(x)$ is transformed to the point $(x - 5, 2y)$ on the graph of $g(x)$. Here is the transformation of some points:

$$(1, 1) \rightarrow (-4, 2), (2, 2) \rightarrow (-3, 4)$$
$$(4, 2) \rightarrow (-1, 4), (5, 3) \rightarrow (0, 6).$$

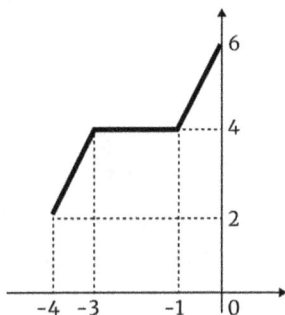

E 11.3. To solve the problem, we suggest plotting the functions $y = x^2 + 1$ and $y = \cos x$. They will clarify the explanation below. The graph of the function $y = x^2 + t$ is a parabola that it is obtained from the "standard" parabola $y = x^2$ by shifting along the y-axis by t units: up, if $t > 0$ and down, if $t < 0$. If we take a look at both graphs, $y = x^2 + t$ and $y = \cos x$, we can see three cases about the number of common points, i.e., the number of solutions of the given equation:

(1) If $t < 1$, there are two common points, so there are two solutions.

(2) If $t = 1$, there is one common point, so one solution $x = 0$.

(3) If $t > 1$, there are no common points, so there are no solutions.

E 11.4. Let's write the quadrantal angles $\pi/2$, π, and $3\pi/2$ as $5\pi/10$, $10\pi/10$, and $15\pi/10$.

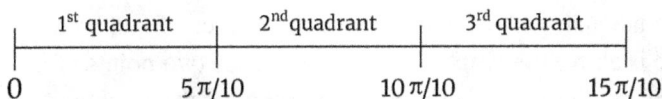

(a) Angles $\pi/10$ and $4\pi/10$ are in the 1st quadrant, in which the sine increases, so, $\sin(\pi/10) < (4\pi/10)$. Angle $7\pi/10$ is in the 2nd quadrant and is symmetric to the angle $3\pi/10$ with respect to quadrantal angle

$5\pi/10$, so sin $(7\pi/10) = \sin(3\pi/10)$. Since the angle $3\pi/10$ is between $\pi/10$ and $4\pi/10$, we have $\sin(\pi/10) < (7\pi/10) < (4\pi/10)$. All three values are positive. Angles $11\pi/10$ and $13\pi/10$ are in the 3rd quadrant, in which the sine decreases, so $\sin(13\pi/10) < \sin(11\pi/10)$. Both values are negative. Finally, we get the arrangement:

$$\sin (13\pi/10) < \sin(11\pi/10) < \sin(\pi/10) < (7\pi/10) < (4\pi/10).$$

(b) Angles, $\pi/10$ and $3\pi/10$ are in the 1st quadrant, in which the cosine decreases, so $\cos(3\pi/10) < \cos(\pi/10)$. Both values are positive. The angles $7\pi/10$ and $9\pi/10$ are in the 2nd quadrant, in which cosine decreases, so $\cos(9\pi/10) < \cos(7\pi/10)$. Angle $12\pi/10$ is in the 3rd quadrant and is symmetric to the angle $8\pi/10$ with respect to the quadrantal angle $10\pi/10$. Therefore, $\cos(12\pi/10) = \cos(8\pi/10)$. Since angle $8\pi/10$ is between $7\pi/10$ and $9\pi/10$, we have $\cos(9\pi/10) < \cos(12\pi/10) < \cos(7\pi/10)$. All three values are negative. Finally, we get the arrangement:

$$\cos(9\pi/10) < \cos(12\pi/10) < \cos(7\pi/10) < \cos(3\pi/10) < \cos(\pi/10).$$

E 11.5. The given interval lies in the 1st quadrant, in which the sine increases and cosine decreases. Also, $\sin(\pi/4) = \cos(\pi/4) = \sqrt{2}/2 \approx 0.7$. Since both α and β are less than $\pi/4$, sin $\alpha < 0.7$. and cos $\beta > 0.7$.

E 11.6. Look at the graphs of sine and cosine:

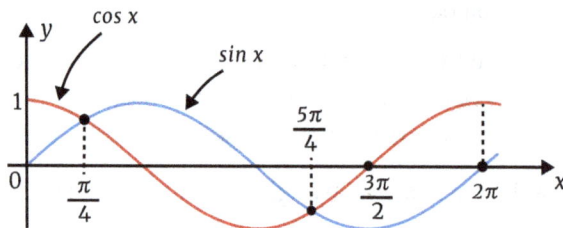

The inequality cos $x > \sin x$ holds for values of x at which the graph of cosine is above the graph of sine. You can see two points of intersection of the graphs. At one of them $x = \pi/4$. We can get the x value of the second point by subtracting $3\pi/2 - \pi/4 = 5\pi/4$. You can also see the same points of intersection of sine and cosine on the unit circle (where the values of functions are equal):

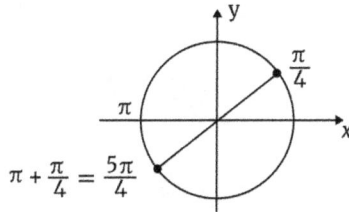

The cosine graph is above the sine graph at two intervals: $(0, \pi/4)$ and $(5\pi/4, 2\pi)$. The union of these intervals is the solution of the problem.

EP 11.1. If we consider all the triples of factors that give 36 as the product, (there are 8 triples in total), then each triple has a unique sum, except two: $(1, 6, 6)$ and $(2, 2, 9)$, the sum of which is 13. Since there is an older child, then the triple $(1, 6, 6)$ is not suitable. Thus, the age of the children is 2, 2, and 9 years old.

Chapter 12

E 12.1.

(a) We have $a = 2$, $b = 3$, $c = \pi/6$. Amplitude is $A = |a| = |2| = 2$, period $P = 2\pi/b = 2\pi/3$, and phase shift $\Phi = -\dfrac{c}{b} = -\dfrac{\pi/6}{3} = -\dfrac{\pi}{18}$. To graph, first plot the basic function $y = \sin x$ on the interval $[0, 2\pi]$ and label the quadrantal angles $0, \pi/2, \pi, 3\pi/2, 2\pi$. Then divide these labels by $b = 3$. The labels become $0, \pi/6, \pi/3, \pi/2, 2\pi/3$. Also, since $A = 2$, mark labels 2 and -2 on the y-axis. We will get the graph of the sinusoid $y = 2 \sin 3t$:

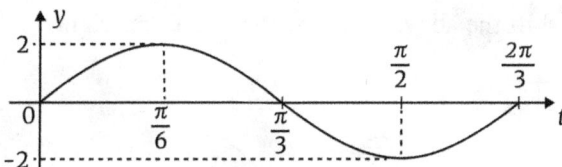

Now, add the phase shift $\Phi = -\pi/18$ to all labels on the t-axis. We get new labels

$$-\pi/18,\ \pi/9,\ 5\pi/18,\ 4\pi/9,\ 11\pi/18.$$

Finally, shift the graph by $\pi/18$ to the left:

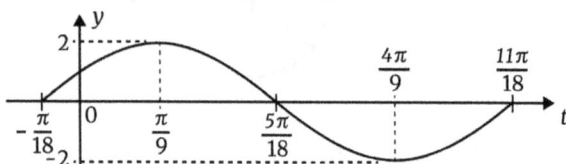

(b) We have $a = -3$, $b = 2$, $c = -\pi/4$. Amplitude is $A = |a| = |-3| = 3$, period $P = 2\pi/b = 2\pi/2 = \pi$, and phase shift $\Phi = -\dfrac{c}{b} = -\dfrac{-\pi/4}{2} = \dfrac{\pi}{8}$.

To graph, plot $y = -\sin x$ on the interval $[0, 2\pi]$ and label the quadrantal angles 0, $\pi/2$, π, $3\pi/2$, 2π on the t-axis. Then divide these labels by $b = 2$. Labels become 0, $\pi/4$, $\pi/2$, $3\pi/4$, π. Also, since $A = 3$, mark labels 3 and -3 on then y-axis. We will get the graph of the sinusoid $y = -3\sin 2t$:

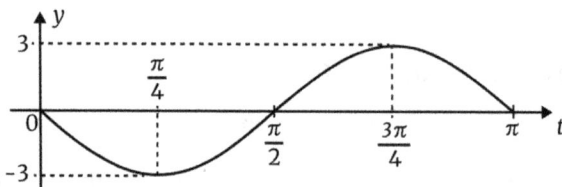

Now add the phase shift $\Phi = \pi/8$ to all labels on the t-axis. We get new labels

$$\pi/8,\ 3\pi/8,\ 5\pi/8,\ 7\pi/8,\ 9\pi/8.$$

Finally, shift the above graph by $\Phi = \pi/8$ to the right:

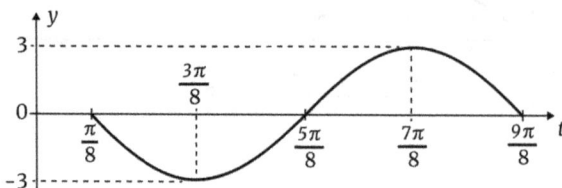

E 12.2. From the formulas for the period P and phase shift Φ, we get $b = 2\pi/P$, $c = -\Phi b$.

(a) $a = 3$, $P = \dfrac{\pi}{3} - \left(-\dfrac{\pi}{6} \right) = \dfrac{\pi}{2}$ $\Rightarrow b = 2\pi \div \dfrac{\pi}{2} = 4$, $\Phi = -\dfrac{\pi}{6}$ $\Rightarrow$

$$c = -\left(-\dfrac{\pi}{6} \right) \cdot 4 = \dfrac{2\pi}{3}.$$

Answer: $y = 3\sin(4x + 2\pi/3)$.

(b) $a = -2$, $P = \dfrac{27\pi}{4} - \dfrac{3\pi}{4} = 6\pi \Rightarrow b = \dfrac{2\pi}{6\pi} = \dfrac{1}{3}$, $\Phi = \dfrac{3\pi}{4}$ $\Rightarrow$

$$c = -\dfrac{3\pi}{4} \cdot \dfrac{1}{3} = -\dfrac{\pi}{4}.$$

Answer: $y = -2 \sin((1/3)x - \pi/4)$.

(c) $a = -4$, $P = \dfrac{14\pi}{15} - \left(-\dfrac{2\pi}{5} \right) = \dfrac{4\pi}{3}$ $\Rightarrow b = 2\pi \div \dfrac{4\pi}{3} = \dfrac{3}{2}$, $\Phi = -\dfrac{2\pi}{5}$ $\Rightarrow$

$$c = -\left(-\dfrac{2\pi}{5} \right) \cdot \dfrac{3}{2} = \dfrac{3\pi}{5}.$$

Answer: $y = -4 \sin((3/2)x + 3\pi/5)$.

(d) $a = 5$, $P = \dfrac{57\pi}{20} - \dfrac{3\pi}{5} = \dfrac{9\pi}{4}$ $\Rightarrow b = 2\pi \div \dfrac{9\pi}{4} = \dfrac{8}{9}$, $\Phi = \dfrac{3\pi}{5}$ $\Rightarrow$

$$c = -\dfrac{3\pi}{5} \cdot \dfrac{8}{9} = -\dfrac{8\pi}{15}.$$

Answer: $y = 5 \sin((8/9)x - 8\pi/15)$.

E 12.3. First, solve the equation for y: $y = -4 \sin(5t - \pi/4) - 24$.

(a) The domain is the entire number line;

(b) Since the largest value of sine is 1 and the smallest is -1, the largest value of y is $4 - 24 = -20$, and the smallest is $-4 - 24 = -28$. Therefore, the range is the interval $[-28, -20]$.

(c) $A = |-4| = 4$; (d) $P = 2\pi/5$; (e) $\Phi = -\left(-\dfrac{\pi}{4} \right) \div 5 = \dfrac{\pi}{20}$, (f) $d = -24$.

E 12.4. We will use the same approach as in Problem 12.5

(a) $\sin x + \cos x = \sqrt{2}\left(\dfrac{\sqrt{2}}{2}\sin x + \dfrac{\sqrt{2}}{2}\cos x\right) = \sqrt{2}\left(\sin x\cos\dfrac{\pi}{4} + \cos x\sin\dfrac{\pi}{4}\right)$

$= \sqrt{2}\sin\left(x + \dfrac{\pi}{4}\right).$

(b) $\sqrt{3}\sin x - \cos x = 2\left(\dfrac{\sqrt{3}}{2}\sin x - \dfrac{1}{2}\cos x\right) = 2\left(\sin x\cos\dfrac{\pi}{6} - \cos x\sin\dfrac{\pi}{6}\right)$

$= 2\sin\left(x - \dfrac{\pi}{6}\right).$

E 12.5. According to Problem 12.5, the expression $a\sin x + b\cos x$ can be written as $a\sin x + b\cos x = \sqrt{a^2 + b^2}\,\sin(x + \varphi)$, where φ is some angle. Since $|\sin(x + \varphi)| \le 1$, we conclude that $|a\sin x + b\cos x| \le \sqrt{a^2 + b^2}$.

E 12.6. We present two methods.
We can use the result the Exercise 12.4 (part a) and modify the equation like this

$$\sqrt{2}\sin\left(x + \frac{\pi}{4}\right) = \sqrt{2} \;\Rightarrow\; \sin\left(x + \frac{\pi}{4}\right) = 1 \;\Rightarrow\; x + \frac{\pi}{4} = \frac{\pi}{2} \;\Rightarrow\; x = \frac{\pi}{4}.$$

2nd method. Square both sides:

$(\sin x + \cos x)^2 = (\sqrt{2})^2 \Rightarrow \sin^2 x + 2\sin x \cos x + \cos^2 x = 2.$

Since $\sin^2 x + \cos^2 x = 1$, we get $2\sin x\cos x = 1$. By the Double Angle formula for sine, the equation becomes $\sin 2x = 1 \;\Rightarrow\; 2x = \pi/2 \;\Rightarrow\; x = \pi/4.$

EP 12.1. Three attempts are enough. You need to try two keys to the same suitcase, and then everything else is obvious.

Chapter 13

E 13.1. We have $y_{min} = 4$ in, $y_{max} = 8$ in. From here, $a = (8 - 4)/2 = 2$, $d = (8 + 4)/2 = 6$. Also, the period of oscillation is $P = 2 \cdot 0.6 = 1.2$ s, and the frequency $f = 1/1.2 = 5/6$. Therefore, the sinusoidal function $y(t) = a \sin(2\pi ft + c) + d$ has the form $y(t) = 2\sin\left(\dfrac{5}{3}\pi t + c\right) + 6$. To find c, we use the initial condition: $y(0) = y_{\min} = 4$.

Thus, $4 = 2\sin c + 6 \implies \sin c = -1 \implies c = 3\pi/2$. We get $y(t) = 2\sin\left(\dfrac{5}{3}\pi t + \dfrac{3\pi}{2}\right) + 6$. Using a reduction formula, we can simplify

this expression: $y(t) = -2\cos\left(\dfrac{5}{3}\pi t\right) + 6$. After 1 s, the height will be

$$y(1) = -2\cos\left(\dfrac{5}{3}\pi\right) + 6 = -2\cdot\dfrac{1}{2} + 6 = 5 \text{ inch.}$$

E 13.2. For the function $y(t) = a \sin(2\pi ft + c) + d$, by the condition of the problem, $f = 75$, $y(0) = d$, $y_{\max} = 124$, $y_{\min} = 72$. From here,

$$a = (124 - 72)/2 = 26,\ d = (124 + 72)/2 = 98,\ y(0) = 98.$$

We get $y(t) = 26\sin(150\pi t + c) + 98$. To find c, we set $t = 0$: $y(0) = 98$ $= 26\sin c + 98 \implies \sin c = 0$. Because the blood pressure decreases, starting from the average value, $c = \pi$. The function takes the form $y(t) = 26\sin(150\pi t + \pi) + 98$. Using a reduction formula, we can simplify the answer: $y(t) = -26\sin(150\pi t) + 98$.

E 13.3. We will use the figure from Problem 12.1, Chapter 12. The height $y(t)$ at time t is defined by the function $y(t) = a\sin(2\pi ft + \varphi) + d$. According to conditions from the problem, $a = 16$ m, $d = 16 + 3 = 19$ m, $\varphi = -\pi/2$, $f = 1/1.5 = 2/3$. We have

$$y(t) = 16\sin\left(\dfrac{4\pi}{3}t - \dfrac{\pi}{2}\right) + 19 = -16\cos\left(\dfrac{4\pi}{3}t\right) + 19.$$

$$y(1) = -16\cos\left(\dfrac{4\pi}{3}\right) + 19 = -16\cos\left(\pi + \dfrac{\pi}{3}\right) + 19$$

$$= 16\cos\left(\dfrac{\pi}{3}\right) + 19 = 16\cdot\dfrac{1}{2} + 19 = 27\,\text{m.}$$

E 13.4. We use a sinusoid as a model for the water level:

$$y(t) = a\sin(2\pi ft + c) + d.$$

The argument of the function is a time t (in hours). We treat 11:00 am as a starting time and assign $t = 0$ to it. We have $y_{max} = 20$ feet, $y_{min} = 4$ feet, $a = (20 - 4)/2 = 8$, $d = (20 + 4)/2 = 12$, $f = 1/12$. The height above the sea level is the function $y(t) = 8\sin\left(\dfrac{\pi}{6}t + c\right) + 12$. To find c, we use the initial condition $y(0) = 20$. We get $20 = 8\sin c + 12$ $\Rightarrow$ $\sin c = 1$ $\Rightarrow$ $c \pi/2$ $\Rightarrow$

$$y(t) = 8\sin\left(\frac{\pi}{6}t + \frac{\pi}{2}\right) + 12, \text{ or } y(t) = 8\cos\left(\frac{\pi}{6}t\right) + 12.$$

What remains is to solve the inequality $y(t) \geq 12$. We have $8\cos\left(\dfrac{\pi}{6}t\right) + 12 \geq 12$ $\Rightarrow$ $\cos\left(\dfrac{\pi}{6}t\right) \geq 0$. Recall that cosine is positive in the 1st and 4th quadrants. The graph of the function $y = \cos\left(\dfrac{\pi}{6}t\right)$ looks like this

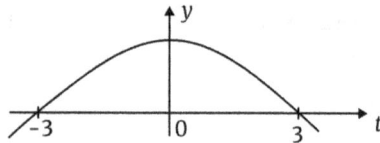

The solution of the inequality $\cos\left(\dfrac{\pi}{6}t\right) \geq 0$ is the interval $[-3, 3]$. Since the value $t = 0$ corresponds to the time 11 am, the values $t = -3$ and $t = 3$ correspond to the times of 8 am and 2 pm. So, to sail safely, the boat should leave the pier between 8 am and 2 pm.

E 13.5. Let t represents the month number, so that $t = 1$ corresponds to January, etc. We denote by $y(t)$ the average temperature in the month t. We assume that $y(t)$ is a sinusoidal function: $y(t) = a\sin(2\pi f t + c) + d$. It has a period of 12 month, i.e. $f = 1/12$. According to conditions of the problem, $y_{max} = 35°C$, $y_{min} = -5°C$. We calculate $a = (35 - (-5))/2 = 20$, $d = (35 + (-5))/2 = 15$. The function $y(t)$ takes the form $y(t) = 20\sin\left(\dfrac{\pi}{6}t + c\right) + 15$. To find c, we use the condition that the maximum average temperature occurs in August ($t = 8$):

$$y(8) = 35 = 20\sin\left(\frac{8\pi}{6} + c\right) + 15 \;\Rightarrow\; \sin\left(\frac{4\pi}{3} + c\right) = 1 \;\Rightarrow\; \frac{4\pi}{3} + c = \frac{\pi}{2} \;\Rightarrow$$

$c = -5\pi/6$. Therefore, $y(t) = 20\sin(\pi t/6 - 5\pi/6) + 15$.

What remains is to solve the inequality $y(t) < 25$. We have $20\sin(\pi t/6 - 5\pi/6) + 15 < 25 \;\Rightarrow\; \sin(\pi t/6 - 5\pi/6) < 1/2$. To solve the last inequality, we can graph $\sin(\pi t/6 - 5\pi/6)$ in the interval $[1, 13]$:

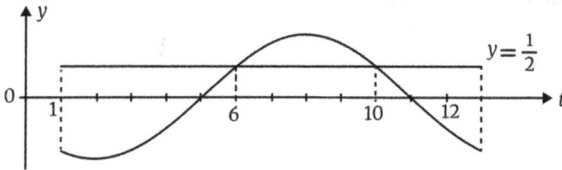

Solutions to the inequality are the values of t for which the graph of sine is below the graph of the line $y = 1/2$. These values are from the intervals $[1, 6)$ and $(10, 12]$. They correspond to the months from January to May and from November to December. We can combine these months to one interval: from November to May.

EP 13.1. Tourists have come up with the following strategy: The first person to be asked will count the number of white hats on all others. If this number is even, then he will reply "White hat," otherwise (if the number of hats is odd), he will reply "Black hat." Based on this information, the other tourists will be able to determine the color of their own hat by looking at others. Only the first tourist will have to rely purely on luck. The surprising thing about this problem is that with only one answer from the first tourist, all other tourists will know the color of their hat.

Chapter 14

P 14.5. There are no such intervals.

P 14.6. Tangent is discontinuous at $x = \pi/2$ in a one-period interval $(-\pi/2, \pi/2]$. Since the period of tangent is π, all points of discontinuity can be described by the formula: $x = \pi/2 + n\pi$, where n is any integer.

P 14.8. In the answers below, n means any integer.

(1) The period is π.
(2) The domain is the union of intervals $(n\pi, (n + 1)\pi)$. Using the notation $\cup$ for the union, we can describe the domain as $\bigcup_{n=-\infty}^{\infty}(n\pi, (n+1)\pi)$.
(3) The range is the entire number line: $(-\infty, \infty)$.
(4) It has point symmetry with respect to the origin.
(5) It takes the positive values in the intervals $(n\pi, \pi/2 + n\pi)$. It takes negative values in the intervals $(\pi/2 + n\pi, (n + 1)\pi)$.
(6) It equals to zero at points $x = \pi/2 + n\pi$.
(7) It does not increase. It decreases at $(n\pi, (n + 1)\pi)$.
(8) It has vertical asymptotes. The equations are $x = n\pi$.

E 14.1. We just need to stretch the graph of $y = \tan x$ twice along the y-axis. In essence, the shape of the graph will be the same as for $y = \tan x$.

E 14.2. To get our graph, we shift the graph of $y = \tan x$ by $\pi/3$ to the left:

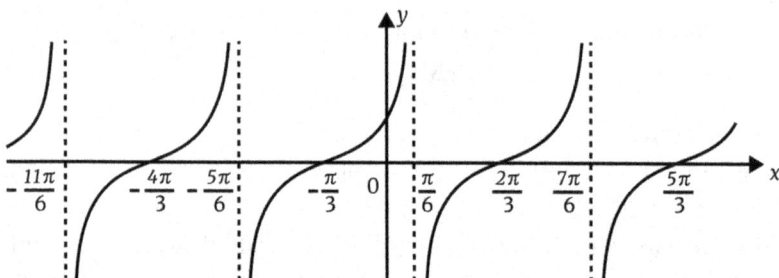

E 14.3. Let P be the period. Then $f(x + P) = f(x)$ $\Rightarrow$ $\tan[3(x + P) + 4]$ $= \tan(3x + 4 + 3P) = \tan(3x + 4)$. Since tangent has a period of π, we conclude that $3P = \pi$ $\Rightarrow$ $P = \pi/3$.

E 14.4. Let's modify the given function:

$$\frac{\sin x / \cos x - \cos x / \sin x}{\sin x / \cos x + \cos x / \sin x} = \frac{(\sin x / \cos x - \cos x / \sin x)\sin x \cos x}{(\sin x / \cos x + \cos x / \sin x)\sin x \cos x}$$

$$= \frac{\sin^2 x - \cos^2 x}{\sin^2 x + \cos^2 x} = -\left(\cos^2 x - \sin^2 x\right)/1 = -\cos 2x.$$

The only difference between graphs of the original function and $y = -\cos 2x$ is that the points $x = n\pi/2$ (n is any integer) must be excluded from the domain of the original function since tangent or cotangent are not defined at these points. The graph looks like this

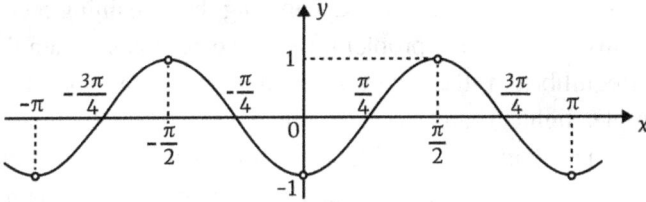

E 14.5. Let's write the quadrantal angles $\pi/2$ and π as $5\pi/10$ and $10\pi/10$.

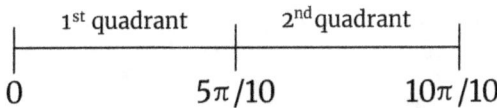

(a) The angles $\pi/10$ and $4\pi/10$ are in the 1st quadrant, in which the tangent increases, so, $\tan(\pi/10) < \tan(4\pi/10)$. Both values are positive. The angles $7\pi/10$ and $9\pi/10$ are in the 2nd quadrant, in which the tangent also increases, so, $\tan(7\pi/10) < \tan(9\pi/10)$. Both values are negative. Also, $\tan(13\pi/10) = \tan(3\pi/10 + \pi) = \tan(3\pi/10)$, which is between $\tan(\pi/10)$ and $\tan(4\pi/10)$. We get the arrangement

$$\tan(7\pi/10) < \tan(9\pi/10) < \tan(\pi/10) < \tan(13\pi/10) < \tan(4\pi/10).$$

(b) In the 1st and 2nd quadrants cotangent decreases. Therefore, $\cot(3\pi/10) < \cot(\pi/10)$ and $\cot(9\pi/10) < \cot(7\pi/10)$. $\cot(12\pi/10) = \cot(2\pi/10 + \pi) = \cot(2\pi)$. Therefore, $\cot(3\pi/10) < \cot(12\pi/10) < \cot(\pi/10)$. We have the arrangement

$$\cot(9\pi/10) < \cot(7\pi/10) < \cot(3\pi/10) < \cot(12\pi/10) < \cot(\pi/10).$$

E 14.6. In the interval $(\pi/4, \pi/2)$ tangent increases and $\tan(\pi/4) = 1$. Therefore, $\tan\theta > 1$.

EP 14.1. Since all the labels are false, then BW is either B or W, but not both. Let's pick a ball from it. Without loss of generality, we can assume that we get a white ball. Thus, this basket is W. Then B cannot be white (white is already taken), and it cannot be B (by condition). Therefore, it is BW, and the other one is B.

EP 14.2. Place 3 coins on each side of the scale. If the scales balance, then all these 6 coins are genuine. Comparing the remaining 5 coins with genuine coins, we solve the problem. But if we get a case when the scales are not in equilibrium, then we take 3 coins from any side of the scale (for example, lighter), and compare them with the remaining genuine ones. So, the problem is solved. The problem can also be solved in general form for any number of coins $N > 2$. The number N should be presented as $N = 3n + r$, where $0 \leq r \leq 2$. In the first weighing, place n coins on each side of the balance. Then act, as we did for 11 coins.

Chapter 15

P 15.2. The value of $-\sqrt{3}/2$ lies on the vertical axis below the x-axis, so angle t is in the 3rd and 4th quadrants. Drawing the unit circle, and keeping in mind that $\sin(\pi/3) = \sqrt{3}/2$, we can find two angles with the given value of sine: $t_1 = \pi + \pi/3 = 4\pi/3$ and $t_2 = 2\pi - \pi/3 = 5\pi/3$. Therefore, all solutions are defined by two infinite sets:

$$t_1 = 4\pi/3 + 2n\pi \text{ and } t_2 = 5\pi/3 + 2n\pi, \text{ where } n \text{ is any integer.}$$

P 15.6. Look at the corresponding figure.

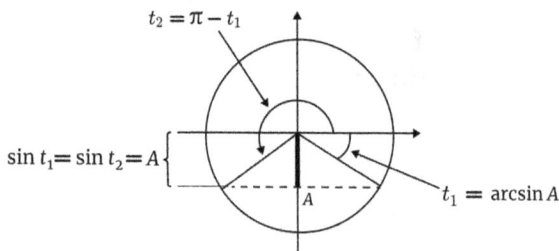

Here, we have the same expressions for t_1 and t_2 as in the previous problem. Note that since arcsin A is negative, we add a positive number $(-t_1)$ to π to get t_2.

E 15.1. Denote x_r as the reference angle for the angle x.

(a) $2\sqrt{2} \sin x - 1 = 1 \Rightarrow \sin x = \sqrt{2}/2 \Rightarrow \sin x_r = \sqrt{2}/2 \Rightarrow x_r = \pi/4$.
Since $\sin x$ is positive, angle x is in the 1st and 2nd quadrants. We have two solutions: $x_1 = \pi/4$ and $x_2 = \pi - \pi/4 = 3\pi/4$.
Answer: $\{\pi/4, 3\pi/4\}$.

(b) $2\sqrt{3} \sin x + 4 = 1 \Rightarrow \sin x = -\sqrt{3}/2 \Rightarrow \sin x_r = \sqrt{3}/2 \Rightarrow x_r = \pi/3$.
Since $\sin x$ is negative, angle x is in the 3rd and 4th quadrants. We have two solutions: $x_1 = \pi + \pi/3 = 4\pi/3$ and $x_2 = 2\pi - \pi/3 = 5\pi/3$.
Answer: $\{4\pi/3, 5\pi/3\}$.

(c) $2\sqrt{3} \sin x - 3 = 1 \Rightarrow \sin x = 2\sqrt{3}/3 \approx 1.15 > 1$.

No solutions.

E 15.2.
(a) $4 \sin x - 1 = 2 \Rightarrow \sin x = 3/4 \Rightarrow \sin x_r = 3/4 \Rightarrow x_r = \sin^{-1}(3/4)$
≈ 0.85. Since $\sin x$ is positive, angle x is in the 1st and 2nd quadrants. We have two solutions: $x_1 = 0.85$ and $x_2 = \pi - 0.85 = 2.29$.
Answer: $\{0.85, 2.29\}$.

(b) $6 \sin x + 5 = 3 \Rightarrow \sin x = -1/3 \Rightarrow \sin x_r = 1/3 \Rightarrow$
$x_r = \sin^{-1}(1/3) \approx 0.34$. Since $\sin x$ is negative, angle x is in the 3rd and 4th quadrants. We have two solutions: $x_1 = \pi + 0.34 = 3.48$ and $x_2 = 2\pi - 0.34 = 5.94$.
Answer: $\{3.48, 5.94\}$.

E 15.3. In all answers, n stands for an arbitrary integer.
(a) $2 \sin x - 3 = -2 \Rightarrow \sin x = 1/2$.
Since $\sin^{-1}(1/2) = \pi/6$, $x = (-1)^n (\pi/6) + n\pi$.
(b) $2\sqrt{3} \sin x + 5 = 2 \Rightarrow \sin x = -\sqrt{3}/2$. Since $\sin^{-1}(-\sqrt{3}/2) = -\pi/3$, we have $x = (-1)^n(-\pi/3) + n\pi$ or $x = (-1)^{n+1} (\pi/3) + n\pi$.
(c) $3\sqrt{2} \sin x + 8 = 2 \Rightarrow \sin x = -\sqrt{2} \approx -1.41 < -1$. No solutions.
(d) $5 \sin x - 2 = 1 \Rightarrow \sin x = 3/5$. Since $\sin^{-1}(3/5) \approx 0.64$,
$x = (-1)^n 0.64 + n\pi$.
(e) $7 \sin x + 6 = 2 \Rightarrow \sin x = -4/7$. Since $\sin^{-1}(-4/7) \approx -0.61$, we have $x = (-1)^n (-0.61) + n\pi$ or $x = (-1)^{n+1} 0.61 + n\pi$.

(f) $2 \sin 3x + 1 = 2 \implies \sin 3x = 1/2 \implies 3x = (-1)^n (\pi/6) + n\pi \implies$
$x = (-1)^n (\pi/18) + n\pi/3$.

(g) $2 \sin(x-\pi/3) = -\sqrt{3} \implies \sin(x-\pi/3) = -\sqrt{3}/2 \implies x-\pi/3 =$
$(-1)^n (-\pi/3) + n\pi$, $x = \pi/3 + (-1)^{n+1} (\pi/3) + n\pi$. It is convenient to split
these solutions into two sets, taking separately even and odd n.

If $n = 2k$ (even n), then $x = \pi/3 + (-1)^{2k+1} (\pi/3) + 2k\pi = \pi/3 - \pi/3 +$
$2k\pi = 2k\pi$.
If $n = 2k + 1$ (odd n), then

$x = \pi/3 + (-1)^{2k+2} (\pi/3) + (2k + 1) \pi = \pi/3 + \pi/3 + 2k\pi + \pi = 5\pi/3 + 2k\pi$.
Answer: There are two infinite sets: $2k\pi$ and $5\pi/3 + 2k\pi$, where k is an
arbitrary integer.

(h) $\sin^3 x = \sin x \implies \sin^3 x - \sin x = 0 \implies \sin x(\sin^2 x - 1) = 0 \implies$
$\sin x (\sin x - 1)(\sin x + 1) = 0$. We can split this equation into three:
$\sin x = 0$, $\sin x - 1 = 0$, and $\sin x + 1 = 0$. The first equation has the
solution $x = n\pi$, the second $x = \pi/2 + 2n\pi$, and the third $x = 3\pi/2 +$
$2n\pi$. All three solutions represent the quadrantal angles, and we can
combine them in one: $x = n\pi/2$.

(i) $2 \sin^2 x - 5 \sin x - 3 = 0$. We can treat this equation as a quadratic one
with respect to $\sin x$. It can be factored $(2 \sin x + 1)(\sin x - 3) = 0$. We
split it into two equations $2 \sin x + 1 = 0$ and $\sin x - 3 = 0$. From the
first one, $\sin x = -1/2 \implies x = (-1)^{n+1} (\pi/6) + n\pi$. From the second
one, $\sin x = 3 > 1$. The last equation does not have solutions.

(j) $3 \sin^2 x - \cos^2 x = 0$. From the Main Identity, $\cos^2 x = 1 - \sin^2 x$.
Substitute this expression into the original equation, and the last
becomes quadratic with respect to $\sin x$: $3 \sin^2 x - (1 - \sin^2 x) = 0 \implies$
$4 \sin^2 x - 1 = 0 \implies (2 \sin x - 1)(2 \sin x + 1) = 0$. We split it into
two equations $2 \sin x - 1 = 0$ and $2 \sin x + 1 = 0$. The first one has the
solutions $x = (-1)^n (\pi/6) + n\pi$ and the second one $x = (-1)^{n+1} (\pi/6) + n\pi$.
We can combine both sets into one: $x = \pm\pi/6 + n\pi$.

(k) $\sin^3 2x + \sin x = 2$. Since sine does not exceed 1, the equation can
have solutions only if $\sin 2x = \sin x = 1$. However, this is not possible. Indeed, $\sin x = 1$ for $x = \pi/2 + 2n\pi$. But $\sin 2x$ for this x equals
to 0. Therefore, the equation does not have solutions.

(l) $\sin^2 2x \cdot \sin x = 1$. For the same reason as in part (k), the equation does
not have solutions.

E 15.4. By the definition of arcsine, the range of this function is the interval $[-\pi/2, \pi/2]$. The number 2 is outside of this interval.

E 15.5. Let $\alpha = \arcsin (\sin \theta)$. By definition, $\sin \alpha = \sin \theta$, and α lies in the interval $[-\pi/2, \pi/2]$ (1st or 4th quadrants).

(a) According to conditions, angle θ is in the 2nd or 3rd quadrants. Let angle θ be in the 2nd quadrant. Then $\sin \theta$ is positive, and the same is true for $\sin \alpha$. Therefore, angle α is in the 1st quadrant, and it is equal to the reference angle of θ, which is $\pi - \theta$.
Let angle θ be in the 3rd quadrant. Then $\sin \theta$ is negative, and the same is true for $\sin \alpha$. Therefore, angle α is the 4th quadrant, and it is equal to the negative reference angle of θ, i.e., $-(\theta - \pi) = \pi - \theta$.

(b) According to conditions, angle θ is in the 4th quadrant. Then $\sin \theta$ is negative, and the same is true for $\sin \alpha$. Therefore, angle α is in the 4th quadrant also, and it is equal to the negative reference angle of θ, i.e., $-(2\pi - \theta) = \theta - 2\pi$.

E 15.6. No. The domain of arcsine is the interval $[-1,1]$ and arcsin 1.2 does not exist.

EP 15.1. Let trains to Eli leave the station, for example, at 1:00, 2:00, etc. And trains to Ben leave at 1:15, 2:15, etc. If Mike comes to the station between 1:00 and 1:15, he will take the train to Ben, (since the train to Eli has already left at 1:00). And if Mike comes to the station between 1:15 and 2:00, then he will take the train to Eli (the train to Ben has already left at 1:15). It is clear that Mike is more often in the 45-min interval than in the 15-min interval. Thus, Mike visits Eli more often than Ben.

Chapter 16

E 16.1.

(a) $2\sqrt{2} \cos x + 3 = 5 \Rightarrow \cos x = \sqrt{2}/2 \Rightarrow x_1 = \cos^{-1}(\sqrt{2}/2) = \pi/4,$
$x_2 = 2\pi - \pi/4 = 7\pi/4.$
Answer: $\{\pi/4, 7\pi/4\}.$

(b) $2\sqrt{3} \cos x + 5 = 2 \Rightarrow \cos x = -\sqrt{3}/2 \Rightarrow x_1 = \cos^{-1}(-\sqrt{3}/2) = 5\pi/6,$
$x_2 = 2\pi - 5\pi/6 = 7\pi/6.$
Answer: $\{5\pi/6, 7\pi/6\}.$

(c) $3\sqrt{2}\cos x - 7 = 2 \Rightarrow \cos x = 3\sqrt{2}/2 > 1.$

No solutions.

(d) $\sqrt{3}\tan x + 2 = 3 \Rightarrow \tan x = \sqrt{3}/3 \Rightarrow x_1 = \tan^{-1}(\sqrt{3}/3) = \pi/6,$
$x_2 = \pi/6 + \pi = 7\pi/6.$

Answer: $\{\pi/6, 7\pi/6\}.$

(e) $\sqrt{3}\tan x + 6 = 3 \Rightarrow \tan x = -\sqrt{3} \Rightarrow x_1 = \tan^{-1}(-\sqrt{3}) + \pi = -\pi/3 + \pi = 2\pi/3, x_2 = -\pi/3 + 2\pi = 5\pi/3.$

Answer: $\{2\pi/3, 5\pi/3\}.$

E 16.2.

(a) $5\cos x - 2 = 1 \Rightarrow \cos x = 3/5 \Rightarrow x_1 = \cos^{-1}(3/5) \approx 0.93,$
$x_2 = 2\pi - 0.93 = 5.35.$
Answer: $\{0.93, 5.35\}.$

(b) $7\cos x + 6 = 2 \Rightarrow \cos x = -4/7 \Rightarrow x_1 = \cos^{-1}(-4/7) \approx 2.18,$
$x_2 = 2\pi - 2.18 = 4.10.$
Answer: $\{2.18, 4.10\}.$

(c) $3\tan x - 1 = 5 \Rightarrow \tan x = 2 \Rightarrow x_1 = \tan^{-1}(2) \approx 1.11, x_2 = 1.11 + \pi = 4.25.$
Answer: $\{1.11, 4.25\}.$

(d) $4\tan x + 7 = 5 \Rightarrow \tan x = -1/2 \Rightarrow x_1 = \tan^{-1}(-1/2) + \pi \approx -0.46 + \pi = 2.68, x_2 = -0.46 + 2\pi = 5.82.$
Answer: $\{2.68, 5.82\}.$

E 16.3. In all answers, n stands for an arbitrary integer.
(a) Factor the given equation as a quadratic equation:
$(2\cos x - 1)(\cos x + 1) = 0 \Rightarrow$
$2\cos x - 1 = 0$ or $\cos x + 1 = 0.$ From the first equation,
$\cos x = 1/2 \Rightarrow x = \pm\pi/3 + 2n\pi.$ From the second equation,
$\cos x = -1 \Rightarrow x = \pi + 2n\pi.$
Answer: Two sets: $\pm\pi/3 + 2n\pi$ and $\pi + 2n\pi.$

(b) Replace: $\sin^2 x = 1 - \cos^2 x$ $\Rightarrow$ $1 - \cos^2 x - 2\cos x + 2 = 0$ $\Rightarrow$ $\cos^2 x + 2\cos x - 3 = 0$ $\Rightarrow$ $(\cos x - 1)(\cos x + 3) = 0$ $\Rightarrow$ $\cos x - 1 = 0$ or $\cos x + 3 = 0$. From the first equation, $\cos x = 1$ $\Rightarrow$ $x = 2n\pi$. From the second equation, $\cos x = -3 < -1$, so no solution.

Answer: $x = 2n\pi$.

(c) $\cos x = \sin 2x$ $\Rightarrow$ $\cos x = 2\sin x\cos x$ $\Rightarrow$ $2\sin x\cos x - \cos x = 0$ $\Rightarrow$ $\cos x(2\sin x - 1) = 0$ $\Rightarrow$ $\cos x = 0$ or $2\sin x - 1 = 0$. From the first equation, $x = \pi/2 + n\pi$. From the second equation, $\sin x = 1/2$ $\Rightarrow$ $x = (-1)^n\, (\pi/6) + n\pi$.

Answer: Two sets: $\pi/2 + n\pi$ and $(-1)^n\, (\pi/6) + n\pi$.

(d) $\tan^2 x = 1$ $\Rightarrow$ $\tan x = 1$ or $\tan x = -1$. From the first equation, $x = \pi/4 + n\pi$. From the second equation, $x = -\pi/4 + n\pi$. It is possible to combine these two sets into one.

Answer: $x = \pi/4 + n\pi/2$.

(e) The range of sine and cosine is the interval $[-1,1]$. Therefore, the given equation is equivalent to the system of two equations: $\sin x = 1$ and $\cos 2x = -1$. The first equation has solutions $x = \pi/2 + 2n\pi$. From the second equation we get $2x = \pi + 2n\pi$ $\Rightarrow$ $x = \pi/2 + n\pi$. Note that the first set of solutions is a subset of the second. Therefore, the solution is the intersection of these sets.

Answer: $x = \pi/2 + 2n\pi$.

E 16.4. When we square both sides of an equation, we may get additional roots that are extraneous to the original equation. In Sofia's case, the equation $(\sin x + \cos x)^2 = 1$ can also be obtained by squaring another equation $\sin x + \cos x = -1$. The last equation provides additional roots which don't belong to the original equation. Therefore, we need to check the roots with the original equation. Let's check the quadrantal angles 0, $\pi/2$, π and $3\pi/2$. You can see that the angles 0 and $\pi/2$ are the roots, but π and $3\pi/2$ are not. So, the correct final answer is two sets: $2n\pi$ and $\pi/2 + 2n\pi$.

E 16.5. By definition, arccosine lies in the 1st or 2nd quadrant where sine is positive. So, the sign is $+$, and the formula becomes $\sin(\arccos x) = \sqrt{1 - x^2}$.

E 16.6. Let $\theta = \arcsin x$. Then $\sin\theta = x$. The problem is to find $\cos\theta$. By the definition of arcsine, angle θ lies in the 1st or 4th quadrants where cosine is positive. Therefore, $\cos(\arcsin x) = \cos\theta = \sqrt{1 - \sin^2\theta} = \sqrt{1 - x^2}$.

E 16.7. Let $\theta = \arctan x$. Then $\tan \theta = x$. First, consider the case when x is a positive number. Then θ lies in the 1st quadrant, i.e., θ is an acute angle. We can draw a right triangle with the angle θ such that $\tan \theta = x$.

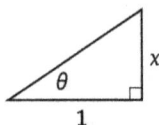

By the Pythagorean theorem, the hypotenuse of this triangle is $\sqrt{1+x^2}$. We have $\sin \theta = \dfrac{x}{\sqrt{1+x^2}}$ and $\cos \theta = \dfrac{1}{\sqrt{1+x^2}}$. If x is nonpositive, we can use odd property of both sine and arctangent, and even property of cosine to get the same results.

E 16.8. Let $\alpha = \arccos(\cos \theta)$. Then α lies in the 1st or 2nd quadrants and $\cos \alpha = \cos \theta$.

(1) If θ lies in the interval $[\pi, 3\pi/2]$ (the 3rd quadrant), then its reference angle $\theta_r = \theta - \pi$. Since $\cos \theta < 0$, angle α lies in the 2nd quadrant and $\alpha = \pi - \theta_r = \pi - (\theta - \pi) = 2\pi - \theta$.

(2) If θ lies in the interval $[3\pi/2, 2\pi]$ (the 4th quadrant), then its reference angle $\theta_r = 2\pi - \theta$. Since $\cos \theta > 0$, α lies in the 1st quadrant and $\alpha = \theta_r = 2\pi - \theta$.

E 16.9. Let $\alpha = \arctan(\tan \theta)$. Then α lies in the 1st or 4th quadrants and $\tan \alpha = \tan \theta$.

(1) If θ lies in the interval $[\pi/2, \pi]$ (the 2nd quadrant), then its reference angle $\theta_r = \pi - \theta$. Since $\tan \theta < 0$, angle α lies in the 4th quadrant and $\alpha = -\theta_r = -(\pi - \theta) = \theta - \pi$.

(2) If θ lies in the interval $[\pi, 3\pi/2]$ (the 3rd quadrant), then its reference angle $\theta_r = \theta - \pi$. Since $\tan \theta > 0$, angle α lies in the 1st quadrant and $\alpha = \theta_r = \theta - \pi$.

EP 16.1. While driving to the center of the bridge, the truck burned over 0.5 pounds of gasoline.

Chapter 17

P 17.7. From the previous problem we saw that $\sin A = a/d$, $\sin B = b/d$, $\sin C = c/d$. Multiplying these expressions, we get $\sin A \sin B \sin C = a/d \cdot b/d \cdot c/d = abc/d^3$.

P 17.11. Denote $\alpha = \arctan(1/2)$, $\beta = \arctan(1/3)$. Then $\tan \alpha = 1/2$, $\tan \beta = 1/3$. Since both tangents are positive, angles α and β are located in the 1st quadrant, and their sum is in the interval $(0, \pi)$. To prove that $\alpha + \beta = \pi/4$, it is enough to show that $\tan(\alpha + \beta) = \tan \pi/4 = 1$. Using the Tangent of the Sum formula (see Exercise 8.4 from Chapter 8), we have $\tan(\alpha + \beta) = (\tan \alpha + \tan \beta)/(1 - \tan \alpha \cdot \tan \beta) = (1/2 + 1/3)/(1 - 1/2 \cdot 1/3) = 1$.

P 17.12. Denote $\alpha = \arctan x$, $\beta = \arctan y$. Then $\tan \alpha = x$, $\tan \beta = y$. Apply tangent to the left part of the identity using the Tangent of the Sum formula:

$$\tan(\alpha + \beta) = (\tan \alpha + \tan \beta)/(1 - \tan \alpha \cdot \tan \beta) = (x + y)/(1 - xy).$$

From here, $\alpha + \beta = \arctan[(x + y)/(1 - xy)]$.

P 17.13. Using the Double Angle Sine formula from Chapter 8,

$$|2\sin x \cos x| = |\sin 2x| \le 1 \quad \Rightarrow \quad |\sin x \cos x| \le 1/2.$$

P 17.19. Denote $m = \tan^n x$. Then $\cot^n x = 1/m$. Since angle x is in the 1st quadrant, $m > 0$. We can use the result of Problem 17.17, (b): $m + 1/m \ge 2 \quad \Rightarrow \quad \tan^n x + \cot^n x \ge 2$.

P 17.26. $\sin^5 x = 1 - \cos^8 x = (1 - \cos^4 x)(1 + \cos^4 x) =$
$(1 - \cos^2 x)(1 + \cos^2 x)(1 + \cos^4 x) = \sin^2 x (1 + \cos^2 x)(1 + \cos^4 x)$.

So, the equation takes the form: $\sin^5 x = \sin^2 x (1 + \cos^2 x)(1 + \cos^4 x)$.

Or, $\sin^2 x[\sin^3 x - (1 + \cos^2 x)(1 + \cos^4 x)] = 0$. This equation splits into two:

(1) $\sin x = 0 \quad \Rightarrow \quad x = n\pi$.
(2) $\sin^3 x - (1 + \cos^2 x)(1 + \cos^4 x) = 0$. Or $\sin^3 x = (1 + \cos^2 x)(1 + \cos^4 x)$.

Since $\sin x \le 1$, we conclude from the last equation that $\cos x = 0$, and $\sin x = 1$. From here $x = \pi/2 + \pi n$. The final answer is the union of two sets: $x = n\pi$ and $x = \pi/2 + \pi n$.

EP 17.1. Consider all possible cases: HH, LL, HL, LH, and what the answer would be for each case. If the answer was "Yes," then the professor could not determine who is who, because in this case, the combinations of two inhabitants are HH, LL, HL. So, the answer must have been "No." Therefore, there is only one case LH, which means that the person who was asked the question is a liar, and the other is honest.

Chapter 18

E 18.1. Using the Law of Cosines for triangle *MDC*, find cosine of angle *M*:

$$\cos M = \frac{a^2 + c^2 - f^2}{2ac} = \frac{60^2 + 30^2 - 40^2}{2 \cdot 60 \cdot 30} = \frac{29}{36}.$$

Using the Law of Cosines for triangle *MAB*, we can calculate distance *e*:

$$e^2 = (a+b)^2 + (c+d)^2 - 2(a+b)(c+d)\cos M =$$

$$(60+110)^2 + (30+90)^2 - 2(60+110)(30+90)\frac{29}{36} = 10,433.$$

From here, $e = \sqrt{10,433} = 102$ miles.

E 18.2. Denote angle *BAD* by θ. Since *AC* is the bisector of the angle θ, angle *CAD* = $\theta/2$. From the right $\triangle CAD$,

$$CD = AC \sin(\theta/2) = c \cdot \sin(\theta/2)[\text{Half-Angle Formula}] = c \cdot \sqrt{\frac{1 - \cos\theta}{2}} \quad (*)$$

From the right triangle *ABE*, $\cos\theta = b/a$. Substitute the last expression into (*):

$$CD = c \cdot \sqrt{\frac{1 - \frac{b}{a}}{2}} = c \cdot \sqrt{\frac{a-b}{2a}}.$$

E 18.3. Denote $CD = x$.

Using the Law of Cosines for $\triangle ABD$, we get $\cos B = \dfrac{c^2 + (a+x)^2 - d^2}{2c(a+x)}$.

Using the Law of Cosines for $\triangle ABC$, we get $\cos B = \dfrac{c^2 + a^2 - b^2}{2ca}$.

Equating expressions for $\cos B$, we get the equation for x:

$$\frac{c^2 + (a+x)^2 - d^2}{2c(a+x)} = \frac{c^2 + a^2 - b^2}{2ca}.$$

By reducing the denominators by $2c$ and cross-multiplying, we get

$$a[c^2 + (a+x)^2 - d^2] = (c^2 + a^2 - b^2)(a+x).$$

We obtained the quadratic equation for x, which can be solved by the standard methods.

E 18.4. Let angle θ be between a and b. Then area S of the triangle is defined by the formula $S = \dfrac{1}{2} ab \sin \theta$. From here we can conclude that maximum S will be reached if $\sin \theta = 1$. From here, $\theta = \pi/2$. Therefore, the required triangle is a right one (as we would expect).

EP 18.1. It happened in 59 days. The next day, the area doubled, and the lake became completely overgrown.

EP 18.2. Blindfolded, place 15 random coins in one pile and the rest in another. Let x be the total number of coins facing heads-up in the first pile. Then there will be $15 - x$ coins facing heads-up in the second pile (since the total number of coins facing heads-up is 15). Let's turn over the coins from the first pile. Now it will also have $15 - x$ coins facing heads-up.

Chapter 19

P 19.1. Using the formulas

$$(a - b)^3 = a^3 - 3a^2b + 3ab^2 - b^3 \text{ and } (a - b)^2 = a^2 - 2ab + b^2,$$

we get $\left(y - \dfrac{a}{3}\right)^3 + a\left(y - \dfrac{a}{3}\right)^2 + b\left(y - \dfrac{a}{3}\right) + c =$

$$y^3 - 3y^2\frac{a}{3} + 3y\left(\frac{a}{3}\right)^2 - \left(\frac{a}{3}\right)^3 + a\left[y^2 - 2y\frac{a}{3} + \left(\frac{a}{3}\right)^2\right] + b\left(y - \frac{a}{3}\right) + c$$

$$= y^3 + y\left(\frac{a^2}{3} - \frac{2a^2}{3} + b\right) + \left(-\frac{a^3}{27} + \frac{a^3}{9} - \frac{ab}{3} + c\right) = y^3 + py + q,$$

where $p = b - \dfrac{a^2}{3}, q = c + \dfrac{2a^3 - 9ab}{27}.$

P 19.4. Expand the left side of the equation:

$$(x - 1)(x - 2)(x + 3) = (x^2 - 3x + 2)(x + 3) = x^3 - 7x + 6 = x^3 + px + q = 0.$$

According to Cardano's formula, the solution x is presented as a sum of two parts u and v: $x = u + v$. Calculate part u:

$$u = \sqrt[3]{-\frac{q}{2} + \sqrt{\frac{q^2}{4} + \frac{p^3}{27}}} = [p = -7, \quad q = 6] = \sqrt[3]{-3 + \sqrt{\frac{36}{4} + \frac{-343}{27}}}$$

$$= \sqrt[3]{-3 + \sqrt{\frac{-100}{27}}} = \sqrt[3]{-3 + \frac{10}{9}\sqrt{-3}}.$$

Similarly, $v = \sqrt[3]{-3 - \dfrac{10}{9}\sqrt{-3}}$. Thus, $x = \sqrt[3]{-3 + \dfrac{10}{9}\sqrt{-3}} + \sqrt[3]{-3 - \dfrac{10}{9}\sqrt{-3}}.$

P 19.5. Reorganize the terms in the sum $1 + 2 + 3 + \cdots + 99 + 100$ by pairing them: $(1 + 100) + (2 + 99) + (3 + 98) + \cdots + (50 + 51)$. We have

50 pairs, and the sum of each pair is 101. Therefore, the total sum is $101 \cdot 50 = 5050$.

EP 19.1. Present the numerator in the form:

$(12 - 2)^2 + (12 - 1)^2 + 12^2 + (12 + 1)^2 + (12 - 2)^2 =$ [double products reduced]

$12^2 + 4 + 12^2 + 1 + 12^2 + 12^2 + 1 + 12^2 + 4 = 12^2 \cdot 5 + 10 = 12(12 \cdot 5) + 10 = 12 \cdot 60 + 10 = 720 + 10 = 730.$

Finally, $730/365 = 2$.

Chapter 20

P 20.2. (a) $i^{100} = (i^2)^{50} = (-1)^{50} = 1$; (b) $i^{150} = (i^2)^{75} = (-1)^{75} = -1$; (c) $i^{27} = i^{26} \cdot i = (i^2)^{13} \cdot i = (-1)^{13} \cdot i = -i$; (d) $i^{37} = i^{36} \cdot i = (i^2)^{18} \cdot i = (-1)^{18} \cdot i = i$.

P 20.5. $(a + bi)(c + di) = ac + adi + bci + bdi^2 = (ac - bd) + (ad + bc)i$.

P 20.6. $(a + bi)(a - bi) = a^2 - (bi)^2 = a^2 - b^2i^2 = a^2 - b^2(-1) = a^2 + b^2$.

P20.8. $\dfrac{a + bi}{c + di} = \dfrac{(a + bi)(c - di)}{(c + di)(c - di)} = \dfrac{ac - adi + bci - bdi^2}{c^2 + d^2}$

$= \dfrac{ac + bd}{c^2 + d^2} + \dfrac{bc - ad}{c^2 + d^2} i.$

P 20.13. Let $n = 2$. According to Problem 20.11, $(\cos \theta + i \sin \theta)^2 = (\cos \theta + i \sin \theta)(\cos \theta + i \sin \theta) = \cos(\theta + \theta) + i \sin(\theta + \theta) = \cos 2\theta + \sin 2\theta$.

Next, for $n = 3$, $(\cos \theta + i \sin \theta)^3 = (\cos \theta + i \sin \theta)^2 (\cos \theta + i \sin \theta) = (\cos 2\theta + \sin 2\theta)(\cos \theta + i \sin \theta) = \cos(2\theta + \theta) + i \sin(2\theta + \theta) = \cos 3\theta + i \sin 3\theta$.

We can proceed[1] for $n = 4, 5, \dots$.

P 20.16. It is enough to show that $z \cdot z^{-1} = 1$. We have $z \cdot z^{-1} = r(\cos \theta + i \sin \theta) \cdot r^{-1} [\cos(-\theta) + i \sin(-\theta)] = rr^{-1} [\cos(\theta - \theta) + i \sin(\theta - \theta)] = \cos 0 + i \sin 0 = 1$.

---

[1] Readers familiar with the mathematical induction, may get a more formal proof.

P 20.17. Using the result of the previous problem, $z_2^{-1} = r_2^{-1}[\cos(-\theta_2) + i\sin(-\theta_2)]$. Then

$$z_1 / z_2 = z_1 \cdot z_2^{-1} = r_1(\cos\theta_1 + i\sin\theta_1) \cdot r_2^{-1}\left[\cos(-\theta_2) + i\sin(-\theta_2)\right] = r_1 / r_2\left[\cos(\theta_1 - \theta_2) + i\sin(\theta_1 - \theta_2)\right].$$

E 20.1. $\dfrac{1}{z} = \dfrac{1}{a+bi} = \dfrac{a-bi}{(a+bi)(a-bi)} = \dfrac{a-bi}{a^2+b^2} = \dfrac{a}{a^2+b^2} - \dfrac{b}{a^2+b^2}i.$

E 20.2.

(a) $r = \sqrt{(-3)^2 + (-\sqrt{3})^2} = \sqrt{12} = 2\sqrt{3}$, $\tan\theta = -\sqrt{3}/(-3) = \sqrt{3}/3 \Rightarrow$

$\arctan(\sqrt{3}/3) = \pi/6$. Since z_1 is in the 3rd quadrant,
$\theta = \pi/6 + \pi = 7\pi/6$.

Answer: $z_1 = 2\sqrt{3}\left[\cos(7\pi/6) + i\sin(7\pi/6)\right].$

(b) $r = \sqrt{4^2 + (-4)^2} = \sqrt{32} = 4\sqrt{2}$, $\tan\theta = -4/4 = -1 \Rightarrow \arctan(-1) = -\pi/4.$
Since z_2 is in the 4th quadrant, $\theta = -\pi/4$. We also can accept
$\theta = 2\pi - \pi/4 = 7\pi/4.$

Answer:
$z_2 = 4\sqrt{2}\left[\cos(-\pi/4) + i\sin(-\pi/4)\right]$ or $z_2 = 4\sqrt{2}\left[\cos(7\pi/4) + i\sin(7\pi/4)\right].$

(c) $r = \sqrt{(-\sqrt{3})^2 + 3^2} = \sqrt{12} = 2\sqrt{3}$, $\tan\theta = 3/(-\sqrt{3}) = -\sqrt{3} \Rightarrow$

$\arctan(-\sqrt{3}) = -\pi/3$. Since z_3 is in the 2nd quadrant, $\theta = -\pi/3 + \pi = 2\pi/3.$

Answer: $z_3 = 2\sqrt{3}\left[\cos(2\pi/3) + i\sin(2\pi/3)\right].$

E 20.3. We convert given complex numbers into the Polar form. Then we apply the de Moivre's formula, and, finally, convert the numbers back to the standard form.

(a) $r = \sqrt{(1/2)^2 + (-\sqrt{3}/2)^2} = \sqrt{1/4 + 3/4} = 1$. Point $(1/2, -\sqrt{3}/2)$ lies
on the unit circle in the 4th quadrant, and $\sin\theta = -\sqrt{3}/2$, $\cos\theta = 1/2$,

$\theta = 300°$. From here, $z_1 = (\cos 300° + i \sin 300°)^{30} = \cos 9000° + i \sin 9000° = \cos 0 + i \sin 0 = 1$.

(b) $r = \sqrt{\left(-\sqrt{2}/2\right)^2 + \left(\sqrt{2}/2\right)^2} = \sqrt{1/2 + 1/2} = 1$, Point $\left(-\sqrt{2}/2, \sqrt{2}/2\right)$ lies on the unit circle in the 2nd quadrant, and $\sin\theta = \sqrt{2}/2$, $\cos\theta = -\sqrt{2}/2$, $\theta = 135°$. From here, $z_2 = (\cos 135° + i \sin 135°)^{20} = \cos 2700° + i \sin 2700° = \cos 180° + i \sin 180° = -1$.

(c) $r = \sqrt{\left(\sqrt{3}/2\right)^2 + \left(1/2\right)^2} = \sqrt{3/4 + 1/4} = 1$, Point $\left(\sqrt{3}/2, 1/2\right)$ lies on the unit circle in the 1st quadrant, $\sin\theta = 1/2, \cos\theta = \sqrt{3}/2, \theta = 30°$. From here, $z_3 = (\cos 30° + i \sin 30°)^{40} = \cos 1200° + i \sin 1200° = \cos 120° + i \sin 120° = -1/2 + \left(\sqrt{3}/2\right)i$.

E 20.4. We can use the result of Problem 20.19. We have $n = 6$, so a given equation has six roots that can be defined by the formula $z_k = \cos(2\pi k/6) + i \sin(2\pi k/6)$, $k = 0, 1, \ldots, 5$. We can reduce fraction inside cosine and sine:

$z_k = \cos(\pi k/3) + i \sin(\pi k/3)$. Let's represent each root in standard form:

$z_0 = \cos\left(\pi \cdot 0/3\right) + i \sin\left(\pi \cdot 0/3\right) = 1$,

$z_1 = \cos\left(\pi/3\right) + i \sin\left(\pi/3\right) = 1/2 + \sqrt{3}/2\, i$,

$z_2 = \cos\left(2\pi/3\right) + i \sin\left(2\pi/3\right) = -1/2 + \sqrt{3}/2\, i$,

$z_3 = \cos\left(\pi\right) + i \sin\left(\pi\right) = -1$,

$z_4 = \cos\left(4\pi/3\right) + i \sin\left(4\pi/3\right) = -1/2 - \sqrt{3}/2\, i$,

$z_5 = \cos\left(5\pi/3\right) + i \sin\left(5\pi/3\right) = 1/2 - \sqrt{3}/2\, i$.

On the complex plane, roots $z_0, z_1, \ldots, z_5$ are located at the vertices of a regular hexagon:

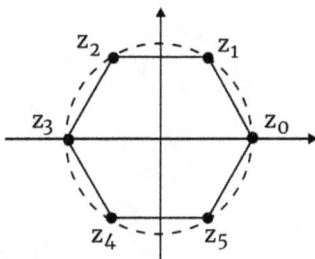

EP 20.1. The third brother will be jailed for one day and set free the next day, with this pattern repeating for the rest of his life. The result would be a reduction of a life sentence in jail by half.

Appendix

A4. Let sides of the Kepler triangle be $1, r, r^2$, and let A, B, and C be angles opposite to these sides respectively. According to the previous problem, $r = \sqrt{\varphi} \approx 1.272$. From here $r^2 \approx 1.618$. Since r^2 is a hypotenuse, we have

$$\sin A = 1/r^2 \approx 0.618 \quad \Rightarrow \quad A = \sin^{-1} 0.618 = 38.17°, \, B = 90° - A = 51.83°.$$

A5. Without loss of generality, we may assume that the sides of the triangle are 1, r, and r^2. We will use the property that the sum of any two sides of the triangle is larger than the third side.

(1) Since 1, r, and r^2 are sides of the triangle, then $1 + r > r^2$. From here, $r^2 - r - 1 < 0$. This is a quadratic inequality. Its solutions are numbers between roots of the equation $r^2 - r - 1 = 0$. One of the roots is a negative number, and the other is the golden ratio φ, which is a positive number. Therefore, $r < \varphi$.

(2) Similarly, $r + r^2 > 1 \quad \Rightarrow \quad r^2 + r - 1 > 0$. The solutions to this inequality are numbers outside the interval whose endpoints are the roots of the equation $r^2 + r - 1 = 0$. One of the roots is a negative number, and the other is a positive number $\dfrac{-1 + \sqrt{5}}{2}$, so r is greater that this root. We can modify this number like this

$$\frac{-1 + \sqrt{5}}{2} = \frac{\left(-1 + \sqrt{5}\right)\left(1 + \sqrt{5}\right)}{2\left(1 + \sqrt{5}\right)} = \frac{-1 + 5}{2\left(1 + \sqrt{5}\right)} = \frac{4}{2\left(1 + \sqrt{5}\right)} = \frac{2}{1 + \sqrt{5}} = \frac{1}{\varphi}.$$

Therefore, $r > 1/\varphi$. Thus, $1/\varphi < r < \varphi$.

Index

A

Abel, Neils Henrik, 270
absolute value, 58
angle
 coterminal, 49
 depression, 21
 elevation, 21
 geometric, 47
 initial side, 48
 negative, 48
 positive, 48
 quadrantal, 56
 reference: definition, 71
 reference: main property, 72
 special, 36
 standard position, 52
 terminal side, 48
astrolabe, 4

B

Barr, James Mark, 298
Bogdanov-Belsky, Nikolay, 277
Bolyai, Janos, 275
Bombelli, Rafael, 272

C

Cardano, Gerolamo, 267
Cardano's formula, 269
Cartesian system of coordinates, 50
Cheops Pyramid, 20

clinometer, 4
complex numbers
 argument, 285
 complex conjugate, 281
 de Moivre's formula, 287
 de Moivre's numbers, 290
 imaginary part, 280
 imaginary unit, 279
 modulus (absolute value), 285
 polar (trigonomic) form, 285
 purely imaginary numbers, 280
 real part, 280
 standard (rectangular) form, 280
complex plane, 283
cosecant
 definition for right triangle, 11
 general definition, 55
cosine
 all solutions of basic equation, 225
 basic (simplest) equation, 205
 definition for right triangles, 12
 difference formula, 107
 double angle formula, 114
 general definition, 52
 graph, 160
 period, 68
 power reduction formula, 124
 principal root of basic equation, 225
 solutions of basic equation for
  specific intervals, 226
 sum formula, 111

cotangent
  definition for right triangle, 11
  general definition, 54
  graph, 201

D
Darwin, Francis, 265
Descartes, Rene, 50
difference of sines formula, 127
distance formula, 106

E
Euler, Leonhard, 108, 273
Euler's quadrilateral theorem, 108

F
Ferrari, Lodovico, 270
Ferro (Ferreo), Scipione, 266
Fibonacci (Leonardo of Pisa), 272
Fiore, Antonio Maria, 266
Forehead Rule, 69
Fourier, Joseph, 178
function
  argument, 6, 149
  cubic parabola, 151
  even, 64, 149
  horizontal asymptote, 192
  hyperbola, 191
  inverse, 209
  line symmetry property, 149
  monotonic, 210
  odd, 64, 149
  parabola, 150
  period, 65
  periodic, 65
  point symmetry property, 150
  value, 6
  vertical asymptote, 192

G
Galois, Evarist, 270
Gauss, Carl Friedrich, 274

golden ratio, 298
Goodrick, John, 186

H
half-angle formulas, 124
Heron's formula, 251
Hertz, Heinrich, 176
Hipparchus of Nicaea, 7

I
inverse cosine function
  definition 1, 224
  definition 2, 224
  domain, 223
  graph, 223
  property, 223
  range, 223
inverse sine function
  definition 1, 213
  definition 2, 213
  graph, 213
  range, 214
inverse tangent function
  definition 1, 230
  definition 2, 230

K
Kepler, Johannes, 301

L
law of cosines, 94
law of sines, 78
  ambiguous case, 84
  geometric interpretation of ratios, 80
Leavitt, Henrietta, 186
Leibniz, Gottfried, 288
Lobachevsky, Nikolai, 275

M
main (Pythagorean) identity, 34
Maxwell, James Clerk, 176
Moivre, Abraham, 287

N

Newton, Isaac, 288

P

Phidias, 298
polar coordinates, 284
Ptolemy, Claudius, 7
Pythagoras, 18
Pythagorean
    main identity, 34, 63
    reciprocal theorem, 27
    theorem, 17
    triple, 27

Q

quadrants, 50

R

radian and degree measures: main
    proportion, 142
radian measure
    definition 1, 135
    definition 2, 138
regular polygons, 35

S

Schiller, Friedrich, 247
secant
    definition for right triangle, 11
    general definition, 55
similar figures, 2
similar triangles, 3
sine
    all solutions of the basic equation,
        216
    basic (simplest) equation, 205
    definition for right triangles, 12
    difference formula, 114
    double angle formula, 114
    general definition, 52
    graph, 156
    period, 65

power reduction formula, 123
principal root of basic equation,
    215
product formula, 122
solutions of the basic equation in
    one period interval, 216
sum formula, 113
term, 9
sinusoidal function
    amplitude, 169
    angular speed, 168
    definition, 168
    frequency, 168
    graphing, 173
    period, 170
    phase shift, 172
special angles in degrees and
    radians, 145
special right triangles, 36
sum of cosines formula, 128

T

tangent
    all solutions of basic equation,
        230
    basic (simplest) equation, 205
    definition for right triangles,
        13
    general definition, 53
    geometric definition, 53
    graph, 196
    graph of inverse function, 228
    period, 66, 197
    solution of basic equation in
        specific intervals, 231
Tartaglia, Nicolo, 266
transcendental equations, 162
triangle
    ambiguous case: how to recognize
        and resolve, 88
    Egyptian, 19
    equilateral, 35
    golden, 297

Kepler, 300
oblique, 77
solving, 77
trig functions
  addition formulas, 116
  basic, 12
  double angle formulas, 117
  even-odd properties, 67
  half-angle formulas, 131
  inverse, 20
  periodic properties, 67
  power reduction formulas, 131
  product formulas, 130
  quadrantal angles, 56
  ranges of additional, 58
  ranges of basic, 58
  reduction formulas, 69
  relations, 42
  signs of basic, 56

special angles, 42
sum and difference formulas, 131
trigonometry: term, 1

U
unit circle, 51
units of angle measurement
  degree, 133
  gradian or grad, 134
  radian, 134
  revolution, 134
  rhumb, 134
  thousandth, 134

V
Volta, Alessandro, 187

Z
Zhukovsky, Nikolay, 294